Smart Business and Education in the Age of VUCA and BANI

G. S. Prakasha
https://orcid.org/0000-0002-1287-7606
Christ University, India

Maria Lapina
https://orcid.org/0000-0001-8117-9142
North-Caucasus Federal University, Russia

Francisco David Guillén Gámez
https://orcid.org/0000-0001-6470-526X
Universidad de Málaga, Spain

Juan J. Leiva-Olivencia
https://orcid.org/0000-0002-0082-1154
University of Málaga, Spain

IGI Global
Scientific Publishing
Publishing Tomorrow's Research Today

Vice President of Editorial	Melissa Wagner
Director of Acquisitions	Mikaela Felty
Director of Book Development	Jocelynn Hessler
Production Manager	Mike Brehm
Cover Design	Jose Rosado

Published in the United States of America by
IGI Global Scientific Publishing
701 East Chocolate Avenue
Hershey, PA, 17033, USA
Tel: 717-533-8845 | Fax: 717-533-7115
Website: https://www.igi-global.com E-mail: cust@igi-global.com

Library of Congress Cataloging-in-Publication Data

Names: Prakasha, G. S. editor | Lapina, Maria, 1981- editor | Guillén Gámez, Francisco David, 1985- editor | Leiva-Olivencia, Juan J., 1979- editor

Title: Smart Business and Education in the Age of VUCA and BANI / Edited by G. S. Prakasha, Maria Lapina, Francisco David Guillén Gámez, and Juan J. Leiva-Olivencia.

Description: Hershey, PA : IGI Global Scientific Publishing, [2026] | Includes bibliographical references. | Summary: "This book explores how smart technologies and data-driven innovation are transforming business and education in the volatile, uncertain, complex, and ambiguous (VUCA) and brittle, anxious, nonlinear, and incomprehensible (BANI) world. It highlights adaptive strategies, digital inclusion, and AI-driven practices shaping sustainable and equitable futures"-- Provided by publisher.

Identifiers: LCCN 2025052871 (print) | LCCN 2025052872 (ebook) | ISBN 9798337379227 hardcover | ISBN 9798337379234 paperback | ISBN 9798337379241 ebook

Subjects: LCSH: Education--Data processing | Business--Data processing

Classification: LCC LB1028.43 .S628 2026 (print) | LCC LB1028.43 (ebook)

LC record available at https://lccn.loc.gov/2025052871

LC ebook record available at https://lccn.loc.gov/2025052872

British Cataloguing in Publication Data
A Cataloguing in Publication record for this book is available from the British Library.

Table of Contents

Detailed Table of Contents

The complexity and unpredictability inherent in efficient markets, intensified by VUCA, and BANI environments, demand intelligent, flexible, and adaptive strategies for effective portfolio selection. Traditional optimization techniques regularly struggle to account for the dynamic behavior and rapid fluctuations inherent in such volatile conditions. In these contexts, computational intelligence techniques integrating evolutionary algorithms, neural networks, fuzzy logic, and hybrid approaches have emerged as an effective tool to tackle the complex portfolio selection and evaluates their performance using both historical and real stock market data. By adapting to actual data and predictive algorithms, the proposed framework, dependent on computational intelligence, supplying enhanced decision-making amid uncertainty. This work contributes to the growing body of intelligent financial analytics literature by presenting a robust, data-driven decision-support system designed to navigate the complexities of modern financial markets shaped by VUCA and BANI forces.

This article examines the information and educational environment (IEE) in its structural and functional representation. The communicative nature of the IEE, its expression, implementation, and representation in its organization and functioning are the subject of this study and form the context for examining the structural components, corresponding properties, and characteristics of the IEE. It is emphasized that the communicative component is inherent in the IEE and subject-based learning, representing a conceptual feature of the IEE and a fundamental condition for its implementation in the context of the digital transformation of learning. Key here is the organization of communicative interaction between educational actors and the creation of effective mechanisms for knowledge transfer within any educational system, regardless of its organizational form: formal, non-formal, or informal education. The relevance of this study stems from the complex problems associated with the formation of this educational continuum.

The deep forests of the Western Ghats are home to the aboriginal tribes Kattunayakans, Categorized under the Particularly Vulnerable Tribal Groups (PVTGs) in India. Despite various government schemes to improve their economic status, the literacy rates remain very low due to geographical isolation and traditional lifestyle. The main objective is to look into designing educational activity centres that exclusively cater to the needs of the Kattunayakan children, integrating indigenous knowledge and methods with teaching-learning sessions to promote lifelong learning. Technological progress and AI can be game changers for accessing quality education, including visual learning, language translation, and personalized methods while preserving cultural heritage. The research is based on a systematic narrative review, critically examining scholarly articles, policy and government reports. The outcome shows that these programs help improve academic skills while keeping them closer to their culture and identity.

This study examines farmers' behavioral intentions toward adopting drone technology in agriculture using a quantitative approach. Data will be collected from farmers familiar with Agri-based technologies through purposive sampling, with responses analyzed via structural equation modeling (IBM SPSS AMOS V21). The study integrates the Technology Acceptance Model (TAM) and the Theory of Planned Behavior (TPB) to understand key factors influencing adoption. Findings highlight that farmers' knowledge of drone operations and economic factors impact their acceptance, while regulations and perceived behavioral control contribute to reluctance. Although the study focuses on farmers, future research can explore perspectives of agricultural specialists, AgriTech start-ups, and policymakers. The insights will help stakeholders enhance the adoption, quality, and flexibility of agricultural technologies. Uniquely, this study addresses the gap in research by examining the role of regulations and behavioral control in farmers' decisions on drone adoption.

This study explores pre-service and in-service teachers' behavioral intention to adopt smart tools by examining their perceptions of usefulness, ease of use, enjoyment, and contextual readiness. Using a quantitative survey design, data was collected from 42 participants across the Delhi-NCR region through purposive sampling. A composite score of 15 Likert-scale items was used to measure behavioural intention, followed by quartile-based analysis, independent samples t-tests (for gender), and one-way ANOVA (for age groups). While results revealed no statistically significant differences across gender and age, the quartile distribution indicated a varied spectrum of intention, with participants broadly categorised into low, moderate, and high-intention groups. The study highlights the need for teacher training programs to go beyond technical instruction. It advocates for a more nuanced, context-aware approach to technology readiness in teacher education that bridges the gap between vision and practice in the evolving ecosystem of Education 5.0.

The text analyses how Artificial Intelligence (AI) can be integrated into education to design contextualized learning situations that increase student motivation and promote equity. Through tools like ChatGPT, content can be personalized, and the curriculum can be connected to students' local realities. The proposed methodology includes context analysis, definition of competencies, the use of prompts, and continuous evaluation with feedback. Additionally, the value of Universal Design for Learning (UDL) is highlighted as an inclusive framework. However, the implementation of AI entails ethical challenges such as data privacy, algorithmic bias, and the need for teacher training. The text concludes that AI should not replace teachers but complement them, and it advocates for inclusive policies and a culture of critical thinking to foster more motivating, personalized, and equitable education.

This study analyzes the role of AI technology, particularly Artificial Intelligence (AI) tools, in fostering multiculturalism and improving English instruction across diverse educational settings.This study employs a 24-item Likert scale survey to examine the perspectives of 115 English as a Foreign Language (EFL) instructors about artificial intelligence and multiculturalism. Results indicate a reasonable consensus on the advantages of artificial intelligence in promoting cultural diversity; yet its effectiveness is constrained by obstacles such as language barriers and cultural disparities.The research emphasizes the necessity of integrating artificial intelligence and offering

(p = 0.006), and years of teaching experience (p = 0.016) on the digital inclusivity of the teachers' pedagogical practices.

The fashion industry is one of the most dynamic industries, with trends that change constantly being moulded by the changing preferences of consumers. In this dynamic environment it is crucial for companies to adjust to the current market demand, failing which companies might plunge into losses and even face shutdowns. Predictive analytics plays an important role in trend prediction, demand forecasting, optimizing the supply chain and inventory management, personalization, dynamic pricing, waste minimization and prediction of sustainable materials. This, of course, is not without challenges. Data privacy concerns and over reliance on AI algorithms leading to inhibition of human creativity stand out as the most concerning challenges that need to be addressed to ensure that predictive analytics is utilised to its fullest potential in the fashion industry. Predictive analytics and AI can ensure that the fashion industry is not just profitable but also consumer centric and sustainable in its operations thus ensuring long term growth and sustenance of businesses in this industry.

Preface

The contemporary world is undergoing transformative shifts marked by rapid technological advancement, global uncertainty, and the emergence of increasingly complex social and economic systems. These dynamics are captured in the frameworks of VUCA (Volatility, Uncertainty, Complexity, Ambiguity) and BANI (Brittleness, Anxiety, Non-linearity, Incomprehensibility), which now define the environments in which business, education, and society operate. In this era, resilience, adaptability, and innovation are no longer optional but essential for sustainable progress. *Reimagining the Future: Smart Business and Education in the Age of VUCA and BANI* brings together a range of scholarly perspectives that highlight the challenges and opportunities of navigating these turbulent times.

The ten chapters in this volume offer a multidisciplinary exploration of future-readiness. The book begins with an analysis of computational intelligence methods for portfolio selection in financial landscapes shaped by volatility and uncertainty. It then shifts toward the intersection of personal development and information subcultures, emphasizing the growing influence of digital learning environments on individual and collective growth. The role of media and information-communication environments in shaping education within the digital society is also critically examined, providing a foundation for understanding systemic transformations.

CHAPTER 1: NAVIGATING COMPUTATIONAL INTELLIGENCE APPROACHES FOR EFFICIENT PORTFOLIO SELECTION IN VUCA AND BANI FINANCIAL LANDSCAPE

The complexity and unpredictability inherent in efficient markets, intensified by VUCA, and BANI environments, demand intelligent, flexible, and adaptive strategies for effective portfolio selection. Traditional optimization techniques regularly struggle to account for the dynamic behavior and rapid fluctuations inherent in such volatile conditions. In these contexts, computational intelligence techniques integrating evolutionary algorithms, neural networks, fuzzy logic, and hybrid approaches have emerged as an effective tool to tackle the complex portfolio selection and evaluates their performance using both historical and real stock market data. By adapting to actual data and predictive algorithms, the proposed framework, dependent on computational intelligence, supplying enhanced decision-making amid uncertainty. This work contributes to the growing body of intelligent financial analytics literature by presenting a robust, data-driven decision-support system designed to navigate the complexities of modern financial markets shaped by VUCA and BANI forces.

CHAPTER 2: CHARACTERISTICS AND PROPERTIES OF THE INFORMATION AND COMMUNICATION ENVIRONMENT AND THE PRINCIPLES OF ITS CONSTRUCTION

This article examines the information and educational environment (IEE) in its structural and functional representation. The communicative nature of the IEE, its expression, implementation, and representation in its organization and functioning are the subject of this study and form the context for examining the structural components, corresponding properties, and characteristics of the IEE. It is emphasized that the communicative component is inherent in the IEE and subject-based learning, representing a conceptual feature of the IEE and a fundamental condition for its implementation in the context of the digital transformation of learning. Key here is the organization of communicative interaction between educational actors and the creation of effective mechanisms for knowledge transfer within any educational system, regardless of its organizational form: formal, non-formal, or informal education. The relevance of this study stems from the complex problems associated with the formation of this educational continuum.

CHAPTER 3: ENHANCING EDUCATIONAL ENGAGEMENT AMONG KATTUNAYAKAN TRIBAL CHILDREN THROUGH CULTURALLY TAILORED EDUCATIONAL ACTIVITY CENTRES

The deep forests of the Western Ghats are home to the aboriginal tribes Kattunayakans, Categorized under the Particularly Vulnerable Tribal Groups (PVTGs) in India. Despite various government schemes to improve their economic status, the literacy rates remain very low due to geographical isolation and traditional lifestyle. The main objective is to look into designing educational activity centres that exclusively cater to the needs of the Kattunayakan children, integrating indigenous knowledge and methods with teaching-learning sessions to promote lifelong learning. Technological progress and AI can be game changers for accessing quality education, including visual learning, language translation, and personalized methods while preserving cultural heritage. The research is based on a systematic narrative review, critically examining scholarly articles, policy and government reports. The outcome shows that these programs help improve academic skills while keeping them closer to their culture and identity.

CHAPTER 4: ANTECEDENTS OF FARMERS' INTENTIONS TO ADOPT DRONE TECHNOLOGY: INTEGRATING TAM AND TPB MODELS IN AGRICULTURE

This study examines farmers' behavioral intentions toward adopting drone technology in agriculture using a quantitative approach. Data will be collected from farmers familiar with Agri-based technologies through purposive sampling, with responses analyzed via structural equation modeling (IBM SPSS AMOS V21). The study integrates the Technology Acceptance Model (TAM) and the Theory of Planned Behavior (TPB) to understand key factors influencing adoption. Findings highlight that farmers' knowledge of drone operations and economic factors impact their acceptance, while regulations and perceived behavioral control contribute to reluctance. Although the study focuses on farmers, future research can explore perspectives of agricultural specialists, AgriTech start-ups, and policymakers. The insights will help stakeholders enhance the adoption, quality, and flexibility of agricultural technologies. Uniquely, this study addresses the gap in research by examining the role of regulations and behavioral control in farmers' decisions on drone adoption.

CHAPTER 5: INVESTIGATING EDUCATORS' BEHAVIOURAL INTENTIONS TOWARDS SMART TOOLS IN EDUCATION

This study explores pre-service and in-service teachers' behavioral intention to adopt smart tools by examining their perceptions of usefulness, ease of use, enjoyment, and contextual readiness. Using a quantitative survey design, data was collected from 42 participants across the Delhi-NCR region through purposive sampling. A composite score of 15 Likert-scale items was used to measure behavioural intention, followed by quartile-based analysis, independent samples t-tests (for gender), and one-way ANOVA (for age groups). While results revealed no statistically significant differences across gender and age, the quartile distribution indicated a varied spectrum of intention, with participants broadly categorised into low, moderate, and high-intention groups. The study highlights the need for teacher training programs to go beyond technical instruction. It advocates for a more nuanced, context-aware approach to technology readiness in teacher education that bridges the gap between vision and practice in the evolving ecosystem of Education 5.0.

CHAPTER 6: AI IN EDUCATION. DESIGNING CONTEXTUALIZED LEARNING TO BOOST MOTIVATION AND ENSURE EQUITY

The text analyses how Artificial Intelligence (AI) can be integrated into education to design contextualized learning situations that increase student motivation and promote equity. Through tools like ChatGPT, content can be personalized, and the curriculum can be connected to students' local realities. The proposed methodology includes context analysis, definition of competencies, the use of prompts, and continuous evaluation with feedback. Additionally, the value of Universal Design for Learning (UDL) is highlighted as an inclusive framework. However, the implementation of AI entails ethical challenges such as data privacy, algorithmic bias, and the need for teacher training. The text concludes that AI should not replace teachers but complement them, and it advocates for inclusive policies and a culture of critical thinking to foster more motivating, personalized, and equitable education.

CHAPTER 7: AI TECHNOLOGIES AND MULTICULTURAL EDUCATION INSIGHTS INTO TEACHERS' ATTITUDES AND CHALLENGES IN HIGHER EDUCATION

This study analyzes the role of AI technology, particularly Artificial Intelligence (AI) tools, in fostering multiculturalism and improving English instruction across diverse educational settings.This study employs a 24-item Likert scale survey to examine the perspectives of 115 English as a Foreign Language (EFL) instructors about artificial intelligence and multiculturalism. Results indicate a reasonable consensus on the advantages of artificial intelligence in promoting cultural diversity; yet its effectiveness is constrained by obstacles such as language barriers and cultural disparities.The research emphasizes the necessity of integrating artificial intelligence and offering opportunities for EFL teachers to improve their intercultural communication skills and cultural competence. It underscores the necessity of inclusion and representation in fostering supportive educational settings by advocating for instructional methods that promote active engagement, collaboration, and culturally pertinent content.

CHAPTER 8: ALIGNING SMART EDUCATION WITH SMART ECONOMY: MULTI-DIMENSIONAL EDUCATIONAL QUALITY ASSESSMENT

This chapter proposes a mathematical framework for assessing educational quality that bridges the gap between educational processes and real-world impacts within a "smart economy," defined as a network of relationships mediated by knowledge. Current educational assessment practices rely on internal metrics that often fail to capture how graduates' capabilities align with evolving economic and societal needs. The proposed framework integrates diverse stakeholder perspectives through two interconnected "fields": the field of student qualities and the field of educational programs. By representing student qualities and programs as vectors in multi-dimensional spaces, the framework formalizes how education transforms capabilities. The chapter presents mathematical formulations from both upward (building from student-program interactions) and downward (deriving from economic system dynamics) perspectives. While acknowledging implementation challenges and limitations, the framework offers a path toward assessment that is holistic, externally aligned, and resistant to metric corruption.

CHAPTER 9: EXPLORING FACTORS INFLUENCING DIGITAL INCLUSION WITHIN MODERN PEDAGOGICAL PRACTICES PREVALENT AMONGST SCHOOLTEACHERS: SMART STUDENTS, SMARTER TEACHERS?

Teachers serve as torchbearers to the vast expanse of knowledge, leaving a lasting impact on the minds of their students through their teaching styles and pedagogical practices. With the advancement of technology, the integration of digital technologies has increasingly been observed in pedagogical approaches. The current study seeks to explore various subjective factors related to teachers, such as their teaching styles, age, classes taught, educational qualifications, and years of experience, in relation to their individual digital competencies and the extent of digital inclusivity in their pedagogies. Three psychometric scales were employed for data collection (TSI-Q, DiPeCoS, and TDiCoS). Data were gathered from 215 school teachers in the United Arab Emirates (UAE) and subsequently analysed. The results revealed a significant influence of teaching styles ($p = 0.045$), classes taught ($p = 0.001$), age ($p = 0.006$), and years of teaching experience ($p = 0.016$) on the digital inclusivity of the teachers' pedagogical practices.

CHAPTER 10: PREDICTIVE ANALYTICS AS A CATALYST FOR DISRUPTION IN THE FASHION ECOSYSTEM

The fashion industry is one of the most dynamic industries, with trends that change constantly being moulded by the changing preferences of consumers. In this dynamic environment it is crucial for companies to adjust to the current market demand, failing which companies might plunge into losses and even face shutdowns. Predictive analytics plays an important role in trend prediction, demand forecasting, optimizing the supply chain and inventory management, personalization, dynamic pricing, waste minimization and prediction of sustainable materials. This, of course, is not without challenges. Data privacy concerns and over reliance on AI algorithms leading to inhibition of human creativity stand out as the most concerning challenges that need to be addressed to ensure that predictive analytics is utilised to its fullest potential in the fashion industry. Predictive analytics and AI can ensure that the fashion industry is not just profitable but also consumer centric and sustainable in its operations thus ensuring long term growth and sustenance of businesses in this industry.

Social and cultural dimensions are equally emphasized. The engagement of Kattunayakan tribal children through culturally responsive educational activity centres illustrates how inclusivity and contextualization are vital in reimagining education. Likewise, the integration of TAM and TPB models in exploring farmers' adoption of drone technology reflects how technological uptake is influenced by behavioral and contextual factors. Issues of mental health and cybervictimization among school students highlight the human side of digital transformation, underscoring the need for emotional resilience in parallel with technological advancement.

The latter chapters address sustainability and preparedness by examining monitoring methods in energy storage systems, management education interventions for VUCA-readiness, and educators' behavioral intentions toward smart tools. Together, these contributions present a holistic vision of how business and education can adapt intelligently in uncertain times.

This volume aims to serve as a guide for researchers, educators, policymakers, and practitioners dedicated to shaping a smarter, more inclusive, and future-ready world.

Chapter 1
Navigating Computational Intelligence Approaches for Efficient Portfolio Selection in VUCA and BANI Financial Landscape

Raghib Noman
Aligarh Muslim University, India

Mohammad Shahid
https://orcid.org/0000-0002-4081-5963
Aligarh Muslim University, India

Mahfooz Alam
https://orcid.org/0000-0003-0668-9796
G.L. Bajaj Institute of Technology and Management, India

Faraz Hasan
https://orcid.org/0000-0001-5666-3813
GITAM University, Hyderabad, India

Maria Lapina
https://orcid.org/0000-0001-8117-9142
North Caucasus Federal University, Russia

DOI: 10.4018/979-8-3373-7922-7.ch001

ABSTRACT

The complexity and unpredictability inherent in efficient markets, intensified by VUCA, and BANI environments, demand intelligent, flexible, and adaptive strategies for effective portfolio selection. Traditional optimization techniques regularly struggle to account for the dynamic behavior and rapid fluctuations inherent in such volatile conditions. In these contexts, computational intelligence techniques integrating evolutionary algorithms, neural networks, fuzzy logic, and hybrid approaches have emerged as an effective tool to tackle the complex portfolio selection and evaluates their performance using both historical and real stock market data. By adapting to actual data and predictive algorithms, the proposed framework, dependent on computational intelligence, supplying enhanced decision-making amid uncertainty. This work contributes to the growing body of intelligent financial analytics literature by presenting a robust, data-driven decision-support system designed to navigate the complexities of modern financial markets shaped by VUCA and BANI forces.

1. INTRODUCTION

The terms VUCA (Volatility, Uncertainty, Complexity, and Ambiguity) and BANI (Brittle, Anxious, Nonlinear, and Incomprehensible) have been used in recent years to characterize the increased unpredictability, complexity, and volatility that have been a defining characteristic of the global financial landscape. These frameworks represent the substantial shifts in market behavior brought about by factors such as geopolitical instability, investor psychology, economic shocks, and quick technical advancements. In such a context, traditional portfolio selection methods, which mostly rely on historical data and static assumptions, are often unable to manage the dynamic nature of stock markets. Portfolio selection, where the goal is to build the best feasible asset mix to maximize returns and minimize risk, is an essential challenge in financial decision-making. In real-world VUCA (Taskan, Junça-Silva, & Caetano, 2022) and BANI (Mieszajkina & Ostapińska, 2024) scenarios, conventional approaches' linearity and market efficiency assumptions, including Markowitz's Modern Portfolio Theory (MPT), usually fall short. Therefore, there is a need for more adaptable, clever, and responsive techniques to analyze high-dimensional data, spot complex patterns, and respond to abrupt changes in market conditions.

considering all of these constraints, researchers and practitioners are using modern CI approaches that supersede conventional techniques. Using computational intelligence (CI) is one promising approach that provides strong tools to handle the challenges of contemporary financial markets. the foundation of financial decision-making is the selection of stock market portfolios, which aims to allocate assets in a way that minimizes risks and maximizes profits. stock market portfolio selection is a cornerstone of financial decision-making, aimed at allocating assets to optimize returns while mitigating risks. traditional methods, such as the Markowitz modern portfolio theory (MPT), often fail to capture the complexity of real-world financial systems, where uncertainties, nonlinearities, and dynamic market conditions prevail. Computational Intelligence (CI) approaches provide a paradigm shift, leveraging techniques such as genetic algorithms, neural networks, and swarm intelligence to handle these complexities effectively. This paper examines the integration of CI techniques in portfolio selection, offering insights into their methodologies, applications, and practical benefits (Markowitz, 1952),

Computational intelligence (CI) is a field of study within artificial intelligence (AI) that develops techniques to enable machines to learn, adapt, and make intelligent decisions. It can be defined as "Computational Intelligence, a set of nature-inspired computational methodologies and approaches to address complex problems associated with real-world applications to which traditional approaches are ineffective or infeasible." (Sugumaran, Geetha, Manjula, & Gopal, 2017). Computational Intelligence (CI) encompasses any computational paradigm that is motivated by biological, natural, or linguistic factors (James Bezdek, IEEE CIS). Examples of such paradigms include neural networks, fuzzy systems, evolutionary computation, swarm intelligence, and nature-inspired metaheuristic methods.

This study aims to offer CI-based techniques for efficient portfolio selection, highlighting versatility and resilience in BANI and VUCA-driven markets. Real-time data processing, adaptive decision-making, and predictive analytics are all featured in the proposed techniques for enhancing risk-adjusted returns. Using real stock market data, this study demonstrates that CI techniques are superior to traditional methods, particularly when it comes to maintaining performance during times of market volatility.

This paper contributes to the growing body of research at the nexus of computational intelligence and portfolio selection by addressing the shortcomings of conventional portfolio selection techniques in VUCA and BANI environments. It offers an in-depth analysis of CI approaches like swarm intelligence, neural networks, and evolutionary algorithms, and suggests incorporating them into frameworks of portfolio selection. The study incorporates CI techniques capable of processing real-time, high-dimensional data and adapting to sudden shifts in the market. The research shows how effective these techniques are at managing volatility and improving risk-adjusted returns by using actual stock market data.

The rest of the paper is organized in the following sections: Background and related work are discussed in Section II. A detailed look into the applications of CI algorithms in the field of finance is provided in Section III. The strengths and advantages of CI are mentioned in Section IV. Challenges and issues are thoroughly covered in Section V. Future research directions are presented in Section VI, and the paper concludes with Section VII.

2. BACKGROUND AND RELATED WORK

Portfolio selection in stock markets is crucial for financial decision-making, aiming to maximize returns while mitigating risks. Traditional methods like Markowitz's Modern Portfolio Theory (MPT), CAPM, and others fail to address portfolio optimization challenges. That is why Computational intelligence techniques are integrated to optimize Portfolio aims to balance risk and return. Computational Intelligence (CI), which is a part of AI, uses nature-inspired methodologies like genetic algorithms, neural networks, and swarm intelligence to solve complex optimization problems, making portfolio optimization easier for investors.

2.1 Portfolio Selection Problem

The portfolio selection problem is concerned with finding the mix of financial assets (stocks, bonds, etc.) that will minimize the risk and maximize the return while accounting for budgetary constraints, risk tolerance, and diversification, which is the challenge. Markowitz's Mean-Variance Optimization is one of the methods commonly used in conventional finance. Although this is no longer the case in the current world, these methods usually assume that markets are stable and predictable.

2.1.1. VUCA framework for portfolio selection problem

The incorporation of the VUCA framework into portfolio selection provides a more resilient and flexible approach to overcoming the drawbacks of conventional methods. Each element of the VUCA model contributes a unique viewpoint that makes portfolio selection more responsive and realistic. VUCA-informed strategies greatly enhance risk management and more robust return optimization by recognizing contemporary marketplaces' complex and uncertain character. The elements of the VUCA framework, volatility, uncertainty, complexity, and ambiguity, can all be addressed using certain techniques and instruments that improve the portfolio selection procedure as a whole.

- **Volatility** refers to regular and erratic changes in stock prices. Real-time volatility metrics like standard deviation, beta, or Value-at-Risk (VaR) can be incorporated into portfolio selection models to address this issue and dynamically modify asset weights. During market volatility, adaptive risk management techniques can be used to reduce the amount of capital allocated to extremely volatile assets, reducing the possibility of drawdowns.

- **Uncertainty** refers to the absence of trustworthy information regarding future occurrences, including shifts in interest rates, earnings announcements, or geopolitical events. This can be addressed by using fuzzy logic systems that accept imprecise or partial input, or by using robust optimization techniques that take into consideration many market conditions. Macroeconomic indicators and sentiment research can also be used to measure and control market uncertainty.
- **Complexity** arises from the interaction of different market elements that affect asset prices, such as sectoral relationships and international economic policy. High-dimensional financial data can be examined for nonlinear relationships and patterns using computational intelligence (CI) approaches, which include swarm optimization, neural networks, and evolutionary algorithms. Moreover, complex inter-asset relationships and systemic risks can be captured and understood through the use of neural networks.
- **Ambiguity** is the problem of contradicting signals or ambiguous data that can cause disparate interpretations of the same data. This can be managed by applying CI approaches such as fuzzy logic systems, which integrate several models or points of view, to produce a more solid and well-rounded investment decision. By providing a more comprehensive assessment of an asset's value, the inclusion of these techniques can reduce ambiguity.

By incorporating the VUCA model into the process of selecting a portfolio, investors could develop strategies that are not only creative and data-driven but also more resilient to the stock market's constantly changing and frequently volatile environment. More resilient, adaptable, and forward-thinking portfolio management systems that can flourish in challenging and unstable financial environments are the result of this strategy (Iryani & Yulianto, 2023).

2.1.2. BANI framework for portfolio selection problem

Integrating the BANI (Brittle, Anxious, Nonlinear, Incomprehensible) framework into a portfolio selection provides a fresh and perceptive perspective for negotiating today's extremely volatile stock market environment. BANI focuses on the emotional and structural weak points in the systems, whereas VUCA emphasizes volatility and ambiguity. This makes it especially essential for developing investment strategies that can withstand systemic shocks and psychological strain.

- **Brittleness** describes how systems that appear robust can break down under pressure in the context of financial markets when choosing a portfolio; this emphasizes the necessity of avoiding over-optimization and over-concentration. A brittle portfolio may seem effective in a typical setting, but it can go horribly wrong when markets change. In order to combat this, investors can use risk-spreading strategies, scenario analysis, and stress testing to create portfolios that emphasize resilience and diversification. Tools for computational intelligence (CI), such as evolutionary algorithms, can be used to model different crisis situations and direct the choice of asset mixes that preserve stability even in the face of extreme circumstances.
- **Anxiety** is a reflection of the overreactions and elevated emotional reactions that are frequently seen in markets, particularly during times of crisis. Irrational trading activity and herd mentality may result from this emotional instability. To solve this, portfolio models can use sentiment analysis and behavioral finance indicators from news, social media, and financial forums. During times of widespread worry, machine learning models trained on these signals can proactively modify the portfolio and identify changes in market sentiment, lowering exposure.
- **Nonlinearity** describes how minor adjustments can have disproportionately significant and unanticipated results. Traditional models frequently presume that the linkages between market forces and assets are linear, but real-world markets do not work like that. Since neural networks and nonlinear regression models can capture intricate, nonlinear correlations in stock prices, risk factors, and macroeconomic variables, they are very helpful in this situation. The portfolio can adjust to nonlinear market behavior and hidden dependencies that might otherwise go unnoticed by including these technologies.
- **Incomprehensibility** describes the overwhelming intricacy and ambiguity of comprehending market movements, particularly when several interconnected factors are at play. This calls for the employment of fuzzy logic systems and explainable AI (XAI), which can handle unclear or insufficient data and still produce outputs that can be understood. Even when conventional analysis fails, investors can still make well-informed judgments by employing models that strike a compromise between predictive strength and transparency (Ponomarenko & Iastremska, 2024).

The BANI framework supports the creation of investment strategies that are not only effective but also structurally sound, psychologically aware, and sensitive to hidden dynamics. It emphasizes how crucial it is to create portfolios that can withstand unexpected shocks, behavioral volatility, nonlinear market reactions, and analytical overload, and even prosper in the face of these challenges. Portfolios that are BANI-aware offer a progressive method of handling investments in the ever-changing and more precarious financial sector. These paradigms make traditional portfolio models inadequate, as they cannot adapt quickly or capture hidden patterns in a nonlinear, chaotic market environment.

2.2 Computational Intelligence Approaches in Finance

CI approaches have gained traction due to their ability to process large datasets, adapt to changing environments, and optimize non-convex objective functions. Common CI methods in finance include:

Neural Networks: Machine learning methods, such as neural networks and support vector machines, have been proven to be more accurate than economic methods in predicting index tracing problems. The forecasting effect of the market is influenced by factors such as market maturity, input variables, forecasting time, and forecasting methods (Holland, 1975). Hybrid models, such as principal component analysis (PCA) and support vector machine (SVM), have shown better performance than single models. Researchers have also used various methods to predict European and American stock markets, with PCA achieving 57% accuracy (Pang, Zhou, Wang, Lin, & Chang, 2020).

Evolutionary Algorithms: An evolutionary algorithm is a type of optimization algorithm that is inspired by the process of natural evolution. It is used to find approximate solutions to optimization problems, such as finding the minimum or maximum of a function. It involves random initialization of potential solutions and modification using rules mimicking natural selection. EAs are adept at handling complex constraints, enabling global optimization, adapting to portfolio optimization problems, and being robust to data noise and uncertainty, making them useful for financial models in portfolio optimization and risk management (Pytel, 2020).

Fuzzy Logic: Engineers and managers often face decision-making challenges due to vague and difficult-to-quantify criteria. Fuzzy logic, a superset of conventional logic, can help in combining these criteria with known quantitative ones to make the best possible alternative. This method is particularly useful when the values of the decision criteria are uncertain. For example, in the selection of a disposal site for hazardous waste material, factors like transportation, surface water quality, groundwater quality, and aesthetics must be combined with known quantitative values to ensure the best possible alternative (Hardy, 1995).

Swarm Intelligence: Swarm intelligence algorithms, such as ACO, PSO, AFS, BFO, and ABC, have revolutionized complex optimization problems. They solve nonlinear design problems in real-world applications across different domains, emphasizing finance in this paper. They are crucial in finance for resource allocation, risk management, and investment strategy enhancement. These algorithms offer fast and reliable solutions for various optimization problems. Traditional methods like linear programming and heuristics have been effective in portfolio optimization and derivative pricing. However, the dynamic nature of financial markets calls for more intelligent methods (Majka, n.d.).

2.3 Related Studies

Numerous studies have demonstrated the efficacy of CI approaches in portfolio optimization. For example, hybrid models combining neural networks and GAs have shown superior performance in capturing nonlinear relationships and generating diversified portfolios. Portfolio Optimization (PO) is a popular method for selecting the best investor portfolio. Heuristic techniques are often used to minimize portfolio risk and maximize expected return. A hybrid approach using a Nonlinear Neural Network and Genetic Algorithm is proposed to solve the Cardinality Constraint Portfolio Optimization Model. Tested on Istanbul Stock Exchange data and publicly available datasets, the hybrid approach offers more viable outcomes than current strategies (Yaman & Dalkılıç, 2021).

Table 1.

References	*Algorithm*	*Idea*	*Strengths*	*Weaknesses*
(Arzanlou & Sardroud, 2024)	Particle Swarm Optimization (PSO)	Uses swarm intelligence to explore solution spaces	-Effective for nonlinear optimization problems	**- May converge prematurely**
(Thakkar & Chaudhari, 2021)	Particle Swarm Optimization (PSO)	A population-based optimization inspired by bird behavior	-Simple to implement	**- Requires fine**
(Zhang, 2020)	Particle Swarm Optimization (PSO)	Mimics social behavior of birds for optimization	- Convergence guaranteed	**-tuning of parameters**
(Chen & Long, 2023)	Particle Swarm Optimization (PSO)	Share information among particles for better search	-Fast initial convergence	-Difficult to escape local optima
(Morris & Comeau, 2020)	Artificial Neural Networks (ANN)	Learns from historical financial data for predictions	Handles nonlinear relationships	- Requires large datasets

continued on following page

Table 1. Continued

References	*Algorithm*	*Idea*	*Strengths*	*Weaknesses*
(Kurani, Doshi, Vakharia, & Shah, 2023)	Artificial Neural Networks (ANN)	Model complex relationships in financial data for forecasting and optimization	- High accuracy in prediction tasks	- High computational cost
(Dube, Nzimande, & Muzindutsi, 2023)	Artificial Neural Networks (ANN)	Mimics the human brain for pattern recognition	- Capable of learning complex patterns	- Prone to overfitting
(Verma, 2024)	Artificial Neural Networks (ANN)	Automatically detects features without manual input	-Performs well in dynamic environments	-Difficult to interpret internal workings (black box)
(López de Prado, Simonian, Fabozzi, & Fabozzi, 2024)	Machine Learning Algorithms	Learns patterns for forecasting and classification	Works well with complex, nonlinear relationships	Requires massive volumes of data
(Gao et al., 2024)	Machine Learning Algorithms	Supports supervised and unsupervised learning modes	Good scalability with big datasets	Interpretation of results can be challenging
(Gogas & Papadimitriou, 2021)	Machine Learning Algorithms	Automates insights from structured and unstructured data	Can handle high-dimensional input features	Models need continuous retraining
(Aziz, Dowling, Hammami, & Piepenbrink, 2022)	Machine Learning Algorithms	Integrates domain knowledge through feature engineering	Useful in real-time decision-making	Prone to bias if the training data is skewed
(Muganda & Kasamani, 2023)	Genetic Algorithms (GAs)	Applies Darwinian principles to evolve solutions	Excellent for global search in large spaces	Computationally demanding due to population updates
(Brabazon, Kampouridis, & O'Neill, 2020)	Genetic Algorithms (GAs)	Works well for discrete and continuous variables	Adapts well to changing environments	Convergence can be slow in complex landscapes
(Contreras et al., 2024)	Genetic Algorithms (GAs)	Uses crossover and mutation to generate new solutions	Effective in multi-objective optimization	May require problem-specific parameter tuning
(Badakhshan & Bahadori, 2024)	Genetic Algorithms (GAs)	Avoids gradient calculations, works with discontinuous functions	Handles multiple constraints flexibly	Randomness can affect the repeatability of results

3. COMPUTATIONAL INTELLIGENCE TECHNIQUES AND THEIR APPLICATIONS IN FINANCE

3.1 Neural Networks for Market Forecasting

Neural networks (NNs), particularly deep learning models, are used to predict stock prices and market trends by analyzing historical data. These predictions serve as inputs for portfolio optimization models. NN is effective in predicting volatility and non-linear stock market prices due to its learning, mapping, generalizing, and self-organizing characteristics. NN is the most common architecture used in stock market forecasting, moving information from input to hidden and output layers. Researchers have used three layers of feedforward NN for IBM's daily stock price prediction (Vui, Soon, On, Alfred, & Anthony, 2013).

3.2 Evolutionary Algorithms for Portfolio Optimization

Evolutionary algorithms (EAs), such as GAs and Particle Swarm Optimization (PSO), are employed to solve multi-objective portfolio optimization problems. PSO has been used in various domains to solve complex optimization problems, emphasizing finance in this study. Its convergence ability allows it to be integrated into a wide range of applications. PSO has been used to optimize and predict various financial products, including technical indicators, stock trading systems, trade credit financing, and generalized type demand. PSO has been used to optimize selling prices, replenishment numbers, and schedules along with fluctuating demand as well as for pricing and time-varying demand and cost (Thakkar & Chaudhari, 2021).

3.3 Fuzzy Logic for Credit Decision-Making

The fuzzy logic theory is based on fuzzy sets, a generalization of classical set theory. A fuzzy set is defined by a membership function that allows every element to represent a different "grade of membership," which is subjective and based on context. fuzzy logic applied to various fields, such as decision making, social policy, and medical sciences, show a wide range of applications. Many researchers have applied fuzzy logic to finance problems, aiming to define the field as applying economic principles to decision-making under uncertainty. Fuzzy logic systems handle imprecise and uncertain market conditions, enabling better decision-making. Fuzzy rule-based systems classify assets into categories based on risk and return profiles (Sanchez-Roger, Oliver-Alfonso, & Sanchís-Pedregosa, 2019).

3.4 Hybrid Approaches for Derivatives Pricing

Traders and investors use derivatives for hedging investments and executing leveraged trading opportunities. Accurate pricing is crucial, with the Black-Scholes formula being the most common technique. Monte Carlo simulation, finite difference methods, and other CI techniques are utilized to enhance the accuracy, efficiency, and reliability in finding out the derivative prices. Deep learning and machine learning are two examples of computational intelligence techniques, mostly, applied in price forecasting (Ashraf et al., 2025).

3.5 Natural Language Processing for Sentiment Analysis

Sentiment analysis is a crucial tool in the digital age, enabling businesses to analyze vast textual data like customer reviews and social media comments. Utilizing natural language processing (NLP) techniques, it categorizes opinions as positive, negative, or neutral, providing valuable feedback on products, services, or brands. This analysis, powered by algorithms like Naive Bayes, SVM, and RNN, is essential for decision-making (Jim et al., 2024).

3.6 Expert Systems for Financial Advisory Services

An expert system is one of the important approaches of CI used in different domains specifically used for financial advisory services. It simulates human expertise and decision-making abilities in specific domains, primarily used in industries like healthcare, finance, manufacturing, etc., to provide reliable, data-driven solutions to complex problems. Financial experts possess valuable knowledge that is not found in literature or other sources. Expert systems in finance provide financial advisory services, including budget suggestions, risk evaluation, parameter addition, and ensuring important financial service parameters (Nedović & Devedžić, 2002).

3.7 Machine Learning for Trading Strategies

Stock trading strategies require machine learning applications for significant commercial yields in finance, drawing the attention of both economists and computer scientists. Machine learning models can analyze large volumes of financial data, identifying patterns and trends that humans might miss. They are able to forecast future market trends, spot potential investments, and assist with profit and loss accounting. Continuous investment strategy optimization is made possible by machine learning models that have been trained on previous data and are able to recognize intricate patterns and adjust to new data (Kwon & Lee, 2024; Jiao & Jakubowicz, 2017).

Figure 1.

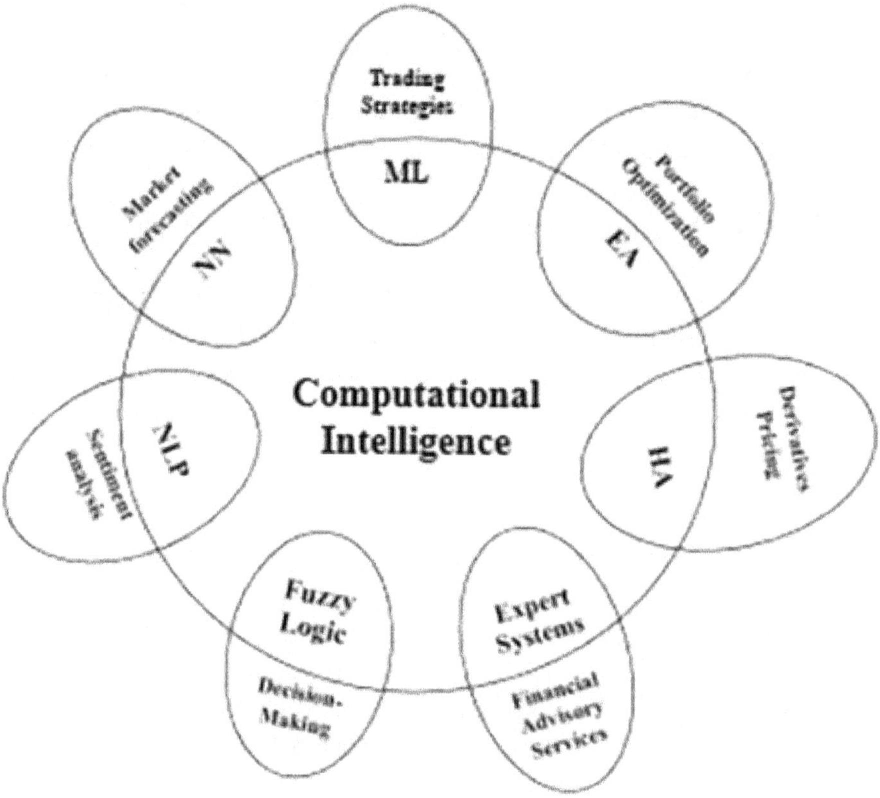

4. STRENGTHS AND ADVANTAGES

Here are some of the advantages of Computational Intelligence (CI) techniques in the field of finance. Let's expand each of them in detail:

Adaptability: CI models are adaptable, making them useful in dynamic environments like stock markets and weather forecasting. They continuously learn from new data, ensuring portfolio optimization under varying market conditions, such as rebalancing portfolios in response to geopolitical events or economic policy changes.

Efficiency: CI techniques optimize multi-objective problems quickly. Efficiency in Computer Intelligence (CI) techniques involves solving complex problems quickly, such as optimizing supply chain management, energy management, and healthcare. Companies like Amazon use CI to manage vast distribution networks, balance supply and demand in smart grids, and suggest accurate treatment plans (Cavalcante, Brasileiro, Souza, Nobrega, & Oliveira, 2016).

Enhance Business Value: CI-powered algorithms can enhance business value by analyzing customer data to tailor financial services, enabling proactive decision-making through predictive analytics, and accurately assessing risk by identifying complex patterns and anomalies in financial data, thereby enhancing the overall value of financial services.

Offers Accuracy: CI algorithms offer accuracy through automated data analysis, precision in financial forecasting, and optimized trading strategies, reducing human error and executing trades with speed and precision because of advanced techniques like neural networks and Evolutionary Algorithms, etc.

Offers Fraud Detection: CI offers fraud detection through anomaly detection and behavioral analytics, identifying unusual patterns in financial transactions and preventing fraudulent activities by analyzing user behavior.

24/7 Customer Interaction: CI-powered chatbots and virtual assistants offer 24/7 customer support and personalized financial advice, leveraging customer data analysis for enhanced customer experience (Agarwal, 2019).

Risk Mitigation: Computational intelligence (CI) has significantly impacted the financial industry by enhancing risk management through techniques like machine learning, neural networks, fuzzy logic, genetic algorithms, and expert systems, thereby improving financial performance and decision-making (Giudici, Centurelli, & Turchetta, 2024).

Portfolio Diversification: Computational Intelligence (CI) techniques, such as machine learning, GWO, and PSO, can significantly enhance portfolio diversification in finance by analyzing vast data, identifying optimal asset allocations, and mitigating risk (Ahmad & Shahid, 2022).

Figure 2.

5. CHALLENGES AND ISSUES

Data Quality: A number of problems, including inconsistency and incompleteness, can affect financial datasets. Getting hold of comprehensive records or restricted access to pertinent databases might make it difficult to obtain high-quality historical financial data. Furthermore, the efficacy of computational intelligence techniques employed for portfolio selection may be impacted by biases or inaccuracies in financial data (Pun, Wang, & Yan, 2023).

Computational Cost: Computational cost refers to the resources required for processing and analyzing large datasets, particularly in financial applications like High-Frequency Trading, Risk Management, Portfolio Optimization, and Fraud Detection, which involve complex calculations and simulations.

Overfitting: Overfitting in financial models can lead to poor trading decisions, underestimating rare events, and inaccurate credit scoring. It can also be prone to overfitting in risk management, causing substantial financial losses. Therefore, it is crucial to avoid overfitting in financial models (Cao, 2022).

Ethical Considerations: AI algorithms can perpetuate biases in training data, leading to unfair outcomes. It's crucial to ensure model transparency and explainability to avoid unintended consequences (Tariq & Sergio, 2025).

Market Volatility and Uncertainty: Financial markets are volatile, requiring robust CI models to adapt to sudden changes. Poorly calibrated models can lead to significant financial losses.

Regulatory Challenges: The rapid development of computational intelligence and machine learning raises regulatory concerns, such as model governance, risk management, and transparency.

Bias in Algorithms: Bias in algorithms can lead to skewed results, damaging financial institutions' reputations, and regulatory issues. To avoid bias, financial institutions should use diverse datasets, regularly test and audit algorithms, and make adjustments for unbiased decisions (Ntoutsi et al., 2020).

Model Drift: A model's performance can be affected when real-world conditions or data patterns have changed, leading to inaccurate results. In finance, dynamic markets can cause models to make poor predictions or decisions due to outdated assumptions. CI techniques might be effective at predicting stock prices but may become less accurate or harmful to decision-making due to changes in investor behavior during an economic crisis (Patchipala, 2023).

Data Privacy and Security: Data privacy is crucial in finance, as it protects sensitive financial and personal information from theft, misuse, or unauthorized access. This is especially important for CI models, which require large amounts of data for accurate predictions and decisions. Inadequate protection can lead to cyberattacks or data breaches, posing risks to customers and financial institutions. Failure to protect data can result in trust issues, legal penalties, and financial losses, such as direct financial theft or damage to a company's reputation (Chanal, Kakkasageri, & Manvi, 2021).

Moral Hazard in Automation: Automated systems can lead to overconfidence in technology, leading to careless or risky behavior. A financial institution using a CI-powered trading algorithm might not closely monitor it, leading to potential financial losses. This problem arises from overconfidence in technology, lack of oversight, and amplified risks due to automated systems operating at scale.

Interpretability and Explainability: Interpretability refers to how well humans can understand how a decision-making model makes certain decisions, which is crucial in finance. Complex models like NN can act like "black boxes," reducing trust, regulatory compliance, and error detection. Poor interpretability can lead to issues like loan rejection, where customers want to know why, and regulatory penalties if the decision-making process is unclear.

Adversarial Attacks: Adversarial attacks trick machine learning models by introducing adversarial examples, reducing their robustness, and raising security concerns. In finance, these attacks include invasion attacks, data poisoning, model inversion, and membership inference. Evasion attacks alter input data, allowing fraudulent activities, while data poisoning manipulates training datasets, and model inversion exposes sensitive financial information (Fursov et al., 2021).

6. FUTURE DIRECTIONS

As financial markets continue to evolve within the frameworks of BANI (Brittle, Anxious, Nonlinear, Incomprehensible) and VUCA (Volatility, Uncertainty, Complexity, and Ambiguity), there is an increasing demand for more resilient and intelligent portfolio optimization strategies. The findings point to several intriguing directions for additional research. Here are some examples of computational intelligence research directions in the finance industry:

Explainable Artificial Intelligence: Explainable Computational Intelligence (CI) can significantly improve the finance sector by enhancing transparency, trust, and decision-making. CI models provide clear reasons for decisions, building trust among clients and investors. They can be used in credit scoring, investment decisions, and regulatory compliance. CI models can also help meet regulatory standards for anti-money laundering (AML) and risk assessment. Trust and accountability are built among clients, investors, and regulators. CI models can also enhance decision-making by providing actionable insights and rationales for predictions, such as fraud detection and market predictions. CI promotes ethical AI use by ensuring fair, accountable, and transparent models. It can help identify and mitigate biases in financial decision-making processes, promoting fairness and inclusivity. Examples of CI applications include loan approval, portfolio management, and fraud detection (Weber, Carl, & Hinz, 2024).

Integration of ESG Factors: Advances in data analytics and CI are improving the integration of ESG factors into investment decisions. Global regulations and growing awareness of sustainability issues are driving the integration of ESG factors into financial reporting and investment practices. This transformation is promoting sustainable and ethical business practices, leading to a more responsible and resilient economy.

Quantum Computing: Quantum computing has the potential to revolutionize the financial sector by offering unprecedented computational power and speed. It can enhance portfolio optimization, risk management, option pricing, high-frequency trading, cryptocurrency and blockchain technology, fraud detection, and machine learning and AI models. Quantum computing can analyze vast amounts of data, identify optimal asset combinations, and provide more accurate simulations and predictions. It can also improve the pricing of financial derivatives, making them more precise and accurate. Quantum algorithms can handle random noise in market data more effectively, leading to better trading strategies. This technology is expected to make the financial sector more efficient and secure (Egger et al., 2020).

Real-Time Decision-Making: Real-time decision-making is revolutionizing the finance sector by enabling quicker, more accurate, and more informed decisions. It is crucial for high-frequency trading, fraud detection, risk management, personalized financial services, portfolio management, and customer support. Real-time analytics help identify opportunities, address suspicious activities, and enable corrective actions. Future prospects include AI and machine learning, IoT integration, and blockchain technology, which will enhance real-time decision-making capabilities and improve efficiency, risk management, and customer experiences (Rangasamy, Rajamohan, Janardhanan, & Manu, 2023).

Real-Time Adaptive Models Using Online Learning: Real-time flexibility is essential for efficient portfolio selection in a volatile, uncertain, complex, and ambiguous VUCA/BANI environment where market circumstances can change quickly. Financial models are made more flexible by including online learning algorithms, which update projections in real time as new data becomes available. Online learning, in contrast to conventional static models, allows for real-time reactivity, enabling investment strategies to dynamically adjust to shifting market patterns. During the times of uncertainty, this constant adjustment helps to maximize profits and minimize risks. Investor success in the quick-paced financial markets can be enhanced by using such adaptive models to make better, more timely judgments.

Hybrid and Ensemble-Based CI Models: In portfolio optimization, hybrid and ensemble-based computer intelligence (CI) models present encouraging developments. Complex market dynamics can be better captured by hybrid systems that combine methods like neural networks, fuzzy logic, and evolutionary algorithms. These models leverage the advantages of each technique: fuzzy logic for decision-making, neural networks for market forecasting, and evolutionary algorithms for portfolio optimization. Furthermore, these techniques increase generalization and decrease the chance of overfitting by integrating outputs from several models, which increases prediction accuracy. Together, these approaches create more robust, adaptive, and flexible systems for portfolio selection in dynamic financial environments.

Robustness under Regime Shifts and Black Swan Events: Traditional portfolio models face serious difficulties as financial markets frequently shift between regimes, such as bull, bear, or crisis times. In order to improve decision-making in diverse market situations, future research can concentrate on creating regime-switching models that dynamically adjust to these changes. Assessing portfolio vulnerabilities in harsh circumstances can also be aided by developing stress-testing frameworks that can replicate infrequent but significant Black Swan occurrences, such as financial crises or geopolitical shocks. By incorporating these tactics into portfolio management, investors can better negotiate volatility and unpredictability and ensure higher resilience. Such models seek to protect assets and preserve stability by taking regime changes and unpredictable disruptions into consideration. This will ultimately result in more resilient and trustworthy financial strategies when unexpected market turbulence occurs.

Decentralized and Blockchain-Enabled Portfolio Management: With the rapid rise of decentralized finance (DeFi), future research in portfolio management may explore the integration of blockchain technology for enhanced transparency, security, and automation. Decentralized and blockchain-enabled portfolio optimization can leverage smart contracts to autonomously execute trades, rebalance portfolios, and manage assets based on triggers generated by CI models. This approach allows for real-time decision-making in a trustless, immutable environment, reducing reliance on intermediaries and enhancing operational efficiency. By embedding CI-driven techniques into blockchain systems, investors can benefit from automated, transparent, and tamper-proof portfolio management. This fusion of DeFi and CI opens new possibilities for creating adaptive, secure, and decentralized investment ecosystems in the evolving financial landscape (Noman, Shahid, & Alam, 2023).

7. CONCLUSION

Traditional portfolio selection strategies frequently fail to provide the flexibility and resilience needed for wise investment decisions in an era dominated by the uncertainty and instability of VUCA (volatility, uncertainty, complexity, and ambiguity) and BANI (Brittle, Anxious, Nonlinear, and Incomprehensible) environments. In order to create flexible, data-driven strategies for portfolio optimization, this study has investigated the revolutionary potential of computational intelligence (CI) techniques such as neural networks, evolutionary algorithms, fuzzy logic, and hybrid models. We have shown how models may learn from non-linear patterns, react dynamically to market swings, and make sound decisions in the face of uncertainty by utilizing CI. These techniques facilitate the integration of various financial indicators, sentiment data, and changing investor behavior, as well as the improved management of intricate risk-return trade-offs. When compared to conventional methods, the proposed CI approach performs better in terms of resilience, flexibility, and risk-adjusted returns.

Finally, this study adds to the expanding corpus of research at the nexus of financial engineering and artificial intelligence. It highlights the significance of smart, adaptable, and forward-looking investing methods. The future of portfolio management systems will be greatly influenced by developments in computational intelligence, especially in the areas of multi-objective optimization, explainable AI, and real-time learning, since markets continue to change in unpredictable ways.

REFERENCES

Agarwal, P. (2019, March). Redefining banking and financial industry through the application of computational intelligence. In *2019 Advances in Science and Engineering Technology International Conferences (ASET)* (pp. 1-5). IEEE. DOI: 10.1109/ICASET.2019.8714305

Ahmad, G., & Shahid, M. (2022, October). Towards portfolio selection in stock markets using grey wolf optimization approach. In *2022 International Conference on Data Analytics for Business and Industry (ICDABI)* (pp. 756-760). IEEE. DOI: 10.1109/ICDABI56818.2022.10041457

Arzanlou, A., & Sardroud, J. M. (2024). Enhancing Project Performance: Particle Swarm Optimization for Optimal Budget Allocation and Maintenance Scheduling. *KSCE Journal of Civil Engineering*, *28*(5), 1635–1644. DOI: 10.1007/s12205-024-1348-1

Ashraf, Z., Shamim, M. S., Noman, R., Shahid, M., Ahmad, F., & Sami, L. (2025). Constrained Portfolio Optimization Using Hybrid Nature-Inspired Metaheuristic Algorithm in Stock Market. In *Nature-Inspired Optimization Algorithms for Cyber-Physical Systems* (pp. 53–74). IGI Global Scientific Publishing.

Aziz, S., Dowling, M., Hammami, H., & Piepenbrink, A. (2022). Machine learning in finance: A topic modeling approach. *European Financial Management*, *28*(3), 744–770. DOI: 10.1111/eufm.12326

Badakhshan, E., & Bahadori, R. (2024). A simulation-based optimization model for balancing economic profitability and working capital efficiency using system dynamics and genetic algorithms. *Decision Analytics Journal*, *12*, 100498. DOI: 10.1016/j.dajour.2024.100498

Brabazon, A., Kampouridis, M., & O'Neill, M. (2020). Applications of genetic programming to finance and economics: Past, present, future. *Genetic Programming and Evolvable Machines*, *21*(1-2), 33–53. DOI: 10.1007/s10710-019-09359-z

Cao, L. (2022). Ai in finance: Challenges, techniques, and opportunities. *ACM Computing Surveys*, *55*(3), 1–38. DOI: 10.1145/3502289

Cavalcante, R. C., Brasileiro, R. C., Souza, V. L., Nobrega, J. P., & Oliveira, A. L. (2016). Computational intelligence and financial markets: A survey and future directions. *Expert Systems with Applications*, *55*, 194–211. DOI: 10.1016/j.eswa.2016.02.006

Chanal, P. M., Kakkasageri, M. S., & Manvi, S. K. S. (2021). Security and privacy in the internet of things: computational intelligent techniques-based approaches. In *Recent trends in computational intelligence enabled research* (pp. 111–127). Academic Press. DOI: 10.1016/B978-0-12-822844-9.00009-8

Chen, X., & Long, Z. (2023). E-commerce enterprises financial risk prediction based on FA-PSO-LSTM neural network deep learning model. *Sustainability (Basel)*, *15*(7), 5882. DOI: 10.3390/su15075882

Contreras, R. C., Xavier da Silva, V. T., Xavier da Silva, I. T., Viana, M. S., Santos, F. L. D., Zanin, R. B., Martins, E. F. O., & Guido, R. C. (2024). Genetic Algorithm for Feature Selection Applied to Financial Time Series Monotonicity Prediction: Experimental Cases in Cryptocurrencies and Brazilian Assets. *Entropy (Basel, Switzerland)*, *26*(3), 177. DOI: 10.3390/e26030177

Dube, F., Nzimande, N., & Muzindutsi, P. F. (2023). Application of artificial neural networks in predicting financial distress in the JSE financial services and manufacturing companies. *Journal of Sustainable Finance & Investment*, *13*(1), 723–743. DOI: 10.1080/20430795.2021.2017257

Egger, D. J., Gambella, C., Marecek, J., McFaddin, S., Mevissen, M., Raymond, R., Simonetto, A., Woerner, S., & Yndurain, E. (2020). Quantum computing for finance: State-of-the-art and future prospects. *IEEE Transactions on Quantum Engineering*, *1*, 1–24. DOI: 10.1109/TQE.2020.3030314

Fursov, I., Morozov, M., Kaploukhaya, N., Kovtun, E., Rivera-Castro, R., Gusev, G., & Burnaev, E. (2021, August). Adversarial attacks on deep models for financial transaction records. In *Proceedings of the 27th ACM SIGKDD Conference on Knowledge Discovery & Data Mining* (pp. 2868-2878). DOI: 10.1145/3447548.3467145

Gao, H., Kou, G., Liang, H., Zhang, H., Chao, X., Li, C. C., & Dong, Y. (2024). Machine learning in business and finance: A literature review and research opportunities. *Financial Innovation*, *10*(1), 86. DOI: 10.1186/s40854-024-00629-z

Giudici, P., Centurelli, M., & Turchetta, S. (2024). Artificial Intelligence risk measurement. *Expert Systems with Applications*, *235*, 121220. DOI: 10.1016/j.eswa.2023.121220

Gogas, P., & Papadimitriou, T. (2021). Machine learning in economics and finance. *Computational Economics*, *57*(1), 1–4. DOI: 10.1007/s10614-021-10094-w

Hardy, T. (1995, March). Multi-objective decision-making under uncertainty-fuzzy logic methods. In *10th Computing in Aerospace Conference* (p. 1000). DOI: 10.2514/6.1995-1000

Holland, J. H. (1975). *Adaptation in Natural and Artificial Systems*. University of Michigan Press.

Iryani, I., & Yulianto, H. (2023). Artificial Intelligence (AI) of Financial in the VUCA Era: A Systematic Mapping Study. *Journal of Computer Networks. Architecture and High Performance Computing*, *5*(2), 398–413. DOI: 10.47709/cnahpc.v5i2.2201

Jiao, Y., & Jakubowicz, J. (2017, December). Predicting stock movement direction with machine learning: An extensive study on S&P 500 stocks. In *2017 IEEE International Conference on Big Data (Big Data)* (pp. 4705-4713). IEEE.

Jim, J. R., Talukder, M. A. R., Malakar, P., Kabir, M. M., Nur, K., & Mridha, M. F. (2024). Recent advancements and challenges of nlp-based sentiment analysis: A state-of-the-art review. *Natural Language Processing Journal*, 100059.

Kurani, A., Doshi, P., Vakharia, A., & Shah, M. (2023). A comprehensive comparative study of artificial neural network (ANN) and support vector machines (SVM) on stock forecasting. *Annals of Data Science*, *10*(1), 183–208. DOI: 10.1007/s40745-021-00344-x

Kwon, Y., & Lee, Z. (2024). A hybrid decision support system for adaptive trading strategies: Combining a rule-based expert system with a deep reinforcement learning strategy. *Decision Support Systems*, *177*, 114100. DOI: 10.1016/j.dss.2023.114100

López de Prado, M., Simonian, J., Fabozzi, F. A., & Fabozzi, F. J. (2024). Enhancing Markowitz's portfolio selection paradigm with machine learning. *Annals of Operations Research*, •••, 1–22.

. Majka, M. (n.d.). Particle Swarm Optimization in Finance.

Markowitz, H. (1952). Portfolio Selection. *The Journal of Finance*, *7*(1), 77–91.

Mieszajkina, E., & Ostapińska, K. (2024). Evaluation of the Impact of the BANI World on Small Business Innovation Activities. *European Research Studies*, *27*(Special Issue B), 272–288. DOI: 10.35808/ersj/3489

Morris, T., & Comeau, J. (2020). Portfolio creation using artificial neural networks and classification probabilities: A Canadian study. *Financial Markets and Portfolio Management*, *34*(2), 133–163. DOI: 10.1007/s11408-020-00350-8

Muganda, B. W., & Kasamani, B. S. (2023, June). Parallel programming for portfolio optimization: A robo-advisor prototype using genetic algorithms with recurrent neural networks. In *2023 International Conference on Intelligent Computing, Communication, Networking and Services (ICCNS)* (pp. 167-176). IEEE. DOI: 10.1109/ICCNS58795.2023.10193396

Nedović, L., & Devedžić, V. (2002). Expert systems in finance—A cross-section of the field. *Expert Systems with Applications*, *23*(1), 49–66. DOI: 10.1016/S0957-4174(02)00027-1

Noman, R., Shahid, M., & Alam, M. (2023, October). The Synergy of Blockchain and Artificial Intelligence in Fintech: Revolutionizing the Financial Landscape. In *2023 4th International Conference on Data Analytics for Business and Industry (ICDABI)* (pp. 425-430). IEEE.

Ntoutsi, E., Fafalios, P., Gadiraju, U., Iosifidis, V., Nejdl, W., Vidal, M. E., Ruggieri, S., Turini, F., Papadopoulos, S., Krasanakis, E., Kompatsiaris, I., Kinder-Kurlanda, K., Wagner, C., Karimi, F., Fernandez, M., Alani, H., Berendt, B., Kruegel, T., Heinze, C., & Staab, S. (2020). Bias in data-driven artificial intelligence systems—An introductory survey. *Wiley Interdisciplinary Reviews. Data Mining and Knowledge Discovery*, *10*(3), e1356. DOI: 10.1002/widm.1356

Pang, X., Zhou, Y., Wang, P., Lin, W., & Chang, V. (2020). An innovative neural network approach for stock market prediction. *The Journal of Supercomputing*, *76*(3), 2098–2118. DOI: 10.1007/s11227-017-2228-y

Patchipala, S.Surya Gangadhar Patchipala. (2023). Tackling data and model drift in AI: Strategies for maintaining accuracy during ML model inference. *International Journal of Science and Research Archive*, *10*(2), 1198–1209. DOI: 10.30574/ijsra.2023.10.2.0855

. Ponomarenko, V., & Iastremska, O. (2024). Determining the impact of Vuca-world and Bani-world on the activities of enterprises in the experience economy.

Pun, C. S., Wang, T., & Yan, Z. (2023). Data-Driven Distributionally Robust CVaR Portfolio Optimization Under A Regime-Switching Ambiguity Set. *Manufacturing & Service Operations Management*, *25*(5), 1779–1795. DOI: 10.1287/msom.2023.1229

Pytel, K. (2020). Hybrid multi-evolutionary algorithm to solve optimization problems. *Applied Artificial Intelligence*, *34*(7), 550–563. DOI: 10.1080/08839514.2020.1730631

Rangasamy, S., Rajamohan, K., Janardhanan, A. K., & Manu, K. S. (2023). The role of financial analytics in decision-making for better firm performance. In *Analytics Enabled Decision Making* (pp. 139–169). Springer Nature Singapore. DOI: 10.1007/978-981-19-9658-0_7

Sanchez-Roger, M., Oliver-Alfonso, M. D., & Sanchís-Pedregosa, C. (2019). Fuzzy logic and its uses in finance: A systematic review exploring its potential to deal with banking crises. *Mathematics*, *7*(11), 1091. DOI: 10.3390/math7111091

Sugumaran, V., Geetha, T. V., Manjula, D., & Gopal, H. (2017). Guest editorial: Computational intelligence and applications. *Information Systems Frontiers*, *19*(5), 969–974. DOI: 10.1007/s10796-017-9794-7

Tariq, M. U., & Sergio, R. P. (2025). Ethical Considerations and Social Impact of AI and Computational Intelligence in Society 5.0. In *Open AI and Computational Intelligence for Society 5.0* (pp. 225-248). IGI Global Scientific Publishing.

Taskan, B., Junça-Silva, A., & Caetano, A. (2022). Clarifying the conceptual map of VUCA: A systematic review. *The International Journal of Organizational Analysis*, *30*(7), 196–217. DOI: 10.1108/IJOA-02-2022-3136

Thakkar, A., & Chaudhari, K. (2021). A comprehensive survey on portfolio optimization, stock price and trend prediction using particle swarm optimization. *Archives of Computational Methods in Engineering*, *28*(4), 2133–2164. DOI: 10.1007/s11831-020-09448-8

Thakkar, A., & Chaudhari, K. (2021). A comprehensive survey on portfolio optimization, stock price and trend prediction using particle swarm optimization. *Archives of Computational Methods in Engineering*, *28*(4), 2133–2164. DOI: 10.1007/s11831-020-09448-8

Verma, R. (2024, June). Revolutionizing Derivative Market Predictions with Hybrid EGARCH-ANN Models: Embracing Machine Learning for Better Forecasting. In *2024 15th International Conference on Computing Communication and Networking Technologies (ICCCNT)* (pp. 1-5). IEEE.

Vui, C. S., Soon, G. K., On, C. K., Alfred, R., & Anthony, P. (2013, November). A review of stock market prediction with Artificial neural network (ANN). In *2013 IEEE international conference on control system, computing and engineering* (pp. 477-482). IEEE.

Weber, P., Carl, K. V., & Hinz, O. (2024). Applications of explainable artificial intelligence in finance—A systematic review of finance, information systems, and computer science literature. *Management Review Quarterly*, *74*(2), 867–907. DOI: 10.1007/s11301-023-00320-0

Yaman, I., & Dalkılıç, T. E. (2021). A hybrid approach to cardinality constraint portfolio selection problem based on nonlinear neural network and genetic algorithm. *Expert Systems with Applications*, *169*, 114517. DOI: 10.1016/j.eswa.2020.114517

Zhang, H. (2020). Optimization of risk control in financial markets based on particle swarm optimization algorithm. *Journal of Computational and Applied Mathematics*, *368*, 112530. DOI: 10.1016/j.cam.2019.112530

Chapter 2

Characteristics and Properties of the Information and Communication Environment and the Principles of Its Construction

Magomed Musaevich Abdurazakov
https://orcid.org/0000-0002-7979-0847
Russian Academy of Education, Russia

Magomed Gosengadzhievich Mukhidinov
https://orcid.org/0000-0003-4627-2995
Dagestan State Pedagogical University, Russia

Marina Georgievna Litvinova
Kuban State Medical University of the Ministry of Healthcare of the Russian Federation, Russia

Abulfat Gulam Oglu Palangov
https://orcid.org/0009-0004-4854-934X
Azerbaijan State Pedagogical University, Azerbaijan

DOI: 10.4018/979-8-3373-7922-7.ch002

ABSTRACT

This article examines the information and educational environment (IEE) in its structural and functional representation. The communicative nature of the IEE, its expression, implementation, and representation in its organization and functioning are the subject of this study and form the context for examining the structural components, corresponding properties, and characteristics of the IEE. It is emphasized that the communicative component is inherent in the IEE and subject-based learning, representing a conceptual feature of the IEE and a fundamental condition for its implementation in the context of the digital transformation of learning. Key here is the organization of communicative interaction between educational actors and the creation of effective mechanisms for knowledge transfer within any educational system, regardless of its organizational form: formal, non-formal, or informal education. The relevance of this study stems from the complex problems associated with the formation of this educational continuum.

INTRODUCTION

The information and educational environment (IEE) is a significant phenomenon in education and a crucial factor in its implementation and development. Consequently, IEE has become a key research focus in pedagogy and subject-specific methods, exploring the issues of its subject identification, formation, and development alongside the development of education itself and its informatization (IE), subject-specific learning in the context of interconnectedness and mutual influence, and the mutual realization of multiple direct and feedback loops. Numerous studies, articles, and monographs have been devoted to the topic of IEE (Elyakov, 2012; Korotenkov Yu. G., 2010; Pisanets, 2020; Studiopedia, 2014), technological developments that ensure the formation and successful functioning of IEE (Zenkina S. V., 2007) and components of the methodological system of subject teaching.

With the advancement of technologies and intelligent systems in the modern information society, as well as the deepening digital transformation of the educational sector itself, the information and educational environment (IEE) is rapidly evolving and expanding its functionality. It is evolving toward creating conditions for the effective organization of students' learning and cognitive activities and ensuring intersystem interaction between all educational entities. A key feature of this evolution is the provision of educational IT tools for interactive dialogue both within the environment and in the "*subject–environment*" format. As modern IEE are increasingly integrated into virtual spaces, where electronic communication (including its remote and telecommunication forms) is the dominant factor in digitalization, *communicativeness* is becoming an attributive property of the environment. This property is *fundamental to its new qualitative characteristic, allowing this environment to be defined as an information and communication educational environment* (ICEE).

Designing an information and educational environment (IEE) is a key element in implementing the Federal State Educational Standard (FSES), as it creates the didactic conditions for implementing its provisions. The task of developing the concept, methodology, and tools for deploying an information and communication educational environment (ICEE) in an educational institution falls to the teacher, who integrates the didactic aspects of the planned activity with the content of the selected model educational program (Kuznetsov et al., 2016 (a); Kuznetsov et al., 2016 (b)).

The digitalization of education and the new generation of Federal State Educational Standards have introduced two very important *new terms into traditional teaching theory and methodology: IEE and distributed content.* In other words, the functioning of an IEE assumes that subject content is concentrated not only in a traditional textbook, but also in IEE components. Moreover, for the convenience of educators, *the principle of structural identity* of the presented subject content is proposed to be used in *distributed content.* What was recently discussed - aspects, functions, means, components – "*this will be, must be*", now already exists and is the technical and technological basis for the new development of the IEE and the expansion of its representation in the information-cultural space, being, at the same time, a new subject of research in conjunction with the "old".

Research Problem

This study aims to address a complex set of interrelated *issues* that have emerged in the field of information and educational environment (IEE) studies. The systemic nature of this problematic field is determined by the following methodological and theoretical "lacuna":

1. **Terminological ambiguity.** The lack of a universal and substantively consistent definition of the concept of "information and educational environment" creates a conceptual barrier to analyzing its systemic relationships with the field of education and the structure of educational and cognitive activities.
2. **Methodological deficiency in IEE design.** Insufficient development of principles for the formation and development of IEEs that guarantee the dialectical unity of its formal characteristics and substantive content. This problem is particularly acute in the context of the lack of identification of invariant aspects of

the selection of content sources and the principles of their didactically sound representation in digital format.

3. **Lack of didactic systematization.** A coherent system of principles regulating the educational functioning and didactic support of IEEs has not been developed. This manifests itself in the absence of clear models for constructive informational and cognitive interaction between subjects and the environment and the unsystematic nature of knowledge about the mechanisms of its positive influence on the personal and professional development of subjects in the educational process.
4. **An underestimated role of communication**. The communicative aspect of IEE remains theoretically fragmented. A comprehensive concept that views communication as an integrated phenomenon fulfilling a threefold function has not been developed: as a means of educational influence, a process for organizing interaction, and a fundamental basis for the formation of subjective relationships within the informational and educational space.

Purpose of the Study

The aim of the work is a multi-faceted analysis of the information and educational environment (IEE), which is implemented through solving the following tasks:

The purpose of this work is a multifaceted analysis of the information and educational environment (IEE), which is realized through solving the following tasks:

1. Conceptual clarification of the definition of the IEE through the study of its systemic interaction with educational practice and the identification of invariant relationships that determine the genesis and evolution of the educational environment.
2. Identification and systemic characterization of the invariant attributes of the IEE, determining its potential for a positive and safe impact on the educational process and its participants. The implementation of this task requires a multilevel analysis – from the informational, psychological and intellectual to the sociocultural aspects of such impact.
3. Systematization of the invariant aspects of the formation and development of the IEE, describing its components and their functional manifestations.
4. Theoretical understanding of information communication as a multidimensional phenomenon, acting as an integral part of the environment, a principle of its organization, as well as a means, method and process of cognitive communication, ensuring the establishment of subjective relations and interrelations.

Methodology and Research Methods

The methodological basis of the study consists of four complementary approaches:

1. *A content-based approach* that ensures the reflection of the content of subject teaching in the components of the information and communication educational environment (ICEE) and the analysis of their didactic relationship.
2. *A formal approach* aimed at identifying and assessing the formal-logical, technological and resource characteristics of the educational system, which determine the effectiveness of interaction with it by educational subjects.
3. *The sociocultural approach*, focusing on the development of the information culture of the individual and society in the context of their positive interaction with the IEE, including aspects of subject-object and interpersonal relationships.
4. *A communicative approach* aimed at organizing productive educational interaction in the context of the IEC and developing competencies for such interaction among educational subjects.

In accordance with the indicated approaches, a set of theoretical methods is used in the work: systematization, formalization, modeling, analysis of communication processes and socio-cultural phenomena in terms of their influence on the organization of the educational an "cognitive process, the mutual influence of approaches in the process of ensuring interaction "subject of training – IEE".

RESEARCH RESULTS

The system environment is part of the overall social environment, exerting *a significant* influence on the corresponding system and, in turn, experiencing a non-trivial feedback effect. This mutual influence is not always positive, which makes the system's task of optimizing it relevant. The specificity of the information and communication educational environment (ICEE) lies in its targeted impact on education through information and educational interaction. The ICEE is formed and functions to improve the quality and effectiveness of achieving educational goals. The feedback effect of education on the ICEE is equally significant and manifests itself in several aspects:

1. **System-wide influence.** Education, as a fundamental structure of society, exerts a decisive influence on the formation and development of the information society, thereby setting the vector for the evolution of the information society itself.

2. **Content and technological development**. The foundations (basic components) of the IKOS are created directly in the field of education through pedagogical developments and their technological implementation. It includes real and virtual objects, either originally intended for educational purposes or adapted for them through the development of appropriate means of educational and cognitive interaction, taking into account didactic principles of teaching.
3. ***Functional integration***. Educational and cognitive interaction between educational entities and the ICEE, aimed at achieving educational outcomes, is an integral component of overall information and educational interaction.

Thus, ICEE not only interacts with the education system but also organically integrates into it, forming a unified whole. Consequently, the feedback from education to ICEE is not only *significant* but also *decisive*.

Another key aspect of the ICEE's identification is the significance of its impact on education. In this context, "significance" implies not only the strength of its positive impact but also its consistency, pedagogically substantiated and enshrined in the tools used. This means that educational tools within the IEC must possess consistent potential, designed for widespread use and reproducibility in order to ensure citizens right to education. In other words, they must be accessible and open to constructive use by all educational institutions. Standard information and educational resources (IER) and electronic educational resources (EER) play a key role here (Zenkina, 2007; Abdurazakov et al., 2019). Their widespread use and accessibility are ensured by their structural organization, the availability of communication tools within the IEE, and the sufficient competence of educational institutions.

Along with standard resources, other informational and cognitive resources common in the social and information environment possess significant potential, albeit less invariant. The value of IER, EER, and other documented objects is determined by their educational *applicability*, guaranteeing a *positive* impact. However, this applicability is often local or personal in nature. Effective work with such resources is ensured by the general *information competence* and *personal culture of* educational subjects, and the interaction itself must be regulated, monitored, and adjusted in accordance with copyright laws (Abdurazakov et al., 2020; Colin & Ursul, 2015; Metaactivity and metaknowledge…, 2022)

Digitalization of the Information and Educational Environment as a Targeted Process

The digitalization of the information and educational environment (IEE) is a targeted process of transforming the education sector, a key element of which is electronic educational resources (EER). Each EER must combine the functions of an object and a learning tool, ensuring a balance between educational content and its presentation for effective learning interaction.

Compared to traditional information and educational resources (IER), EER have a significant didactic advantage, since they allow the transformation of *quantitative parameters of information* (volume, transmission speed, multimedia) into *a qualitatively new impact on the subject of learning*.

Didactic functions and advantages of electronic educational resources:

1. **Structure-forming function:** Additional information in the electronic educational resource (metadata, navigation, multimedia elements) forms a didactically sound structure that facilitates productive interaction with the content.
2. **Catalytic function:** Multimedia in electronic educational resources acts as a catalyst for the learning process, improving perception, understanding, and assimilation of material through the use of various perception channels. This corresponds to the trend toward the development of a visual type of thinking, synthesizing logical-verbal and visual-figurative components (Korpan, 2013, Chernyakova, 2013).
3. **Intelligent environment:** The most promising are electronic educational resources with integrated artificial intelligence elements, implemented on platforms of intelligent electronic-digital systems (SMART devices and technologies). Such resources not only reproduce information, but are also capable of processing it, providing the user with unlimited contextual access to knowledge (Tsybulsky & Levin, 2019) and shaping a new digital reality (Floridi, 2014).

Problematization and research context. Despite its prevalence, the basic concept of "EER" is often identified with electronic textbooks (ET), integrated educational resources (IER), or digital educational resources (DER) and remains the subject of academic debate (Korotenkov Yu. G.; Korpan, 2013; Grigor`ev & Grinshkyn, 2006; Zenkina, 2010). Existing studies provide a general understanding of the potential of EER, but their didactic classification and effectiveness criteria require further clarification in the context of digitalization. The needs for EER can be deduced from an analysis of their didactic capabilities, such as visualization, interactivity, access to relevant knowledge, and operational control, which are studied in detail in the psychological, pedagogical, and methodological literature (Zenkina, 2007; Zenkina et al., 2015; Suvorova, 2014).

The relationship between the ICEE and the education sector. It is important to emphasize that the ICE is not generated solely by the education sector. From a pragmatic perspective, such autonomy is rational and contributes to its effectiveness. Semantically, the ICEE is not a component of the educational system, but rather an independent environment enriched with material and technological infrastructure, information technology products, and informational, educational, and media resources. Thus, the *functioning and development of the ICEE is interdisciplinary, encompassing pedagogical, technological, scientific, applied, and sociocultural aspects*.

The Concept and Purpose of the Information and Communication Educational Environment

The interpretation of the concept and role of the information and communication educational environment (ICEE), characterized by terminological ambiguity, is directly linked to the assessment of the degree of its interaction with the education system. The effectiveness of this interaction is determined by the depth of understanding and successful implementation of processes such as differentiation, alignment, combination, and integration of key aspects of the functioning and development of the ICEE.

The purpose of the ICEE to improve the efficiency and quality of education by promptly providing educational entities (teachers and learners) with information and educational resources (IER), documents, and educational services (organizational, methodological, reference) in an online interaction mode.

The strategic goal of the ICEE is to ensure the transition of the education system to a state adequate to the level of development of the information society, and its subsequent associated development. As Yu. G. Korotenkov notes, the information society is characterized by a high level of development of knowledge, the IT industry, legal regulation, social and information culture, which determines the corresponding level of education (Korotenkov, 2010). This level is defined as *noospheric* . Since the education system functions in close connection with the processes of its informatization and the ICEE, their common strategic goal is ”en”uring the transition of society to an informational state, and education itself to a noospheric state” (*ibid*: Korotenkov, 2010).

Consequently, an information and communication educational environment (ICEE) must meet the parameters of the information society in terms of the intellectual, material, technical, and sociocultural development of all its components, including subjects and objects, and ensure a high level of informational and cognitive interaction between them.

In the context of progressive informatization and global digital transformation, the traditional information and educational environment is integrating with the virtual educational environment, presented in electronic format using telecommunications channels, gradually replacing a number of its functions. Virtual IEE are becoming increasingly important both in terms of their universality and global reach, and in the context of the possibility of creating specialized and localized solutions. In accordance with this trend, the range of information and educational resources (IER) is expanding to include electronic educational resources (EER), and educational interaction in physical space is complemented by communication in a virtual environment.

On the one hand, a virtual environment seems more accessible, visible, and understandable. And in many ways, this is indeed true: it is more mobile, easily modified and expanded, and allows one to create one's own personal virtual world and a sphere of productive cognitive interaction with it. On the other hand, implementing such an approach requires differentiation and appropriate structuring of this environment in the context of both specification and unification for subsequent integration and systematization. For a virtual IEE with pronounced manifestations of specification and differentiation, these factors are especially important.

Thus, *the purpose of* the information and communication educational environment and the virtual IEE is *to ensure* the implementation of educational goals and *improve* the quality of educational results through the information and educational resources contained therein, electronic communication tools, intelligent systems of educational interaction and its productive subject implementation in accordance with the requirements and needs of the modern information society.

As already noted, the definition of the IEE is ambiguous; there are numerous definitions and interpretations. In addition to the above-mentioned reasons, this can also be explained by the fact that the IEC is a multifaceted phenomenon. Therefore, each definition only captures some of its aspects, those most important for a specific study and the priorities of the researchers themselves. General concepts of the IEC as an objective factor directly related to education are provided in the Federal State Educational Standard: "The information and educational environment of an educational institution includes: a set of information educational resources, including digital educational resources, a set of technological means of information and communication technologies (ICT): computers, other ICT equipment, communication channels, a system of modern pedagogical technologies that ensure learning in a modern information and educational environment" (Federal… (FSES), 2012).

Clearly, this is a purely formal, quantitative definition–identification by *form.* It is stated that the IEE should "include", without denying or specifying its content– the purpose, goals, relationships, interrelations, and other components of the IEE. That is, this definition can only be relied upon by "infusing" it with the content of subject-based learning.

Aspects of the Methodology of Pedagogy in the Information Society

In this study, the content of the information and communication educational environment (ICEE) is analyzed from the perspective of its mutual influence with the goals of education, as well as its significance and importance in the context discussed above.

Pedagogy explores the abstract potential of the positive impact of ICEE on the education system, presenting its results in the form of a system of knowledge, models, methods, pedagogical systems and technologies.

The subject area of modern pedagogy includes:

- the field of education and its digital representation in cyberspace;
- educational process;
- informational educational interaction and communication.

The methodological aspect requires special attention. Pedagogic methodology is the study of the logical structure and principles of organizing pedagogical activity. According to the existing approach, it presupposes "the identification of lawful relationships between the subject and the subject of education, and principles for organizing the processes of implementing methodological knowledge based on these laws" (Abdurazakov et al, 2016). As scientific knowledge develops, methodological foundations are also subject to revision and refinement. Thus, the development of pedagogical science and its methodology are mutually dependent and mutually influential.

In the era of digital technologies, when the methods and means of informatics become general scientific concepts, they also become part of pedagogy and subject-specific methods: "The methodology of education becomes informational, pedagogical systems are realized in pedagogical technologies, independent educational goals" (Abdurazakov et al, 2016). Consequently, the focus of modern pedagogy is on two interrelated subject areas: on the one hand, the information culture of the individual, integrating competencies in the field of information security, media competence, and media culture; on the other hand, the representation of the information and educational environment in educational cyberspace, including all the resources of public cyberspace that have the potential for a positive impact (scientific and educational, socio-cultural, ideological, psychological) on the subject of education.

Electronic technologies, intelligent educational systems (Nurmukhamedov, 2010; Frolov, 2017), and a multitude of IOR and EOR created in the informatization of education also contain the abstract potential for constructive and productive educational interaction with educational subjects and the corresponding positive impact on them.

In the field of education, in educational processes and in real educational interactions, this potential is realized in its multiple (unlimited in time and space) reproduction.

The means of the IEE infrastructure and virtual IEE, computer systems, and telecommunications facilities ensure this implementation.

Thus, *the content of the ICEE* is the relationships and connections of its material, embodied, abstract and virtual components, abstract processes implemented in real educational interaction.

The significant impact of the information and communication educational environment (ICEE) on the education system is determined by the qualitative characteristics of its content and their quantitative expression. Therefore, to ensure the effective functioning of the formal structure of the IEE, targeted actions are necessary to shape and develop its content. This process requires consideration of the criterion of significance and the search for optimal formal solutions expressed in the architecture of the environment, its component composition, procedural organization, and communication mechanisms.

More relevant for understanding the complex nature of the IEE is the definition proposed by O. I. Sokolova: "The information and educational environment is a pedagogical system plus its support, i.e. financial and economic, material and technical, regulatory and marketing, and management subsystems" (Sokolova, 2001).

In the context of this study, "pedagogical system" refers to the conceptual and practical embodiment of the pedagogical aspect of the ICEE, including its theoretical foundations, development methods, and direct pedagogical implementation in the educational space. Thus, this concept serves as a collective image of the internal component of the ICEE, generated by its "*pedagogical dimension*". Pedagogical dimensions are the area of pedagogy that deals with issues of testing and assessing students' knowledge, skills, and abilities. Their main goal is to obtain and provide objective information on the actual level of academic achievement based on these same pedagogical dimensions (qualimetric, motivational, research, expert, assessment, diagnostic, etc.).

In the context of the digital transformation of education, the ICEE simultaneously acts as its source and recipient, creator and consumer of pedagogical content. It is an abstract model representing the potential of the environment and incorporating a system of knowledge about it, basic information and educational content, information technology tools, and mechanisms for information and methodological support of education and management of the educational process.

Moreover, in accordance with this definition, the structural elements of the pedagogical system can legitimately be classified as the entire set of educational and pedagogical activities, as well as an ordered system of goals, requirements, trends, patterns, and principles that form the basis of the educational process. The concept of "pedagogical system" accumulates within itself the systematized professional creativity of teachers, which is reflected in the following definition: "A pedagogical system is a set of interconnected structural and functional components subordinated to the goals of education and acting in interaction with the environment as a holistic phenomenon" (Sivashinskaya, 2010). Moreover, its practical implementation is carried out through the interactive interaction of all educational subjects - both students and teachers, including both theoretical and practicing teachers. Thus, pedagogical systems are created and implemented through the entire diversity of forms of pedagogical creativity, the educational process, and educational interaction. A meaningful interpretation of this definition presupposes the presence of certain initial ideas about the content of the IEE and requires the researcher to conceptually understand its components.

Based on the analysis carried out, it seems appropriate to carry out a semantic integration of the definitions considered, the result of which is the following synthesized definition:

The Social Information and Communication Educational Environment (SICEE) is a specific component of the overall social and information environment, existing as a holistic unity of the educational system and the entire set of conditions for its implementation. This unity is embodied through substantive, communicative, procedural, technological, and other necessary components that ensure the functioning and development of information and educational interaction.

The purpose of the SICEE the prompt provision, including *online*, *of* the educational resources it contains (IOR, EOR, DOR), as well as educational services through digital devices and technologies in accordance with the goals of education, the requirements and conditions of the modern information society.

In fact, the modern educational process is inextricably linked to the information and communication educational environment (ICEE), which is defined as a universal pedagogical system. Its infrastructure, resource base, technological solutions, and intelligent learning systems are becoming integral components of the educational space. At the same time, the ICEE is not completely identified with the formal educational sphere regulated by state standards, is not absorbed by it, and does not represent a simple extension of it. The ICEE functions as an autonomous information and cognitive ecosystem, whose scope of application encompasses not only formal education but also areas such as additional education, personal self-education, and socio-cognitive self-development. This environment represents a universal and systematically organized information and cognitive resource for a wide range of purposes, constantly open to development and scaling in both pedagogical and material-technological aspects.

The concept of ICEE formulated in this way has the following essential characteristics:

1. *Activity-developmental aspect.* The ICEE should be studied exclusively through the prism of the development of the pedagogical system as an activity, which implies the effective activation of internal factors in the educational sphere and their subsequent systemic expansion.
2. *Hierarchical priority.* The consideration of the ICEE in the paradigm of the "pedagogical system" naturally determines the primacy of the study of its highest level of systemic organization - the general educational IKOS, which sets the basic principles and standards for all lower levels.
3. *Creative-managerial dichotomy.* The analysis of the ICEE should be carried out not only from an organizational and managerial perspective (management of information and educational resources), but also from a creative perspective, which includes a methodological solution to the problems of generating, preparing, and contextualizing these resources within the integrated pedagogical education system.

Aspects of the Information and Communication Educational Environment

ICEE is a multifaceted and multifunctional phenomenon of the modern social and information-cognitive environment.

An aspect of an ICEE is its specific properties relative to the qualitative attribute applicable to it or its main focus. The universal aspect of an ICEE is its implementation based on a basic methodological approach to subject-specific teaching, expressed in the principles, methods, content, forms, and means of instruction. Since all methodological approaches in the scientific and educational sphere implement a fundamental *systems-information* approach, all of these approaches share a significant *characteristic of homogeneity*: the expression of systemicity and informational content.

The basic aspects and functions of the social information and communication educational environment (SICE) are formed at the macro level of the general education SCE within the framework of its universal pedagogical system. At the local level, they are concretized and implemented through direct information and educational interaction, expressed in a system of ordered subject and subject-object relations. Thus, within the structure of the general SCE pedagogical system, two interconnected levels of implementation are clearly differentiated: system-wide and local.

The essential (basic) aspects and functional characteristics of the social information and communication educational environment (SIEE) are defined and shaped at the macro level – the level of the general educational ICEE, embodied in its universal pedagogical system. At the local level, their operationalization and practical implementation occur within the framework of direct information and educational interaction, structured by an ordered set of subject and subject-object relations. Thus, within the structure of the general SIE pedagogical system, it is relevant to distinguish two interconnected levels of implementation: *the macro* (system-wide) and *the micro* (local) levels.

The most significant aspects that determine the formation and development of the general IKOS include the following:

The information aspect is directly linked to the processes of informatization and the development of the information society. The information system is a direct product of the informatization of education, simultaneously developing in pedagogical, technological, and sociocultural dimensions. Its status and maturity serve as a key indicator of the development of the information society as a whole.

Sociocultural and regulatory aspects are also determined by the context of the information society. These aspects "encompass" the formation and development of *the information culture* of educational institutions, as well as its specialized forms, such as *information-legal* culture and *information security culture.*

The psychological aspect relates to the complex impact of the ICEE on the individual's cognitive and emotional-volitional spheres. This aspect includes an analysis of the informational-psychological influence of the environment and its intellectualized impact during educational interactions.

The communicative aspect encompasses the entire range of communication processes within the educational system, as well as the interaction between the environment itself and the educational sphere. Its essence lies in the communicative realization of the corresponding subjective and subject-objective relationships that form the basis of educational activity.

Communication in the modern information and computer environment has become more than just an expression of the processes of transmitting information–remote and local, internal and external to its sources and recipients. It has long been a process of *communication*, interconnection, and effective interaction in this environment, using a specialized *communicative language*, vocabulary, word formation, grammar, syntax, and semantics. This communicative language forms complex expressions–website addresses, resources, and documents–and formulates search queries for the desired information–by form. Moreover, searches can also be organized by *content* using search engines as intelligent communication tools. In this aspect of communication, we can speak of an invariant educational "dialect" in the communicative language, reflecting the specification of informational and cognitive interaction in the ICEE.

The networked form of information presentation and communication requires *web- based word formation* and corresponding syntax. The semantics of such lexical units (communicative forms) in the ICEE is determined by the semantics of educational concepts and terms. "The essence of the communication model... is that a person simultaneously and constantly plays the role of both the source and the recipient of a message or information. The development of this model reflects the transformation of ideas about communication: from a linear model to a process model with an emphasis on internal and external factors of the process environment" (Studiopedia, 2014). In this regard, we can speak of the formation, under the influence of the ICEE, *of a unique model* of educational communication, expressing the purposefulness of education and the specifics of educational interaction with the environment.

The educational aspect of ICEE, considered at three levels:

- training in educational interaction with the ICEE, communicative language and practical proficiency in it;
- development of information competence and socio-culture of students in interaction with the IKOS, media competence and personal media culture (Zhilavskaya, 2013; Korotenkova & Korotenkov, 2021);
- implementation of subject-based educational training in constructive and targeted interaction with the ICEE.

We agree with the authors' opinion that "Communication is becoming not only the main factor of education, a process, a means, but also *a goal* – to teach the subject of education effective, productive and reproductive communication in the information and cognitive environment. ... The introduction of digital technologies into education changes the concept of dialogue: it can be *virtual*, networked and only simulating real interactive communication" (Korotenkova & Korotenkov, 2021).

The Function of the Information and Communication Educational Environment

In fact, the universal aspects *of social ICEE* are also its ***principles*** – the principles of its purposeful construction, formation, development and the principles of its productive application and functioning.

*The **function of the ICEE*** is the means of its active educational interaction and influence on the sphere and subjects of education, expressed in a pedagogical system, information model or abstract process. Universal *functions of the general ICEE (social ICEE)* are the following:

The managerial and organizational function of the ICEE expresses its psychological and regulatory aspects through informational and psychological influence.

The ideological function of the ICEE, expressed in the formation of the informational worldview of educational subjects.

The educational function of ICEE, expressed in the formation and development of universal human and civic qualities of the subject of education with the aim of *the harmonious development of his personality*.

The developmental function of ICEE, expressed in the focus on the personal development of educational subjects and ensuring their self-development.

Educational function of ICEE, expressed in the education of educational subjects regarding the general directions and forms of development of the world and the life of society, the problems, trends and principles of this development.

Communicative function expressing the communicative and informational aspects of the ICEE in the semantics of the concept of communication in the modern information environment:

– network, media transmission of information, differentiated information objects, display of educational information, media information on standard means of perception of their video and audio images;

- identification of information and educational objects by means of the communication language of the educational specification, allowing the organization, ordering and secure storage of a set of objects identified in this way;
- search for necessary information and educational objects by their identifiers and search queries formulated in terms of a given communication language;
- organization and management of the implementation of the educational process and subject-based educational training using the ICEE system through the formation and operational maintenance of a multitude of standard documents reflecting the current status and statistics of training.

The educational function ensures the implementation of the educational and communication aspects of the ICEE in conjunction with the implementation of its other aspects.

Each function of the ICEE can be multi-aspect, and each of its aspects can be expressed by several of its functions.

At local levels of the ICEE (at the level of the ICEE of the educational institution), continuity is achieved with respect to universal aspects and the functions that implement them, their productive use of the results of this implementation, and their implementation in specific conditions in accordance with existing specifics. At the same time, local information and communication educational environments of educational institutions are autonomous systemic entities characterized by stable internal invariant interconnections. This systemic independence determines the presence of specific aspects of functioning and development in local ICEE, which, in turn, determines the development of their own unique functions reflecting their essential characteristics.

CONCLUSION

Thus, the requirements, purpose, and main goals of the ICEE are outlined, its aspects and functions are highlighted, and its fundamental, qualitative properties are described. An identification definition of the ICEE is provided, combining form and content, in the context of its purposefulness and the significant impact it has on education. The following basic *principles* for the development and functioning of the ICEE are formulated:

1. Materiality is understood as *the invariance* of impact, pedagogically determined and recorded in the means of educational influence. These means must possess the corresponding *invariant potential,* designed for *mass* application and reproduction.
2. The effectiveness of the mutual influence of education and the ICEE is determined by the implementation of processes of differentiation, compliance, combination, and integration of socio-cultural, pedagogical, technological, and scientific-applied aspects of the functioning and development of the ICEE.
3. The following are fundamental, invariant *aspects* of the formation and development *of the ICEE*:

Information aspect related to informatization and information society.

Sociocultural and regulatory aspects, related to the information subculture, information security subculture.

Psychological aspect related to the informational, psychological and intellectual impact of ICEE.

The communication aspect is related to communication in the ICEE and the interaction with it of the educational sphere, the communication implementation of the corresponding subjective and subject-object relations.

Communication is understood as the process of communication, interconnection, and effective interaction in an information-cognitive environment using a specialized communication language, vocabulary, word formation, grammar, syntax, and semantics. Under the influence of the IKOS, *a unique model of educational communication* is formed, expressing the purpose of education and the specifics of its interaction with the environment.

The educational aspect of ICEE, considered at three levels:

- training in educational interaction with the ICEE, communicative language and practical proficiency in it;
- development of information competence and socio-culture of students in interaction with the IKOS;

- implementation of subject-based learning in cooperation with ICEE.

4. Universal *functions are defined* that implement invariant aspects of the formation and functioning of the ICEE, which are also *invariant.*

Each function of the ICEE can be multi-aspect - express several aspects, and each of its aspects can be expressed by several of its functions.

ACKNOWLEDGEMENTS

The authors are grateful to the anonymous reviewers who read the article and made valuable comments that allowed us to improve its quality.

REFERENCES

Abdurazakov, M., Gadjiev, D., Volkova, S., & Lyaginova, O. (2019). The Concept Of Electronic Learning With The Application Of Digital Technologies. // In: Joˇze Rugelj, Maria Lapina (eds.): Proceedings of SLET-2019 – International Scientific Conference Innovative Approaches to the Application of Digital Technologies in Education and Research, Stavropol – Dombay, Russia, 20-23 May 2019, published at https://ceur-ws.org/Vol-2494/

Abdurazakov, M., Volkova, S., Vasil'eva, P., Matveeva, E., & Tyutyunkova, M. (2020). Electronic educational resources as a means of digital education development. // CEUR Workshop Proceedings: "SLET 2020 – Proceedings of the International Scientific Conference on Innovative Approaches to the Application of Digital Technologies in Education". P p. 188–202. http://ceur-ws. org/Vol-2861/paper21.pdf

Abdurazakov, M. M., Korotenkov, Yu. G., & Mukhidinov, M. G. (2016). Educational space represent a tion in cyberspace. SHS Web of Conferences. T. 29. P. 01001. http://dx.doi.org/DOI: 10.1051/shsconf/20162901001 /20162901001

Chernyakova, N. S. (2013). Digitized embodiment of meanings as a subject of cultural interpretation. Cultural sciences in the perspective of "digital humanites". St. Petersburg. Materials of the International Conference - St. Petersburg: Asterion

Colin, K. K., & Ursul, A. D. (2015). Information and culture. Introduction to Information Cultural Studies. Moscow: "Strategic - priorities"– 300 p.

Elyakov A. D. (2012). Homo Informaticus and the modern information environment. Philosophy and Society. Issue #3 (67)

Federal State Educational Standards. (2012). Order of the Ministry of Education and Science of Russia dated 05/17/2012 N 413 (as amended on 12/11/2020). https://fgos.ru/ Access mode: 02/27/2025.

Floridi, L. (2014). *The fourth revolution: How the infosphere is reshaping human reality*. OUP.

Frolov, A. A. (2017). Technology of intellectual education. Access mode: 04/15/2025. http://ncrao.rsvpu.ru/sites/default/files/library/tehnologiya_intellektualnogo_obrazovaniya.pdf

Grigor'ev, S. G., & Grinshkyn, V. V. (2006). *Educational electronic editions and resources: study guide. manual for stud. pedagogical universities and students of the system of advanced training of educational workers*. KSU.

Korotenkov, Yu. G. (2010). *Information educational environment of the basic school: a teaching aid // Yu. G.Korotenkov. – M*. Academy of IT.

Korotenkova, V. V., & Korotenkov, Yu. G. (2021). *Methodological aspects of modern education monograph, M*. Pero.

Korpan, L. M. (2013). Graphic design and design education in the discourse of information society culture. St. Petersburg. Materials of the International Conference - St. Petersburg: Asterion.

Kuznetsov, A. A., Monakhov V. M., & Abdurazakov M. M. (2016a). Methodological recommendations for the teacher on designing the main educational program in informatics in accordance with the requirements of the second generation of the Federal State Educational Standards. (Text). *Informatics and education.*, *10*(279), 9–17.

Kuznetsov, A. A., Monakhov V. M., & Abdurazakov M. M. (2016b). Research activities of an informatics teacher in new didactic conditions for the functioning of the Federal State Educational Standard. (Text). *Informatics and education.*, *6*(275), 4–16.

Metaactivity and metaknowledge. methodological advice. (2022). Access mode: 02/27/2025. https://metodsovet.ru/metadeyatelnost_metaznaniya

Nurmukhamedov, G. M. (2010). Multimedia textbook universal pedagogical teaching aid in modern education. *Informatics and education.*, (6), 29–34.

Pisanets, N. G. (2020). Functional purpose of the information educational environment in the work of an educational institution. Access mode: 04/15/2025. https://www.prodlenka.org/metodicheskie-razrabotki/435430-funkcionalnoe-prednaznachenie-informacionnoj-

Sivashinskaya, E. F. (2010). *Pedagogical systems and technologies: a course of lectures for students of pedagogical specialties of universities*. Ecoperspective.

Sokolova, O. I. (2001). Modern information and educational environments. Access mode: 02/26/2025. https://search.rsl.ru/ru/record/01003220700

Suvorova T. N. (2014). Didactic functions, opportunities and properties of educational resources. Standards and electronic monitoring in education. No. 2.Pp. 27–35.

Studiopedia. (2014). System of information communications. Access mode: 02/27/2025. https://studopedia.ru/1_87877_sistema-informatsionnih-kommunikatsiy .html

Tsybulsky, D., & Levin, I. (2019). Science teachers' worldviews in the age of the digital revolution: Structural and content analysis. *Teaching and Teacher Education*, *86*, 102921. DOI: 10.1016/j.tate.2019.102921

Zenkina, S. V. (2007). *Information and communication environment focused on new educational results (monograph)*. Education.

Zenkina, S. V. (2010). *Methodology for the development and assessment of electronic educational resources: textbook. Manual for students of the advanced training system, educators and students ped. Universities*. Izvestia Publishing House.

Zenkina, S. V., Suvorova, T. N., & Nikolaev, M. V. (2015). Electronic educational resources as part of the information and educational environment: educational and methodological manual for students of pedagogical universities and students of the system of advanced training of educators. Moscow – Kirov: "Publishing house Raduga – PRESS", – 99 p.

Zhilavskaya, I. V. (2013). Media education of youth: monograph / Moscow: MGGU, – 243 p.

Chapter 3

Enhancing Educational Engagement Among Kattunayakan Tribal Children Through Culturally Tailored Educational Activity Centres

Emmanuel Daniel
https://orcid.org/0000-0002-2222-9012
Christ University, India

Vigneshwaran Subbiah Akkayasamy
https://orcid.org/0000-0002-5167-2443
Christ University, India

Grace Anu Backia
https://orcid.org/0009-0005-5329-2451
Christ University, India

Nisha Varghese
https://orcid.org/0000-0003-4379-9644
Christ University, India

S. Pooja
https://orcid.org/0000-0002-8831-8552
Indian Institute of Management, Rohtak, India

Greta D. Souza
Christ University, India

ABSTRACT

The deep forests of the Western Ghats are home to the aboriginal tribes Kattunayakans, Categorized under the Particularly Vulnerable Tribal Groups (PVTGs) in India. Despite various government schemes to improve their economic status, the literacy rates remain very low due to geographical isolation and traditional lifestyle. The main

DOI: 10.4018/979-8-3373-7922-7.ch003

objective is to look into designing educational activity centres that exclusively cater to the needs of the Kattunayakan children, integrating indigenous knowledge and methods with teaching-learning sessions to promote lifelong learning. Technological progress and AI can be game changers for accessing quality education, including visual learning, language translation, and personalized methods while preserving cultural heritage. The research is based on a systematic narrative review, critically examining scholarly articles, policy and government reports. The outcome shows that these programs help improve academic skills while keeping them closer to their culture and identity.

1. INTRODUCTION

In the age of globalisation, where there are new innovations every now and then, many nations have supported providing equitable access to services to their citizens. However, in some pockets of the globe, the tribal population seems to be distant from the so-called mainstream society for various reasons. They face a lot of challenges in their lifetime, very specifically with education, which reflects in their socio-economic disparities and cultural dilution, which results in marginalisation. In spite of many government initiatives, there is a massive gap in understanding the implementation of educational programs and the impact it has created among the tribal communities. The literature review highlighted challenges due to a lack of infrastructure and student enrollment (Nanda & Samanta, 2017a; Xaxa, 2016). However, very few studies have focused on the importance of culturally relevant education (Bindhani, 2021).

In the VUCA (Volatile, Uncertain, Complex, and Ambiguous) and BANI (Brittle, Anxious, Nonlinear, and Incomprehensible) era, progress in the technological aspects can be bridge to reduce disparities among the indigenous people in terms of education. With the support of mobile technologies and E-learning applications, civil society organizations can reach the Kattunayakan children in the remote tribal areas, bringing inclusivity through quality education. Furthermore, the career aspirations of tribal children often remain underexplored. This results in them not being able to enjoy the benefits of development, so they are only seen as a mere vote bank. During the election campaigns and in the election manifesto, these marginalised people are brainwashed and later neglected by the elected representatives. (Rupavath, 2021). The sustainable Development Goal-4 Quality Education aims to provide inclusive quality education for all including the indigenous population. It also paves way to integrate traditions of Kattunayakan knowledge base into the curriculum (United Nations, 2015).

India, being a global powerhouse, is a pluralist country with a very rich legacy, diversity, a multitude of cultures, a land in which some of the major religions of the world had their origin, a land of languages, castes, and racial stocks which has made the social functioning highly stratified and hierarchical. As an outcome, the economic and social opportunities are differently distributed in terms of caste, religion and other social affiliations. India has a lot of dense forests, hilly terrain, and fertile coastal belts, besides the Indus, Indo-Gangetic, and Brahmaputra plains. Such a vast scape of ecology has ensured spatial and occupational differences to a greater extent. Even in the 21st century, India is one of the most developing countries in the global south; the majority of the population's predominant occupation is agriculture. On the global front, India has the most significant tribal population, which exceeds 10.4 crores, forming 8.6% of the country's entire population. Even though they are a tiny proportion, they are highly ethnically diverse (Xaxa, 2016).

Despite the rich traditions that the tribes have had for a very long time, most of the so-called mainstream educational systems have failed to accept and accommodate their traditional wisdom, language, and ways of learning. Several studies emphasise that the majority of the tribes in India still live in poverty, engaged in ancient agricultural practices which do not yet have results and, most importantly, with poor literacy rates (Bindhani, 2021). Inequalities are more common across the globe and have been a typical attribute in many societies since its inception. Though the scenarios are changing, the pace at which they are swift is very slow compared to other global advancements. This situation is very intense and worst in many tribal societies as they are mostly educationally backward (Nanda & Samanta, 2017b).

The deep forests of the Western Ghats are home to many aboriginal tribes, such as Kattunayakans. The tribe lives closely with their traditional lifestyle, which keeps their culture alive. Despite many government and civil society interventions, the literacy rate remains very low and access to quality education is still a dream.

1.1 Objectives

- To study the specific educational challenges faced by Kattunayakan Tribal Children in the Nilgiris.
- To explore how a culturally relevant educational activity centre can create a lasting impact on the learning process.
- To analyse the available literature on the role of technology and the tribal educational models.
- To recommend interventions to improve quality education, with the support of technology-driven educational activity centres.

2. METHODS

The research is based on a systematic narrative review, critically examining scholarly articles, policy reports, government reports, and other academic materials. A structured method ensures careful and detailed exploration of the literature. A narrative review approach was used to integrate and critically examine available research on the topic (Snyder, 2019). The research seeks to highlight major debates, gaps, and an integrated knowledge of the topic. Sources were chosen from reputable academic databases, such as Google Scholar, PubMed, Scopus, JSTOR, and ResearchGate. Non-academic sources like government reports, policy briefs, and material from international organizations were also integrated to depict the various viewpoints (Boote & Beile, 2005). The selection criteria gave preference to contemporary studies of the last decade, complemented where applicable by landmark works. Inclusion Criteria: Peer-reviewed articles in journals, books, policy reports, and research papers pertinent to themes of the study (Cooper, 2010). Exclusion Criteria: Non-academic literature like blog sites, opinion pieces, and non-peer-reviewed papers, only when providing critical contextual information.

Critical evaluation of the literature was conducted to compare, contrast, and integrate findings across studies. A comparative framework was applied to bring out key differences and similarities to gain an in-depth understanding of the knowledge base (Machi & McEvoy, 2016). Being a review-based research, results are inextricably bound by the quality and prevalence of available research. Biases within the literature used might shape conclusions (Tranfield, Denyer, & Smart, 2003). Attempts were made to use various viewpoints and make a fair-minded critical assessment.

2.1 Western Ghats

The Western Ghats which is found in the southern part of India are much older than the greater Himalayan mountain range, which has a unique biophysical and ecological outlook (UNESCO World Heritage Centre, n.d.). The Western Ghats are the mountain range that traverses the states of Tamil Nadu, Kerala, Karnataka, and Goa. They cover around 1,80,000 square kilometres, which is roughly 6% of the total landmass of the country. It has a minimal interruption in between in the northern Kerala state, the Palghat Gap (Molur et al., 2011).

The Western Ghats, otherwise known as Sahyadri, is a mountain range which is home to some of the world's unique flora, fauna and fungi (Molur, 2009). The mountain range is home to more than 30% of the Indian mammal species. The Lion-tailed macaque and the Nilgiri tahr are only found in this region. This region is also a very important sphere for a variety of large animals. Around 17% of the world's tigers and 30% of the Asian elephants live in the Western Ghats. It also has a lot of dense forests and a lot of medicinal plants. Western Ghats is a very crucial source of water inside the southern mainland as most of the rivers in the peninsula originate from here, providing water to around 245 million people (*About the Western Ghats*, n.d.). Compared to the other hotspots in India, the western ghats have the highest population per unit, more than 300 individuals/km2, which poses a more significant challenge to conserving it. (Molur, 2009). The loss of forest in the Western Ghats is very rapid. Out of the original land mass for primary vegetation, only 6.8% remains (Myers et al., 2000).

There have been evident changes seen ever since the first mankind started settling there. However, massive changes took place in the early 1800s, with the British initiation of coffee and tea plantations and open land cultivation by the locals. The evergreen forest habitats have undergone a serious decline of 65-91% since the early 1900s (Gadgil and Meher-Homji, 1986). A humongous 12% decline was evident in the primary forests of the Western Ghats inside Karnataka (Ramesh et al., 2010). Nearly 40% of the primary forests were converted either for plantation or agriculture between the years 1920 and 1990 (Menon and Bawa, 1997).

The forests in the Western Ghats in India have been a home for many indigenous tribes such as Kattunayakans, Paniyas, Kotas, Kurumbas, Irulas and Todas. Over the years, each community has acquired a unique set of conditions that has helped them shape their culture, language, religion and their lifestyle (Jayakumar et al., 2024). The tribal population faces constraints in access to resources, technology and quality education, which categories them as Particularly Vulnerable Tribal Groups (PVTGs) (Behera et al., 2006). Out of the mentioned tribes the Kattunaakayans and the Paniyas who live in the Gudalur region are more marginalised due to several factors (Jayakumar et al., 2024).

2.2 Kattunayankans- The Invisible Voices of the Forest

The Kattunayakans reside in the Western Ghats of South India in the states of Tamil Nadu, Kerala and Karnataka. In South India, the Kattunayakans are known as Jenu Kuruman, Naickan, Jenu Kurumba, Kathu Nayakkan, Ten Kurumba, Ten Kurumban, and Jenu Koyyo Sholanayakas. Kattunayakan means "King of the Jungle" in Tamil and Malayalam. Kattunayakans are considered to be one of the early inhabitants of the Western Ghats, and traditionally, they were involved in gathering and collecting wild honey and wax (Parthasarathy, 2013). Kattunayakans were once known for their expertise in bird trapping, hunting, and fishing. Currently, honey is very important in their daily lives as they consume it, use it for preparing their medicines and trade it to outsiders to make some livelihood. The Kattunayakans can very easily climb a tree that is 30 meters in search of honey. During the process, they also get so many bee stings that their bodies easily develop immunity. There were times when they had to compete with the bears to collect honey (Project, n.d.).

Due to a lot of human-animal conflict and restrictions from the Tamil Nadu Forest Department, formerly the Madras Forest Department, in the recent past, the Kattunayakans are now employed in estates and other plantations. Kattunayakans in the Western Ghats live in hamlets, and traditionally, their houses are made of cow dung, mud and charcoal. There is an elder for the whole of the community who is consulted for different matters (Ramachandran et al., n.d.). Mostly, the houses of Kattunayakans are very small huts in a cluster, just very similar to the Paniyan Tribes. The houses are usually one-roomed in the measurements of either 8x6 or 8x8, respectively. The doors are usually smaller, and they are made of bamboo thatties with no provision for ventilation at all. The temple house, which they call Ambalam, originated from the Malayalam word and is a very unique settlement for the tribe. This resembles a small hut located at the entrance of their hamlet. Their temple is locked for most of the time in the year. No idols are placed inside the temple house, but some assorted ancient items are found inside (Natarajan, 1985). The Kattunayakans give a lot of importance to their family and hut. They tend to protect the belongings of their dead ancestors. To date, they have a practice of abandoning their house if anyone dies inside the house. In such cases, the immediate family members will construct a new house next to the abandoned house. The Hill Area Development Programme (HADP) has constructed many houses for the Kattunayakans in their old settlements (Sasikumar, 1999).

The Kattunayakans have a unique way of dressing. In a traditional sense, women wear long clothes that are wound around their bodies, covering their upper bodies and tied under the right shoulder, keeping the other arms open. Some women also drape a strip of cloth, tightening their waist. Usually, the women carry a small cloth having lime, betel leaves and areca nuts. Most of the women still follow this practice. On the other hand, the men wear dhoti and a small towel on their shoulders (Sasikumar, 1999). The Kattunayakan women had a practice of having a tattoo on their shoulders. Both women and men wear brass rings. Married women wear anklets and toe-rings (Parthasarathy, 2013).

Music and dance are very integral to Kattunayaan's cultural heritage. They are monogamous tribes, and a lot of girls are pushed to marriage soon after they attain puberty (Project, n.d.). Black magic is more prevalent among the women in the tribe, and everyone cannot access it. They have a very strong belief in animal sacrifices, prayers and herbal medicines. However, currently, they do not practice animal sacrifice as a religious practice due to government restrictions. The other tribes in the Nilgiris, such as Kurumbas and Paniyans, visit Kattunayakan Magicians, who are known for black magic for problems in family life, childbirth and future predictions (Parthasarathy, 2013). There is no society in this world without the means for production, distribution, consumption, and some sort of exchange (Herskovits, 1955). Kattunayaakans, once hunters and gatherers of the forest produce, have now entered other activities, such as farming or working in the plantations in the Nilgiris District (Luiz, 1962). The Kattunayakans are currently in their economic transformation, trying to become a self-sustained community involved in agriculture (Sasikumar, 1999).

The wisdom passed on from generation to generation, the ability to solve problems and work as a team have been some of the strengths of the Kattunayakans. Several factors, such as observation, meticulously planning to gather honey, and teamwork, enhance their social and cognitive skills. Nevertheless, these skills are not taught in formal education. (Reddy & Reddy, 2020).

2.3 The Paradox of Education- A Dream Yet a Distant Reality

During the British administration in India, the Government adopted a few policies that would isolate the tribal communities (Desai, 2010). Soon after the independence from the Britishers, the Government of India took many steps to uplift the tribal communities by providing education. It has become essential for the tribals to be equipped with new knowledge and skills to cope with massive economic shifts. Formal education acts as the bridge connecting the traditional practices of the tribals and the modern world (Brahmanandam & Bosu Babu, 2016).

The Government of India initiated several educational programs and schemes that reserve seats for them in academic establishments such as universities and colleges for their educational and social advancement. Though the policy for reservation has been in place for more than seven decades, the population of the tribes has grown significantly over the years. Filling up the quota based on government policies has always been a problem, even today (Xaxa, 2001). The state also tries its best to give the tribes legal recognition. Also, they are encouraged to use their language for their education and other purposes, which enables them to grow economically and socially on their own terms (Xaxa, 2005).

In all the states in India, the schools teach the children in the language of the state. A few decades back in Bihar, the education department came with the publication of books in the language of some tribal groups to help children learn in their heart's language. However, the officials never used those resources and are kept in government repositories. However, in Tamil Nadu, the tribal children are taught in Tamil, the state's official language (Xaxa, 2005). The tribes had no option but to learn an alien language, tradition, culture and history to identify themselves with the mainstream society. Though the constitution of India had many provisions to protect the indigenous culture and language, the individual states had no interest in implementing them (Xaxa, 2016). The tribes are very different from each other in manifold ways. Some are more wealthy, literate, and politically empowered than others—factors or influences from the outside cause these differences rather than internal ones. The tribes in India have suffered more and more with poor health status and lack of quality education, which has questioned the nation-building process after the Indian independence. In recent times, India has emerged as one of the superpowers in the world with many advancements, but most of the tribes still lack basic things. Looking at this scenario raises a question about India's national identity (Xaxa, 2005).

Women have very low literacy rates compared to men. Therefore, there is a need for some special programs from the government and effective interventions from civil society organisations to address the issue. For example, the literacy rate among the Kattunayakan community in Kerala is 59.37%, which is much below that the National or State level. Out of 17,436 Kattunayakan children who are above the age of 5, a significant number of 8,857 have not progressed beyond 10th standard. Only 515 have completed 10 standard (SSLC). Furthermore, only 16 have completed their graduation. These statistics calls for a very urgent need for tailored interventions from the government and civil societies to work on these disparities (Anusree & Umajyothi, 2013). One of the key elements for the low enrolment of tribal girls in school is the lack of relevance to their culture. If education could cater to the felt needs and aspirations of the tribal population, the enrolment of girls would substantially improve. Even though Kattunayakans do not hesitate to send their female children to schools dominated by boys, the presence of a Lady teacher in the school can make a humongous difference. Another factor in the low enrolment of boys and girls in school is that the parents are least interested in education and feel it is only for mainstream society (Parthasarathy, 2013).

2.4 Understanding the Barriers: Why Mainstream Education Fails Tribal Children

The history of education in the Nilgiris Hills dates to 1881, when there was only one school attended by 20 children, and their community status was not reported. This school for Indigenous children gave some kind of basic education. In these schools, basic Canarese and Tamil were taught. This school was supported by some people, and the school teachers were paid in kind with a small annual fee in cash during the Dussehra festival. The classes used to take place in different places based on the availability, such as temples, residences of influential people or other public places. The children were also taught some basic arithmetic; however, they left the school before attaining adequate knowledge (Gopalakrishnan, 1995). As per the documents in the government repository, the British government initiated many schools in the Nilgiris district. As per the statistics from the Census 1901, the people of Nilgiris were well-educated compared to any other district in the Madras Presidency, barring the Madras Town. In 1899, the British Government, based on the report of their educational authorities in India, established a few schools to cater to the indigenous people of Nilgiris. In 1908, 40 indigenous schools were functioning with a strength of 1270 students (Francis, 1908).

Kattunayakans, seen as the friends of wildlife, found interest in formal education for the first time when they encountered some outsiders in 1958. These outsiders came from a place called Srangapattinam, Mysore region to explore Mudumalai and Bandipur Wildlife sanctuaries. The visitors sought the help of the Kattunayakans to navigate the forest area and identify some medicinal plants and trees. After they completed their work, they taught the Kattunayakan children a few Kannada alphabets and numbers. This small act ignited a spark among the entire tribe, for which, the Government of Tamil Nadu in the Nilgiris district established schools for the tribal children. These schools cater to the Kattunayakan children and neighbouring tribes such as Paniyans, Jenu, and Betta Kurumbas (Parthasarathy, 2013).

Education is seen as the driving factor in the new realm of tribal development, covering all the essential aspects of the community. Undoubtedly, mainstream education has empowered the Kattunayakans to face extreme exploitation and boldly fight for their rights (Parthasarathy, 2013). Despite these advancements, the Kattunayakan children face many challenges regarding their education. Most of the Kattunayakans were in their traditional occupations, resulting in frequent travel, which disrupted their children's education. Furthermore, a considerable gap between their cultural aspects and mainstream education, has dented their progress (Singh, 2019). Moreover, the formal education discounted the indigenous knowledge base of Kattunayakans, which made things worse for the children. Along with this, poor infrastructure inside the deep forests and a lack of culturally sensitive pedagogy for teachers have contributed to high dropout rates among Kattunayakan children (Rao, 2021).

2.5 Barriers to Tribal Education:

The Kattunayakan Tribes are spread over Tamilnadu, Karnataka and Kerala (Ramakrishnan et al., 2024). The study among the tribal residential schools in Kerala brings out the fact that students find difficulty in connecting the lessons in class with their everyday life, and report difficulty in reading the text books and talking with the teachers (Muneer V, 2015). Figure 1 representing the educational status of the Kattunayakan Tribal Men and Women. The number of literates is very less compared to Non-Literates and few percentage passed schooling and no one graduated as per the statistics. The literacy rate is indirectly proportional to the age as represented in Figure 2.

Figure 1. Educational Status of Kattunayakans

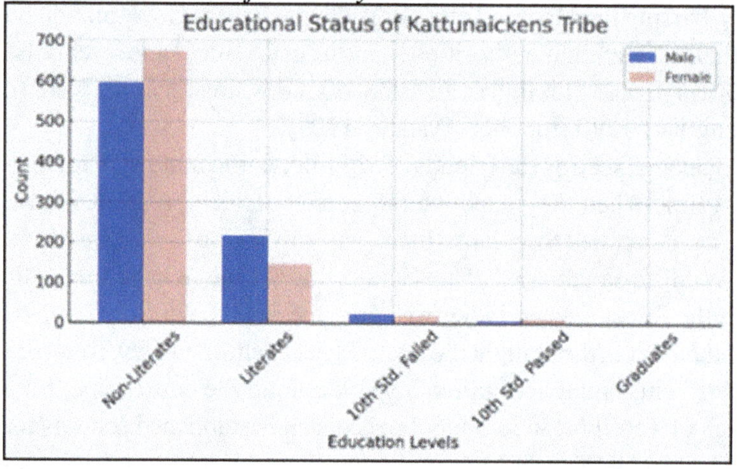

Source: Educational Status of Kattunayakans in Nilgiris District, Tamil Nadu (Parthasarathy, 2013)

Figure 2. Literacy and Non Literacy rate of Kattunayakans

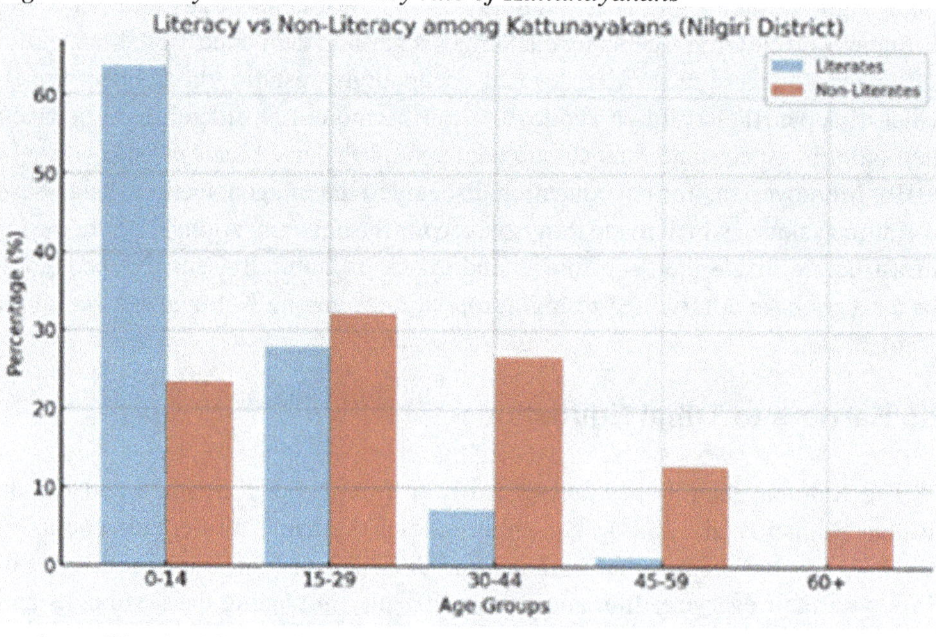

Source: Educational Status of Kattunayakans in Nilgiris District, Tamil Nadu (Parthasarathy, 2013)

T. Brahmanandam and T. Bosu Babu (2018) also cites language as one of the main barriers for education among the tribals as their vernacular is different from the state language in which they are taught. Their study also suggests introducing tribal art forms and culture in the school curriculum to retain their place at school. A study in Koraput district of Odisha discusses the problems of lack of interest in studies and difficulties in reaching school among many others which also includes the aspect of language (Bindhani,2021).

Although the government of India has come up with various interventions such as scholarship for the pre and the post matric and the higher education, hostels for tribal students, Ashram schools and vocational training institutes (Kaur & Kaur, 2022), the level of education is still very low among the tribals. Despite the provision of schemes like the Post-matric scholarship for ST students, hostels for ST boys and girls, Ashram school, National overseas scholarship for ST students, some of the aforementioned challenges continue to persist, keeping the tribal children from accessing education (Daripa, 2017).

3. MODELS OF EVENING TUITION CENTRES FOR TRIBAL STUDENTS IN TAMIL NADU

The evening tuition centres by The Centre for Tribal and Rural Development (CTRD) Trust in partnership with a German Organisation SODI, support the tribal children from the hamlets in Nilgiris for education. The centre has appointed qualified teachers who assist the students in learning, doing their homework and in enhancing their skills. Additionally, these centres also provide food and a comfortable environment to learn (*Evening Coaching Centers (ECC) – CTRD Trust*, 2025). The CORE (Children Organization for Relief and Education) NGO, based out of Tiruvannamalai has initiated the project called Bridge School Concept that regularly conducts evening tuitions for Irular Tribal children in the district. They teach life skills, in addition to helping them learn the subjects. One of the main objectives of the project is to prevent dropouts and help the children be regular to schools (*TDH Core*, 2016). One another initiative was by the forest department of Sathyamangalam in Tamilnadu, catering to the children from around 10 Tribal hamlets. This initiative is no longer functioning. But when it began in 2017, they focused on bridging the gap in education as these were tuition centres where the subjects they study at school were taught in a way for students to better understand and learn (Saravanan, 2024). This is said to have reduced the number of dropouts.

In comparing the Indian and the global models, the Indian models are more of assimilation into the mainstream education without much focus on the traditional lifestyles of the tribals while the global models majorly incorporate the tribal ways of life in their curriculum. This paper talks about bridging the gap between formal education and cultural practices of Kattunayakans, leading to a holistic development of children.

3.1 Theories and Models of Tribal Education: Kattunayakans

The Kattunayakan community, possesses a rich cultural heritage and well cultivated knowledge and survival system which remains ignored in absence of informed policy and socio-economic and socio-political barriers. There have been many such instances across the world, historically where tribal groups were identified and their indigenous knowledge system was revived which helped both the community and the larger society (Bose, 2021). Educational engagement and mainstream cross connections are particularly helpful to preserve what has organically developed (Nakata, 2007). Advocating educational participation for Kattunayakans can not merely happen by making schools available but by giving them the confidence of their learning and participation leaving an indelible mark of identity at a bigger canvas (Corntassel, 2012). Since there has been a fractional educational participation of PVTG, Kattunayakans, it is not possible to include them in mainstream education without no conditions of preservation of all that makes them a unique pool (Vowel, 2016).

Some of the indigenous educational models and theories like Tribal Critical Race Theory (TribalCrit), Indigenous Standpoint Theory, Transculturation Theory, Culturally Relevant and Sustaining Pedagogy (CRSP), Land-Based Education, Two-Eyed Seeing, and Indian initiatives of digital technology can guide us to to create and continuously develop a meaningful and engaging educational experience for Kattunayakan children (Alfred, 2014; Bartlett et al., 2012; Carter & Hollinsworth, 2017; Redvers, 2020).

3.2 Tribal Critical Race Theory and the Kattunayakan Context

The Tribal Critical Race Theory highlights how colonialism still affects tribal education and stresses the importance of tribals in their decision making pertaining to curriculum development (Brayboy, 2005). This means actively recognising and incorporating their own lived experiences along with their relevant cultural knowledge into their educational aspects such as curriculum, teaching and learning. Educational marginalisation among Kattunayakan children is usually due to priority given to mainstream approaches. A TribalCrit-informed strategy involves designing culturally relevant teaching-learning strategies which reflect their environmental, festival and traditional knowledge (Brayboy, 2005). For example, story based pedagogies can include some aspects like ecological knowledge, folklore, myths, etc. The mainstream connection of this approach can be by publishing stories of everyday Kattunayakan life. Such initiatives are taken by National Book Trust of India by publishing graphically engaging children's books and short stories. However, language is the most important area to focus upon. Localisation, a form of translation, is booming with the advent of online communications and large language models or AI. It will make a larger number of people aware of the tribe's strengths and challenges (García & Wei, 2014). When Kattunayakan myths, folklore, and environmental knowledge are presented through story based learning programs, the children will be filled with a sense of pride about their ethnic identity. Eze and Nnia (2022) in their study bring out the ways of integrating African knowledge in Nigeria's formal education system by incorporating the same through oral literature such as myths, legends and through music.

Moreton-Robinson (2021) stated that indigenous knowledge must be recognised as valid and central to educational research and practice. Remembering that Kattunayakan children are inseparable from forests, in their everyday knowledge and practices, the forest is a character and not just a habitat. Their education needs regard and reverence for forest related discourses. Their being is entrenched in the wild-life and educational practices cannot overlook the same according to Indigenous Standpoint Theory which talks about reclaiming knowledge systems, such values need a particular attention in science and ecology education. Primary, lived experiences in the foremost school years may also influence the liberal scientific studies as even young children can exercise their presence effectively as a stakeholder. The Kalinga Institute of Social Sciences in the state of Odisha, India is an educational institution run exclusively for the tribal children of the state which includes students from 64 tribes and 11 PVTGs (Particularly Vulnerable Tribal Groups) is a model on the same. Among the many ways adopted, one particular method is where the school comes up with planting saplings and each sapling is named after a child who should be taking care of it. Tribals being those who live in forest and have close contact with nature, the school aims to impart this in the institution along with giving them education, while the curriculum is also designed in a way that makes them ready to be part of the mainstream (Kanti, 2015).

From a cultural lens, when our country is experiencing the influence of LPG reforms, Kattunayakans are much underrepresented, a mainstream education may force them to assimilate their culture. It may blur the meaning of their struggle to survive. As language is a tool to exercise cultural identity therefore while printing textbooks and other materials the dialect is what the policy makers need to pay attention (Ortiz, 1947). A study by Magga et al., (2005) emphasises the medium of instruction for the tribal children to be in their mother tongue. They do not condone placing the tribal children in boarding schools as these would keep them from communicating in their mother tongue and practicing their own culture. Elek et al., (2020) in their study highlight that the focus of the Early Childhood Education and Care programs for the children from indigenous communities focuses more on language and "school readiness." In terms of language, sometimes, the script becomes a challenge, in such situations the multimedia approach with translations can support the indigenous worldview.

4. DESIGNING PEDAGOGICAL TOOLS FOR KATTUNAYAKAN COMMUNITY CHILDREN: INTEGRATING INDIGENOUS EDUCATIONAL MODELS

The Kattunayakans possess a rich cultural heritage and well-cultivated knowledge and survival system, which remains ignored in the absence of informed policy and socio-economic and socio-political barriers (Anusree & Umajyothi, 2021). There have been many such instances across the world, historically, where tribal groups were identified, and their indigenous knowledge system was revived, which helped both the community and the larger society (Brayboy, 2005).

Educational engagement and mainstream cross-connections are constructive in preserving what is organically developed. Advocating educational participation for Kattunayakans can make schools available by giving them confidence in their learning and participation, leaving an indelible identity mark on a bigger canvas (Anusree & Umajyothi, 2021). Since there has been a fractional educational participation of PVTG, Kattunayakans, it is impossible to include them in mainstream education without any conditions to preserve all that makes them a unique pool. Some of the indigenous educational models and theories like Tribal Critical Race Theory (TribalCrit), Indigenous Standpoint Theory, Transculturation Theory, Culturally Relevant and Sustaining Pedagogy (McCarty & Lee, 2014), Land-Based Education, Two-Eyed Seeing (Saulis, 2018), and Indian initiatives of digital technology can guide us to to create and continuously develop a meaningful and engaging educational experience for Kattunayakan children.

Designing a pedagogy for the upcoming generation of Kattunayakans will not be easy because they must immediately be aligned with contemporary practices. Ladson-Billings (1995) offered Culturally Relevant and Sustaining Pedagogy (CRSP), which focuses on a community-based learning approach. Following this approach, the children will be given such projects as a part of their learning that offer tangible and beneficial results for the community, infusing a sense of value while preparing them for upward mobility. Such a project can be forest conservation, linking it to classroom and community engagement. Children will be motivated to seek help and inform the elders about it. Hare, (2011) in the research brings out the different methods that indigenous knowledge is imparted to the children from the first Nations in Canada through a programme called AHSOR (Aboriginal Head Start on

Reserve). One of the learning centres of AHSOR takes children outdoors, teaching them to fish, gather plants and hunt, as the children come back to the centres and write or draw about what they have seen and learnt. He also observes elders from the community coming to the centres to teach the children the preparation of traditional medicines, thus passing on the knowledge from one generation to another. Beaulieu (2019) and Goforth et al., (2022) suggest a model where the education for the native Americans would bring in elders from the community in order to teach the children culture and the concept of family. The native studies program teaches them the history and culture. He also models art and crafts and training teachers from the local community. Consequently, a mutual exchange will help the two generations. Initially, the government stakeholders can appoint formal educators, which may reinforce the cross-cultural dissemination of knowledge resources.

CRSP also focuses on land-based education (Simpson, 2014), which is always a concern for any tribal community. A land-based mathematics curriculum can integrate practical learning experiences such as measuring tree growth, estimating water sources, and tracking animal movement patterns. Bringing children at par with national educational standards while maintaining their cultural identity can be thought about by linking concepts to seasons. This approach is called the seasonal learning module. The children then learn core concepts with daily life situations like harvest cycles or wildlife migration patterns. This ensures that all the subjects are taught in ways that are meaningful and applicable to their daily lives.

Indeed, to maintain harmony between cultures and ensure that students get representation beyond borders, the Mi'kmaq Nation advocates for Two-Eyed Seeing. As Bartlett et al. (2012) put it, integrating Indigenous and Western knowledge offers a practical and long-term view. Shava (2016) puts forth a case study among the indigenous people in Southern Africa where science is taught through culturally relevant methods to the students from the indigenous community. He highlights a situation where the researchers in teaching fermentation methods, replaced the modern methods with those in the culture which also led to them identifying the higher nutritional values of the traditional fermentation method of the community. Science concepts such as winnowing, post-harvest pest control practices are also modelled from the traditional methods thus helping the children relate and learn better. Kattunayakans have their own system of health and medicine. Applying Two-Eyed seeing is particularly useful by combining knowledge of scientific learning with their existing understanding of health and wellness. In such discourses, the technological advance globally also needs to be mentioned. Field visits or experiential learning for young children can be highly stimulating (Kolb, 1984).

Rowe and Shawnda (2023) argue that technological integration requires digital equity and inclusion, which requires governmental attention and support. It also comes at the risk of cultural ignorance if technology is not integrated with cultural sensitivity. With governmental support, too, there is a chance of underutilized resources in far-flung areas. Open and Distance Learning Systems through mobile applications with easy multimedia interfaces can help students study while keeping them engaged. The government can partner with edtech enterprises, as Nag (2024) suggests, that virtual reality with indigenously informed games and activities can allocate resources promptly without wastage. For example, there can be dedicated times of years when 3-D movies inspired by the daily lives of the community are screened in their language, and community participants are invited to voice-over (Nayak et al., 2020 & Tuttle, 2013). Envisioning the educational participation of indigenous tribal groups in India is challenging. While we learn from global experiences, there is no guarantee that the existing measures will be highly effective. However, implementing continuous monitoring and evaluation and redefining and redesigning the programs will ensure participation and enhanced knowledge of the mainstream about the Indigenous ethos and the community with cultural insemination.

5. RECOMMENDATIONS

Based on this review, the following recommendations are suggested to enhance the teaching-learning experience of Kattunayakan children. These suggestions aim to make learning engaging and meaningful, by incorporating technology while valuing their cultural heritage.

5.1 Culturally relevant Curriculum

The activity centres should focus on developing a curriculum that would give importance to the culturally rich heritage of the Kattunayakan tribes. The learning process should integrate traditional elements: a deep understanding of their forest area, wildlife tracking, and medicinal herbs. For instance, the teaching-learning may include some transect walks to explore biodiversity and wildlife, hands-on activities on gathering forest materials, and sessions led by the tribal elders on cultural practices, meaning, and relevance in the 21st century. Some of the pedagogy may include project-based learning, such as identifying medical plants and creating herbal gardens, which will add to their knowledge base and enhance their practical skills. This type of teaching-learning will help the children respect their heritage and make education a joyful experience.

5.2 Community-Run Educational Activity Centres

Community-Run activity centres are one of the sustainable models of development as it involves the community taking ownership of the development. This includes setting up evening coaching centres. Such centres ought to be focused on contributing to the holistic development of a child rather than the mere repetition of lessons from schools. These can also address the issues of dropout rates, parental hesitation and mobility. Such centres can also act as bridge schools for children who drop out of school due to the issues of mobility. While these centres would primarily focus on education for children, it can also be made interesting through unconventional methods such as involving tribal leaders and local youth as facilitators. Such initiatives could help bridge the intergenerational knowledge gaps. Involving elders from the community also provides space for healthy mentoring that is culturally sensitive. When people from the are involved, it gives an increased sense of belonging and hence for the initiative to be taken forward with enthusiasm.

5.3 Technology with Cultural Sensitivity

The activity centres should bridge culture and technology to make education a meaningful and joyful experience for the Kattunayakan children. The NGOs can partner with technological companies and design a learning application that can work offline, which can also work in remote areas and areas where there is no internet penetration. These apps can use AI technology to analyse the children using these applications and customise the learning based on the child's interest levels and understanding. In addition, including virtual and augmented reality in the teaching-learning process of the Kattunayakan children can be a game-changing opportunity that can facilitate their exploration of their rich culture and forests. The stories and the voices of the Kattunayakan elders can be integrated into applications that will support future generations' access to ancestral knowledge. This approach will create interest in learning and a deep sense of attachment to their own culture. This idea shows how the usage of technology which is deeply rooted in tribal culture and tradition, can transform the learning process.

5.4 Mother Tongue-Based Multilingual Education

Learning in their mother tongue is the best way any child gets to understand the content in its entirety. The primary problem with the children of Kattukayakans is their language being different from the formal Tamil that is usually taught at school. It is essential that children learn their lessons in their heart language in order to reap the benefits of learning. Assimilating into another language while making it difficult for the children in one hand, also leads to the endangerment of their native tongue. Lack of scripts may be another challenge in the area of bilingual or multilingual education. This can be dealt with by the use of technology and making education available to the children in the mother tongue. Promoting a multilingual education proves beneficial in a way that children would be made ready to experience the best of education in their native tongue while also exposed to the commonly spoken language of the state. The responsibility rests on the government and the department of education of the state to bring about such a change in collaboration with various stakeholders in the field of education and the community in terms of providing lessons to the children in their native or hear language.

5.5 Capacity building for Educators

Teachers must be recruited from the Kattunayakan tribal community and trained to design meaningful learning experiences for Kattunayakan tribal children. Educators from the community naturally know the children's language, traditions and way of thinking, so the learning becomes joyful and relatable. Teachers from non-tribal backgrounds who sincerely want to help also should be given culturally immersive experiences paired with mentoring and hand-holding by tribal teachers. Lastly, sensitization programs should be frequent so that non-tribal teachers can appreciate and embrace their students, who will be trained for a different worldview. Such programs may concentrate on language subtleties, the rites of the culture and the respectful art of teaching. Basic and advanced technology training will empower teachers, help bridge the learning gaps and make digital education inclusive and accessible. This way the students will have better learning outcomes and feel a deep sense of meaningful connection with the teachers in the classroom.

5.6 Intersectoral Collaboration

Any development has in its core the role of stakeholder involvement and this one is no different from that. Designing culturally tailored centres involves collaborating with various stakeholders like the educators, social workers, anthropologists, technologists and the tribal leaders themselves in order to provide a holistic approach towards improving the access to and the quality of education. While educators can step in with their expertise, the involvement of the tribal leaders gives a foundation on which various pedagogical approaches can be built. The collaborative efforts of various stakeholders provide a better outcome. Apart from individual stakeholders, institutions like the NGOs, educational institutions, forest departments can step in to make the experience more holistic for the tribal children. Collaborations of various kinds also enable in developing education models that combine aspects of health, livelihood and culture, along with education.

5.7 Gender-Sensitive Approaches

Access to education for girls is a challenge especially in tribal communities. This calls for an increased awareness on the role of women in the areas of education and subsequently in the areas of career. Such tuition centres can be forerunners in the area of gender equality through appointing women educators who can be rolemodels for girls to study. These can also be spaces of gender equality among students and of creating awareness among parents. Targeted efforts for the same could include campaigns on creating awareness on the importance of education of girls, community gatherings on gender sensitization to parents with the assurance of a safe space making sure girls can go to school will also help prevent early marriages. This will encourage girls and also help boys and men think more positively about gender equality.

5.8 Participatory Monitoring and Evaluation

Using technology for monitoring and evaluation can help NGOs understand whether the educational activity centres are making a difference in the lives of kattunayakan children. Local people can support to make sure everything is running properly. From mobile apps to digital surveys, simple tools can gather real-time feedback, ensuring their input is always part of the conversation. There is also a strong connection between the community and the teachers by using the Participatory Action Research (PAR) model. These are the facilities that the community can count on to provide them with assistance, and they are more likely to trust these centres if they feel included and respected, which could be a shared path to finding meaningful goals and realistic solutions that benefit their children. This approach is not merely about bringing quality education but about restoring pride, dignity, honour and ownership to the Kattunayakans. Such a community members-driven model can inspire the globe.

6. CONCLUSION

Lack of essential resources and inadequate infrastructure have been constraints for tribal education. However, reckless exclusion from mainstream society has been the most pressing issue for a long time. A tailored educational activity centre would not just accelerate progress but also address the educational gap by creating inclusivity and belongingness. Mother tongue-based education allows children to learn in the language of their heart, not being forced to give up their roots in the name of development or success. It will equip Kattunayakan children with their language, make education a joyful experience, and foster inclusive development. Moreover, the educational activity centres will forge a passing community-centred approach to create a sustainable teaching-learning environment where knowledge can flow from generation to generation. Kattunayakan elders would be the Human Libraries, producing and preserving the knowledge base. When used appropriately, technology will magnify the voices of the most marginalised communities instead of silencing them. One of the most advanced technological tools in the 21st century such as Artificial intelligence or virtual classrooms can provide quality education for the Kattunayakan children ensuring equitable teaching-learning opportunities in the VUCA (Volatile, Uncertain, Complex, and Ambiguous) and BANI (Brittle, Anxious, Non-linear, and Incomprehensible) world. Furthermore, educational activity centres and technology aligned with the Sustainable Development Goal 4 will be a game changer creating impact and promoting lifelong learning opportunities.

The educational models for the Indigenous people, which are found in different corners of the globe, offer noteworthy insights into preserving cultural identity without discounting academic growth. By integrating Participatory Action Research (PAR) models, the NGOs can ensure the interventions are not imposed but co-created for the Kattunayakans. Bridging the bridge by closing the information divide and holding the space for meaningful interactions between social workers, educators, technologists, and tribal leaders will help develop holistic support systems to nurture the intellectual and emotional well-being of tribal children. This approach is not just a demand for temporary measures but a strong insistence on permanent systemic changes. Integrating indigenous knowledge into the teaching-learning advocates for a community and adds layers of richness to their collective human experience.

REFERENCES

About the Western Ghats. (n.d.). WWF India. https://www.wwfindia.org/about_wwf/critical_regions/western_ghats2/about_the_western_ghats/

Alfred, T. (2014). The Akwesasne cultural restoration program: A Mohawk approach to land-based education. *Decolonization, 3*(3), 134–144.

Anusree, G. I., & Umajyothi, V. (2021). A case study of the educational and socio-economic status of the Kattunayakan community in Kerala. *Journal of the Achutha Menon Centre for Health Science Studies, 15*(4), 1–10. https://achuthamenon foundation.org/images/publications/vol15-4.pdf

Anusree, & Umajyothi. (2013). A case study of the Educational and Socio-Economic status of the Kattunayakan community in Kerala. https://achuthamenonfoundation .org/images/publications/vol15-4.pdf

Bartlett, C., Marshall, M., & Marshall, A. (2012). Two-eyed seeing and other lessons learned within a co-learning journey of bringing together Indigenous and mainstream knowledges and ways of knowing. *Journal of Environmental Studies and Sciences, 2*(4), 331–340. DOI: 10.1007/s13412-012-0086-8

Beaulieu, D. (2019). A Survey and Assessment of Culturally Based Education Programs for Native American Students in the United States. https://www .semanticscholar.org/paper/A-Survey-and-Assessment-of-Culturally-Based-for-in -Beaulieu/c08108f30cf49353ac4f01a3c4cdf4accfb1e46b

Behera, K. K., Mandal, P., & Mahapatra, D. (2006). Green Leaves for Diarrhoeal Diseases Used by the Tribals of Kenojhar and Mayurbhanj District of Orissa, India. Ethnobotanical Leaflets, 2006(1), Article 34. https://opensiuc.lib.siu.edu/ebl/vol2006/iss1/34

Bindhani, B. (2021). *Tribal education in India: Issues and challenges*. Concept Publishing Company.

Bindhani, B. K. (2021). Status and problems of educational scenario among the tribals in Koraput district, Odisha. *The Oriental Anthropologist a Bi-annual International Journal of the Science of Man, 21*(1), 158–163. DOI: 10.1177/0972558X21990627

Boote, D. N., & Beile, P. (2005). Scholars Before Researchers: On the Centrality of the Dissertation Literature Review in Research Preparation. Educational Researcher, 34(6), 3-15.

Bose, P. (2021). *Indigenous knowledge and sustainable development: Lessons from tribal communities*. Routledge.

Brahmanandam, T., & Babu, B. T. (2016). Educational status among the Scheduled Tribes: Issues and challenges. The NEHU Journal, 14(2), 55–70. https://nehu.ac.in/public/downloads/Journals/NEHU-Journal-July-Dec-2016-A5.pdf

Brayboy, B. M. J. (2005). Toward a Tribal Critical Race Theory in Education. *The Urban Review*, *37*(5), 425–446. DOI: 10.1007/s11256-005-0018-y

Carter, J., & Hollinsworth, D. (2017). Teaching Indigenous geography in a neo-colonial world. *Journal of Geography in Higher Education*, *41*(2), 174–187. DOI: 10.1080/03098265.2017.1290591

Cooper, H. (2010). *Research Synthesis and Meta-Analysis: A Step-by-Step Approach*. Sage publications.

Core, T. D. H. (2016, September 22). Terre Des Hommes CORE Trust. https://tdhcore.org/6540-2/

Corntassel, J. (2012). Re-envisioning resurgence: Indigenous pathways to decolonization and sustainable self-determination. *Decolonization*, *1*(1), 86–101.

Daripa, S. (2017). Tribal Education in India: Government Initiative and Challenges. *International Journal of Research in Social Sciences*, *7*(11), 2249–2496. https://www.ijmra.us/project%20doc/2017/IJRSS_OCTOBER2017/IJMRA-12325.pdf

Desai, A. R. (2010). *Tribes and Indian society*. Rawat Publications.

Elek, C., Gubhaju, L., Lloyd-Johnsen, C., Eades, S., & Goldfeld, S. (2020). Can early childhood education programs support positive outcomes for indigenous children? A systematic review of the international literature. *Educational Research Review*, *31*, 100363. DOI: 10.1016/j.edurev.2020.100363

Evening Coaching Centers (ECC) – CTRD Trust. (2025). Ctrd.in. https://ctrd.in/evening-coaching-centers-ecc/

Eze, U. T., & Nnia, I. (2022). Integrating African Indigenous Knowledge In Nigeria's Formal Education System: It's Potential For Sustainable Development. Journal of Education and Practice. https://www.semanticscholar.org/paper/Integrating-African-Indigenous-Knowledge-In-Formal-Eze-Nnia/d20712b209d23c5775894d5726dd32da2240788b

Francis, W. (1908). *Madras District Gazetteers: The Nilgiris*. Logos Press.

Gadgil, M., & Meher-Homji, V. M. (1986). Role of protected areas in conservation. In Chopra, V. L., & Khoshoo, T. N. (Eds.), *Conservation for Productive Agriculture* (pp. 143–159). Indian Council of Agricultural Research.

García, O., & Wei, L. (2014). *Translanguaging: Language, Bilingualism and Education*. Palgrave Macmillan. DOI: 10.1057/9781137385765

Goforth, A. N., Nichols, L. M., Sun, J., Violante, A. E., Brooke, E., Kusumaningsih, S., Howlett, R., Hogenson, D., & Graham, N. (2022). Cultural Adaptation of an Educator Social–Emotional Learning Program to Support Indigenous Students. *School Psychology Review*, *53*(4), 1–17. DOI: 10.1080/2372966X.2022.2144091 PMID: 39100141

Gopalakrishnan, M. (1995). Gazetteers of India, Tamil Nadu State, The Nilgiris District. Madras Press.

Hare, J. (2011). "They tell a story and there's meaning behind that story": Indigenous knowledge and young indigenous children's literacy learning. *Journal of Early Childhood Literacy*, *12*(4), 389–414. DOI: 10.1177/1468798411417378

Herskovits, M. F. (1955). Cultural anthropology. Oxford & IBH Publishing Co.

Indigenous People's Plan Kattunayakans. (n.d.). https://www.cepf.net/sites/default/files/56154-ipp-kattunayakan.pdf

Jayakumar, C., Sankaran, S., & Gangadharan, P. (2024). Looking inward, looking forward: Articulating alternatives to the current education system for adivasis, by Adivasis. *Contemporary Education Dialogue*, *21*(1), 108–124. DOI: 10.1177/09731849231212111

Kanti, A. (2015). Holistic Tribal Education: A Case Study of Kalinga Institute of Social Sciences in Odisha, India. https://www.semanticscholar.org/paper/Holistic-Tribal-Education%3A-A-Case-Study-of-Kalinga-Kanti/b510a16c24faa918bbcdb649befc18e5ac46563c

Kaur, N., & Kaur, G. (2022). Educating the Tribals: Government Interventions. GHG Journal of Sixth Thought, 9(1). https://www.ghgkce.org/files/education/journal/j_6_3_2022.pdf

Kolb, D. A. (1984). *Experiential learning: Experience as the source of learning and development*. Prentice Hall.

Ladson-Billings, G. (1995). Toward a theory of culturally relevant pedagogy. *American Educational Research Journal*, *32*(3), 465–491. DOI: 10.3102/00028312032003465

Luiz, A. A. D. (1962). *The tribes of Kerala*. Bharatiya Adimjati Sevak Sangh.

Machi, L. A., & McEvoy, B. T. (2016). *The Literature Review: Six Steps to Success.* Corwin Press. DOI: 10.4135/9781071939031

McCarty, T. L., & Lee, T. S. (2014). Critical culturally sustaining/revitalizing pedagogy and Indigenous education sovereignty. *Harvard Educational Review*, *84*(1), 101–124. DOI: 10.17763/haer.84.1.q83746nl5pj34216

Menon, S., & Bawa, K. S. (1997). Applications of Geographic Information System, Remote-Sensing, and Landscape Ecology Approach to Biodiversity Conservation in the Western Ghats. *Current Science*, *73*, 134–145. https://www.scirp.org/reference/referencespapers?referenceid=3010812

Molur, S. (2009). Habitat and status assessment of mammals with special reference to rodents and bats in Western Ghats of Karnataka [PhD Thesis]. University of Mysore.

Molur, S., Smith, K. G., Daniel, B. A., Darwall, W. R. T., & Zoo Outreach Organisation. (2011). The status and distribution of freshwater biodiversity in the Western Ghats, India. IUCN.

Moreton-Robinson, A. (2021). *Talkin' up to the white woman: Indigenous women and feminism*. University of Minnesota Press.

Muneer, V. (2015). Analysis of status of tribal students in selected aspects and functioning of tribal residential schools in Kerala. [Doctoral dissertation]. http://hdl.handle.net/10603/349029

Myers, N., Mittermeier, R. A., Mittermeier, C. G., Da Fonseca, G. A. B., & Kent, J. (2000). Biodiversity hotspots for conservation priorities. *Nature*, *403*(6772), 853–858. DOI: 10.1038/35002501 PMID: 10706275

Nag, D. (2024). When Technology Meets Tradition. https://doi.org/DOI: 10.1002/9781394200498.ch2

Nakata, M. (2007). *Disciplining the Savages: Savaging the Disciplines*. Aboriginal Studies Press.

Nanda, A. K., & Samanta, D. (2017). *Education and social change among Indian tribes*. Rawat Publications.

Nanda, A. K., & Samanta, S. (2017). Mainstreaming tribals through financial literacy – a review of literature. *International Journal of Social Economics*, *45*(2), 437–444. DOI: 10.1108/IJSE-12-2016-0371

Natarajan, T. S. (1985). *Tribal habitats of Nilgiri District: A profile*. Tribal Research Centre.

Nayak, S. R., Kant, N., & Anjali, K. (2020). Strategy of using ICT in ODL to disseminate higher education in tribal communities: a case of MP, India. Asian Association of Open Universities Journal, 15(2), 189-206.

Ole Henrik Magga. Nicolaisen, I., Trask, M., & Dunbar, R. (2005). Indigenous Children's Education and Indigenous Languages. ResearchGate. https://www.researchgate.net/publication/255571823_Indigenous_Children

Ortiz, F. (1947). *Transculturation and Cuba*. University of Miami Press.

Parthasarathy, J. (2013). *Education and Development Among The Tribes. Tribal Research Center (TRC) & Hill Area Development Programme (HADP)*. Government of Tamil Nadu.

Parthasarathy, J. (2013). Kattunayakans of Nilgiris District, Tamil Nadu. Tribal Research Center (TRC) & Hill Area Development Programme (HADP), Government of Tamil Nadu.

Project, J. (n.d.). Kattunayakan in India. https://joshuaproject.net/people_groups/17305/IN

Ramachandran, B., Zvelebil, K. V., David, N. B., Buchanan, F., Dubois, J. A., & Aiyappan, A. Census of India, Tribal Research Center, Ooty, Keystone Foundation, & Seetha Kakkoth. (n.d.). Indigenous People's Plan Kattunayakans. https://www.cepf.net/sites/default/files/56154-ipp-kattunayakan.pdf

Ramakrishnan, R., Rajendrakumar, S., & Kothurkar, N. K. (2024). Regional sustainability of the Kattunayakan tribe in Kerala, India through the enhancement of agricultural, livestock, and livelihood options. *Agricultural Systems*, *217*, 103929. DOI: 10.1016/j.agsy.2024.103929

Ramesh, B. R., Venugopal, P. D., Pélissier, R., Patil, S. V., Swaminath, M. H., & Couteron, P. (2010). Mesoscale patterns in the floristic composition of forests in the central western Ghats of Karnataka, India. *Biotropica*, *42*(4), 435–443. DOI: 10.1111/j.1744-7429.2009.00621.x

Reddy, S. K., & Reddy, M. (2020). *Indigenous knowledge systems and sustainable practices: The case of the Kattunayakan tribe*. Indigenous Studies Press.

Redvers, J. (2020). "The land is a healer": Perspectives on land-based healing from Indigenous practitioners in northern Canada. *International Journal of Indigenous Health*, *15*(1), 90–107. DOI: 10.32799/ijih.v15i1.34046

Rowe, R. K., & Shawanda, A. (2023). Indigenous Pedagogies and the Implications of EdTech, Data, and AI in the Classroom. In Perspectives on Indigenous Pedagogy in Education: Learning From One Another (pp. 215-247). IGI Global.

Rupavath, R. (2021). Poverty and Education: Attainments and challenges for tribal communities. *Contemporary Voice of Dalit, 15*(2), 230–247. DOI: 10.1177/2455328X211042718

Saravanan, S. P. (2024, November 18). Evening coaching classes enhance skills of tribal students in Sathyamangalam Tiger Reserve. The Hindu. https://www.thehindu.com/news/cities/Coimbatore/evening-coaching-classes-enhance-skills-of-tribal-students-in-sathyamangalam-tiger-reserve/article68881434.ece

Sasikumar, M. (1999). *Kattunaickans: An ethnographic report*. International School of Dravidian Linguistics.

Saulis, P. (2018). Indigenous knowledge and Western science [Video]. YouTube. https://www.youtube.com/watch?v=cwCfsAiERyk

Shava, S. (2016). The Application/role of Indigenous Knowledges in Transforming the Formal education Curriculum: Cases from Southern Africa. In V. Msila & M. T. Gumbo (Eds.), Africanising the Curriculum: Indigenous Perspectives and Theories (pp. 121–137). AFRICAN SUN MeDIA.

Simpson, L. B. (2014). Land as pedagogy: Nishnaabeg intelligence and rebellious transformation. *Decolonization, 3*(3).

Snyder, H. (2019). Literature review as a research methodology: An overview and guidelines. *Journal of Business Research, 104*, 333–339. DOI: 10.1016/j.jbusres.2019.07.039

Tranfield, D., Denyer, D., & Smart, P. (2003). Towards a methodology for developing evidence-informed management knowledge by means of systematic review. *British Journal of Management, 14*(3), 207–222. DOI: 10.1111/1467-8551.00375

Tuttle, H. (2013). Technology-Based Curriculum Development in Native Language Learning: How Computer-Assisted Cultural Education Supports Contemporary Native Youths.

UNESCO World Heritage Centre. (n.d.). Western Ghats. https://whc.unesco.org/en/list/1342/

United Nations. (2015). Transforming our world: The 2030 Agenda for Sustainable Development. United Nations. https://sdgs.un.org/2030agenda

Vowel, C. (2016). *Indigenous Writes: A Guide to First Nations, Métis, and Inuit Issues in Canada*. Highwater Press.

Xaxa, V. (2001). Protective Discrimination: Why Scheduled Tribes Lag behind Scheduled Castes. *Economic and Political Weekly*, *36*(29), 2765–2772. https://www.jstor.org/stable/4410888

Xaxa, V. (2005). Politics of Language, Religion and Identity: Tribes in India. *Economic and Political Weekly*, *40*(13), 1363–1370. https://www.jstor.org/stable/4416402

Xaxa, V. (2016). Tribes and Indian national identity. *The Brown Journal of World Affairs*, *23*(1), 223–237. https://www.jstor.org/stable/26534720

Chapter 4

Antecedents of Farmers' Intentions to Adopt Drone Technology:

Integrating TAM and TPB Models in Agriculture

M. P. Sunil
https://orcid.org/0000-0002-7781-450X
Christ University, India

Akanksha Khanna
https://orcid.org/0000-0003-2156-4852
Symbiosis International University, UAE

ABSTRACT

This study examines farmers' behavioral intentions toward adopting drone technology in agriculture using a quantitative approach. Data will be collected from farmers familiar with Agri-based technologies through purposive sampling, with responses analyzed via structural equation modeling (IBM SPSS AMOS V21). The study integrates the Technology Acceptance Model (TAM) and the Theory of Planned Behavior (TPB) to understand key factors influencing adoption. Findings highlight that farmers' knowledge of drone operations and economic factors impact their acceptance, while regulations and perceived behavioral control contribute to reluctance. Although the study focuses on farmers, future research can explore perspectives of agricultural specialists, AgriTech start-ups, and policymakers. The insights will help stakeholders enhance the adoption, quality, and flexibility of

DOI: 10.4018/979-8-3373-7922-7.ch004

agricultural technologies. Uniquely, this study addresses the gap in research by examining the role of regulations and behavioral control in farmers' decisions on drone adoption.

1. INTRODUCTION

Agriculture is a dynamic system and a primary sector of the world economy. There is a high dependency on agriculture for food, raw materials, employment, and infrastructural development (Vincent et al., 2019; Cheruku & Katekar, 2023). In many countries, agricultural growth significantly impacts national income and economic development (Pingali, 2007). Likewise, in India, the Economic Survey (2019-20) states that more than 65% of the population depends on agriculture for their livelihood, contributing 16.5% to GDP. Despite being the largest sector of the Indian economy, farmers still struggle to achieve adequate agricultural yield due to several challenges, including traditional cultivation methods, insufficient knowledge, lack of innovation, limited technological awareness, changing climatic conditions, environmental pollution, poor soil productivity, labor shortages, trade disruptions, inadequate irrigation, drought, and flooding (Mariappan & Zhou, 2019; Daponte et al., 2019; Selvan et al., 2021).

Digital technology has the potential to transform agriculture by improving efficiency and sustainability. Emerging technologies such as Artificial Intelligence (AI), the Internet of Things (IoT), Big Data, and Blockchain technology are key drivers in modernizing agricultural practices (Liu & Liu, 2023; Arhin et al., 2024). Water and wind erosion are major issues that digital technologies can help mitigate (Selvan et al., 2021). Additionally, AI plays a crucial role in providing farmers with key information about weather, soil conditions, pest infestations, and crop health, enabling better decision-making (Joshi et al., 2020; Sood et al., 2022). Farmer-centric innovation ensures that technological advancements align with their needs and practical applications (Madureira et al., 2022). Technology adoption has been found to increase efficiency and production (Cao et al., 2025; Klerkx et al., 2019).

Among digital innovations, drone technology, also known as Unmanned Aerial Vehicles (UAVs), has emerged as a game-changer in modern agriculture. Drones provide accurate, timely, and efficient data collection for field analysis, planting, crop spraying, crop monitoring, irrigation management, crop health assessment, crop insurance, weed identification, and geofencing (Rani et al., 2019; Cheruku & Katekar, 2023). AI-driven drones optimize resource usage, reduce environmental impact, and enhance agricultural sustainability (Canicattì & Vallone, 2024). Learning ease and trialability are primary factors influencing farmers' adoption of drones (Takagi et al., 2020). However, many Indian farmers remain hesitant to adopt drone technology due to regulatory concerns and a lack of awareness. Several countries are implementing regulations to manage risks associated with drone adoption in agriculture (Ayamga et al., 2020). If widely accepted, drone technology could significantly transform Indian agriculture with governmental support and policy enhancements (Hua & Wang, 2019; Ouko et al., 2022).

To understand and explore farmers' behavioral intentions toward drone adoption, this study integrates the Technology Acceptance Model (TAM) and the Theory of Planned Behavior (TPB). TAM, proposed by Davis (1989), examines the acceptance of various technologies based on perceived usefulness and ease of use. TPB, developed by Fishbein and Ajzen (1975), provides a framework for understanding behavioral intentions by analyzing attitudes, subjective norms, and perceived behavioral control. These models are widely used in information technology and behavioral studies (Zhang et al., 2019; Akudugu et al., 2023). The research aims to identify key determinants influencing farmers' decisions to adopt drones and assess how behavioral factors impact their final acceptance of this technology (Mukherjee et al., 2017; Cao et al., 2025).

By integrating TAM and TPB, this study seeks to provide insights into the factors driving drone adoption in agriculture. Given the increasing role of digital technology in farming, understanding farmers' perspectives can help develop policies and strategies to encourage widespread adoption. Additionally, government initiatives and regulatory frameworks must align with farmers' needs to facilitate the seamless integration of drone technology into agricultural practices (Noor & Noel, 2023; Mack et al., 2024).

2. THEORETICAL BACKGROUND

2.1 Technology Acceptance Model (TAM) and Theory of Planned Behaviour (TPB)

The Technology Acceptance Model (TAM), originally proposed by Davis in 1989, identifies four key variables: perceived usefulness, perceived ease of use, attitude towards usage, and intention to use. These variables help in understanding the acceptance of technology by users (Davis, 1989). TAM was initially designed to explain the general determinants of computer acceptance but has since evolved into a widely used framework in the fields of information technology and information systems (Zhang et al., 2019; Ouko et al., 2022).

The Theory of Planned Behaviour (TPB), proposed by Fishbein and Ajzen in 1975, was derived from the Theory of Reasoned Action (TRA). TPB includes three important variables: attitude, subjective norm, and perceived behavioural control, which collectively influence behavioural intention (Fishbein & Ajzen, 1975; Akudugu et al., 2023). TPB is widely applied to study human behaviour in specific circumstances (Hua & Wang, 2019; Bhuiyan & Mahato, 2024).

This study emphasizes the application of drone technology in agricultural activities and explores its potential to enhance performance and productivity. Given the necessity for farmers to adopt such technology, TAM and TPB serve as foundational theories to understand the factors influencing their acceptance. The research aims to examine the determinants of farmers' intention to use drone technology by focusing on user acceptance and the behavioral influences shaping their decisions.

2.2 Perceived Usefulness

Perceived usefulness refers to an individual's belief that a specific system will enhance performance (Davis, 1989). Drones are recognized for their ability to improve crop health monitoring, optimize resource usage, and promote sustainability by reducing chemical overuse (P et al., 2024; Cheruku & Katekar, 2023). According to the TAM framework, perceived usefulness significantly impacts behavioural intention to use a technology (Cheruku & Katekar, 2023).

In the agricultural sector, farmers prioritize methods that require minimal effort and time while ensuring high accuracy. The perceived ease of use of drone technology directly influences its perceived usefulness. If drone technology simplifies agricultural tasks and improves efficiency, farmers are more likely to adopt it (G. C et al., 2023). Thus, we hypothesize that farmers will accept drone technology if they find it beneficial in enhancing effectiveness and quality in agricultural activities (Hua & Wang, 2019; Shang et al., 2021).

2.3 Perceived Ease of Use

Perceived ease of use is defined as the extent to which an individual believes that using a specific system will be simple and convenient (Davis, 1989). It is a key determinant of behavioural intention in the TAM framework. When a technology reduces effort and simplifies tasks, users are more likely to adopt it.

In the context of agriculture, drones represent a shift from traditional farming methods. Their ease of use depends on factors such as learning simplicity, operational clarity, and flexibility in meeting agricultural needs (Venkatesh, 2000; P et al., 2024). Farmers' willingness to adopt drone technology is influenced by how intuitive and user-friendly they perceive it to be (Mahasneh, 2024). Consequently, perceived ease of use has a positive impact on attitudes toward using drone technology in agriculture.

2.4 Attitude Towards Usage

Attitude is an individual's positive or negative evaluation of a particular behaviour (Ajzen & Fishbein, 1980). Attitudes are shaped by behavioural beliefs, which involve evaluating the consequences of an action and judging its favourability (Ajzen, 1991; Shang et al., 2021).

In this study, attitude reflects farmers' perceptions of drone technology adoption, which can be positive or negative. Perceived usefulness and perceived ease of use significantly influence attitudes toward drone technology, as farmers value its benefits in improving agricultural practices (Hua & Wang, 2019; Sundar et al., 2023). Attitudes toward drone usage positively impact farmers' intention to adopt the technology (Verma & Sinha, 2018; Noor & Noel, 2023).

2.5 Intention to Use

Intention to use refers to an individual's tendency to adopt a specific technology (Phillips et al., 1994; Subramanian et al., 2021). Both TAM and TPB emphasize intention to use as a predictor of actual adoption.

Attitude, regulation, and perceived behavioural control influence farmers' intention to adopt drone technology. Regulations and policies can either facilitate or hinder adoption, depending on their flexibility. Additionally, farmers with higher perceived behavioural control, such as economic resources and confidence in technology use, exhibit stronger intentions to adopt drones (C et al., 2024; Suvittawat, 2024).

2.6 Social Influence

Social influence is the extent to which an individual perceives that their environment affects their behaviour (Venkatesh et al., 2003; Barathkumar et al., 2024). It plays a crucial role in technology adoption, particularly in early adoption stages.

Farmers' decision to use drone technology is influenced by opinions from their social circles, including peers, family members, and agricultural experts (Suvittawat, 2024). Social influence can shape farmers' perceptions of the usefulness and ease of using drones, ultimately impacting their adoption decisions (Kamble et al., 2024).

2.7 Compatibility

Compatibility is the extent to which an innovation aligns with users' past experiences, values, and needs (Rogers, 1995). Farmers are less likely to adopt technology that does not integrate well with their existing agricultural practices (Kamble et al., 2024).

For drone technology to be widely adopted, it must be compatible with agricultural activities, financial conditions, and farmers' knowledge levels. Compatibility directly influences perceived usefulness and serves as a critical factor in farmers' final decisions on adoption (Subramanian et al., 2021; Suvittawat, 2024).

2.8 Self-Efficacy

Self-efficacy refers to an individual's belief in their ability to successfully perform a task (Bandura, 1977; McCarthy et al., 2023). It impacts confidence, skills, and willingness to overcome obstacles (Rezaei et al., 2020; Barathkumar et al., 2024).

Farmers with high self-efficacy are more likely to find drone technology useful and easy to use. These individuals may also provide guidance to others, further promoting drone adoption in agricultural communities.

2.9 Regulation

Regulation consists of official rules established by governing authorities to control the use of technology (Levi-Faur, 2011). It plays a crucial role in farmers' decisions regarding drone adoption.

Lack of awareness about drone-related regulations may discourage farmers from using the technology. Additionally, rigid policies may create skepticism. The study highlights the need for flexible government policies that encourage drone adoption in agriculture (Mack et al., 2024). Implementing supportive regulations can transform the agricultural sector and improve farmers' lives (Cheruku & Katekar, 2023).

2.10 Perceived Behavioural Control

Perceived behavioural control refers to an individual's judgment of how easy or difficult it is to perform a specific action (Ajzen & Fishbein, 1980; Bhuiyan & Mahato, 2024). It encompasses factors such as financial resources, time, and technological confidence.

Farmers with sufficient resources and opportunities exhibit higher perceived behavioural control and greater willingness to adopt drone technology (Hua & Wang, 2019; McCarthy et al., 2023). Additionally, drones are perceived as a solution to agricultural challenges, such as water scarcity (Noor & Noel, 2023). Thus, perceived behavioural control positively influences farmers' intention to use drone technology in agriculture.

Table 1. Definition of the variables

Variable	Definition	References
Perceived Usefulness	The degree to which an individual farmer is convinced that accepting Drone technology will help in achieving great performance in agriculture.	(Davis, 1989)
Perceived Ease of Use	The degree to which an individual farmer believes that using Drone technology would be simple and convenient for operation in agriculture.	(Davis, 1989)
Attitude Towards Usage	An individual farmer's positive or negative feeling and evaluation of performance about accepting drone technology in agriculture.	(Ajzen, I., & Fishbein, 1980)
Intention to Use	Specialist's Tendency of Intention to extension Drone Technology in agriculture among farmers.	(Phillips et al., 1994)
Social Influence	The degree to which a farmer feels that his behaviour is influenced by surrounding environment important individual who believes in utilizing a drone technology in agriculture.	(Venkatesh et al., 2003)
Compatibility	The degree to which an innovation is considered to be reliable and consistent along with past experience, needs and existing values with the drone technology in agriculture.	(Rogers, 1995)
Self-Efficacy	Farmer's belief of capability in performing and accomplishing agriculture task successfully with the use of Drone Technology.	(Bandura, 1977)
Regulation	The official rules which is made exclusively by the government or the concerned authority in order to have control which is enforced by the law pertaining to usage of Drone Technology in agriculture.	(Levi-Faur, 2011)
Perceived Behavioural Control	Farmers judgement on the degree of difficulty in the way of carrying out drone operations in agricultural activities.	(Ajzen, I., & Fishbein, 1980)

Source: Author Compiled

3. RESEARCH PROBLEM AND HYPOTHESIS DEVELOPMENT

The study sought to establish the influence of farmer's behaviour on the Intension to use Drone Technology. The study assists in identifying various types of factors influencing the farmers behaviour on acceptance of Drone technology in agricultural activities. It also explores which factors would be causing transformation in acceptance of Drone Technology. Farmers think differently based on their experience, opinion, influence, methods, knowledge and awareness, particularly in decision-making. Considering all the factors, we have framed major variables affecting the influencing behaviour of farmers in acceptance of drone technology.

The theoretical model of our study is based on the Technology Acceptance Model (TAM) and the Theory of Planned Behaviour (TBH). The collective basis of Social influence, Compatibility and Self-Efficacy has a significant effect on both perceived usefulness and perceived ease of use and they together affect the farmer's attitude towards accepting Drone Technology in agricultural activities. At the same time, regulation and perceived behavioural control directly impact farmer's behavioural intentions.

Based on the theoretical framework and the presented model, the following hypotheses are proposed.

H1: Social Influence, Compatibility and Self-Efficacy collectively are the basis for Perceived Usefulness

H2: Social Influence, Compatibility and Self-Efficacy collectively are the basis for Perceived Ease of Use

H3: Perceived Ease of Use has a Positive impact on Perceived Usefulness

H4: Perceived Usefulness has a positive impact on Attitude Towards Usage

H5: Perceived Ease of Use has a Positive impact on Attitude Towards Usage

H6: Attitude Towards Usage has a Positive impact on Farmer's Intention to Use

H7: Regulation has a positive impact on Farmer's Intention to Use

H8: Perceived Behavioural Control has a positive impact on Farmer's intention to Use.

4. METHODS AND MATERIALS

Data was collected using online questionnaire and offline questionnaire by adopting a random sampling technique to select 430 respondents. Respondents included farmers who have a little awareness of technologies like drone technology. The study used Random sampling which is a technique where a random portion of the population represents the entire set of data and has an equal probability of being chosen. The data was collected during the period of May 2024 till December 2024. The questionnaire had Seven-Point Likert Scale ranging from (1) Strongly Disagree to (7) Strongly Agree. The Measure of constructs of questionnaire are referred to Table 2. The Questionnaire was divided into two parts; the first part was about demographic factors, and the second with a measuring range of variables in constructs.

Table 2. Constructs and Measurement Scale Variables

Constructs	Question Tag	Items	References
Perceived Usefulness	PU1 PU2 PU3 PU4 PU5 PU6	Using Drones in Agriculture would accomplish tasks more quickly Using Drones would enhance Agricultural performance Using Drones in Agriculture would increase productivity Using Drones would enhance effectiveness of the agricultural activities Using Drones would make it easier to perform the agricultural activities Drones found to be useful in Agriculture	Adopted from (Davis, 1989)
Perceived Ease of Use	PEOU1 PEOU2 PEOU3 PEOU4 PEOU5 PEOU6	Learning to operate Drones in Agriculture would be easy Drones would make work easy as per requirements in Agriculture Operating with drones would be clear and understandable Drones in Agriculture would be flexible to operate with It is easy to become skilful in Agriculture by using Drones Drones are simple to use in Agriculture	Adopted from (Davis, 1989)
Attitude Towards Usage	ATU1 ATU2 ATU3 ATU4	Using Drones in Agriculture would be beneficial Using Drones would be pleasant in Agricultural practices Using Drones in Agriculture would be good Using Drones in the Agricultural practices are valuable	Adopted from (Ajzen, 2002)
Intension to Use	ITU1 ITU2 ITU3 ITU4	There is high possibility to use Drones in Agriculture within the foreseeable future Intend to use Drones in the Agricultural activities within the foreseeable future Complete willingness in using Drones in Agriculture within the foreseeable future Using Drones in Agriculture would be important in the foreseeable future	Adopted from (Venkatesh et al., 2008)
Social Influence	SI1 SI2 SI3	People who are important to you would think that you should use Drones in agriculture People would influence the behaviour of yours to use Drones in agriculture People's whose opinions are valued by you would prefer usage of Drones in agriculture	Adopted from (Venkatesh et al., 2012)
Compatibility	COMP1 COMP2 COMP3 COMP4	Using Drones would be compatible with most conditions of the Agriculture Drones would fit well with required production practices and agricultural activities Using Drones in Agriculture would be consistent with the financial situations Using Drones in agriculture would be consistent with information and knowledge acquired along with prior experience	Adopted from (Moore & Benbasat, 1991), (Abdollahzadeh et al., 2016), (Sharifzadeh et al., 2017)
Self-Efficacy	SE1 SE2 SE3	You can be confident in using Drones for different agricultural practices Skills, Experience and knowledge are essential to use Drones in agriculture By using Drones in Agriculture, you can give technical advice to others	Adopted from (Ajzen, 1991), (Francis et al., 2004), (Sharifzadeh et al., 2017)
Regulation	RE1 RE2	You are well aware of all rules and regulation pertaining to the flying of Drones for agricultural activities The Effective Policy Registration process pertaining to use of Drones in agriculture are simple and user-friendly	Adopted from (Rajagopalan, R. P., & Krishna, R. 2018), (Singh et al., 2019)
Perceived Behavioural Control	PBC1 PBC2 PBC3	Whether you would purchase the Drone or not is completely up to your decision making The resources, time and opportunities are the requirements to purchase Drones in Agriculture You can be confident, to perform the required activities and purchase Drones at your convenience	Adopted from (Kim & Han, 2010)

5. RESULTS AND DATA ANALYSIS

5.1 Proposed Model

Figure 1. Proposed Model

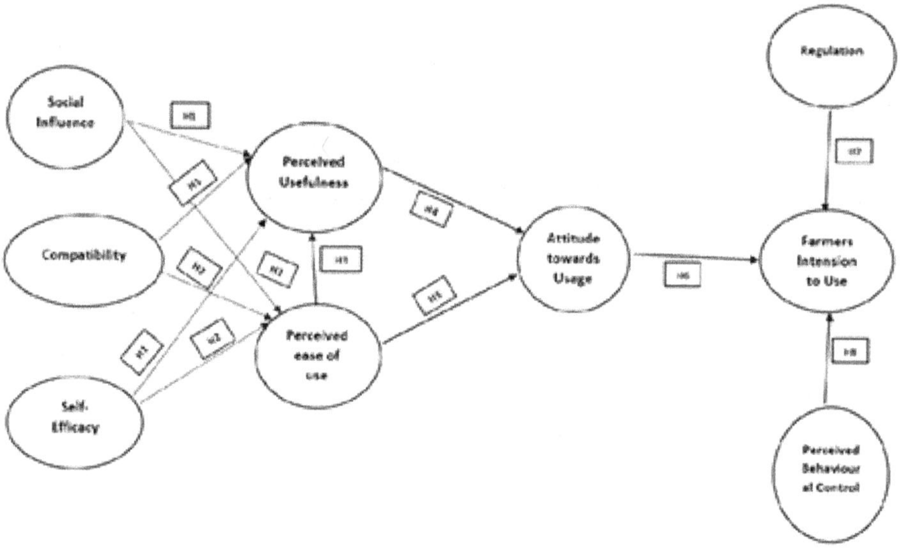

Figure 1 shows the proposed model developed using the theoretical background and hypotheses development. With the implementation of TAM and TPB, 9 variables were developed, and a questionnaire was devised as a measurement scale for the study. Figure 1: The proposed Model clearly mentions the variables used and identified for this study. Perceived Usefulness and Perceived Ease of Use is adopted from (Davis, 1989), Attitude towards usage from (Ajzen, 2002), Intension to Use from (Venkatesh et al., 2008), Social Influence from (Venkatesh et al., 2012), Compatibility from (Moore & Benbasat, 1991), (Abdollahzadeh et al., 2016), (Sharifzadeh et al., 2017), Self-Efficacy from (Ajzen, 1991), (Francis et al., 2004), (Sharifzadeh et al., 2017), Regulation from (Rajagopalan, R. P., & Krishna, R. 2018), (Singh et al., 2019) and Perceived Behavioural Control from (Kim & Han, 2010). The new variables which is introduced to the research is Regulation. The structured questionnaire is constructed with the items, and a 7-point Likert scale is used to measure the level of agreement with each item. Here, '1' signified the least agreement and '7' signified the highest level of agreement with the given statement. A total of 296 questionnaires were collected. The majority of the respondents were in the age group of 31-60 and above 60. The majority of respondents have at least a school education and are aware of different technologies to use in agriculture.

5.2 Demographic Profile

Table 3 shows the demographic profile. It describes that 64.2% are males, 34.9% are females and 0.9% of transgender in the agricultural activity. 20.2% are in the age group of 18-30, 37.7% are in 31-60 age group and in the age group of more than 60 there are 42.1% respondents. Educational Qualification discloses that 48.1% of farmers have minimum school education, 23.5% are qualified in Intermediate College/Diploma, and 17.9% are qualified as undergraduates, 1.9% with post-graduation where in 8.6% with the other qualifications. Annual income disclosed that 61.4% farmers fall in the group of less than 2.5 lakhs of annual income, 31.2% in 2.5-5 lakhs, 6.3% in the annual income group of 5-10 lakhs and lastly 1.2% of farmers are having the annual income of more than 10 Lakhs. 36% farmers are having a landholding of less than 4 acres, 22.6% having between 4-10 acres of land holding, whereas 41.4% owns more than 10acres of land holdings. Agricultural speciality is in horticulture of about 29.1%, farmers involved in irrigation is about 24.7%, Plantation 16.0%, agronomy of about 25.1% and other agricultural activity is about 5.1%.

Table 3. Demographic Profile

Demographic Details	N	Valid percentage
AGE	87	20.2
18-30	162	37.7
31-60	181	42.1
More than 60		
LAND HOLDING	155	36.0
Less than 4 acres	97	22.6
Between 4-10 acres	178	41.4
More than 10 acres		
GENDER	150	34.9
Male	276	64.2
Female	4	0.9
Transgender		
EDUCATION QUALIFICATION	207	48.1
School Education	101	23.5
Intermediate College/Diploma	77	17.9
UG	8	1.9
PG	35	8.6
Others		

continued on following page

Table 3. Continued

Demographic Details	N	Valid percentage
ANNUAL INCOME	264	61.4
Less than 2.5 lakhs	134	31.2
2.5 – 5 lakhs	27	6.3
5 – 10 lakhs	5	1.2
More than 10 lakhs		
AGRICULTURE SPECIALITY	125	29.1
Horticulture	106	24.7
Irrigation	69	16.0
Plantation	108	25.1
Agronomy	22	5.1
Others		

Source: survey data

5.3 Measurement model

Figure 2. Measurement model

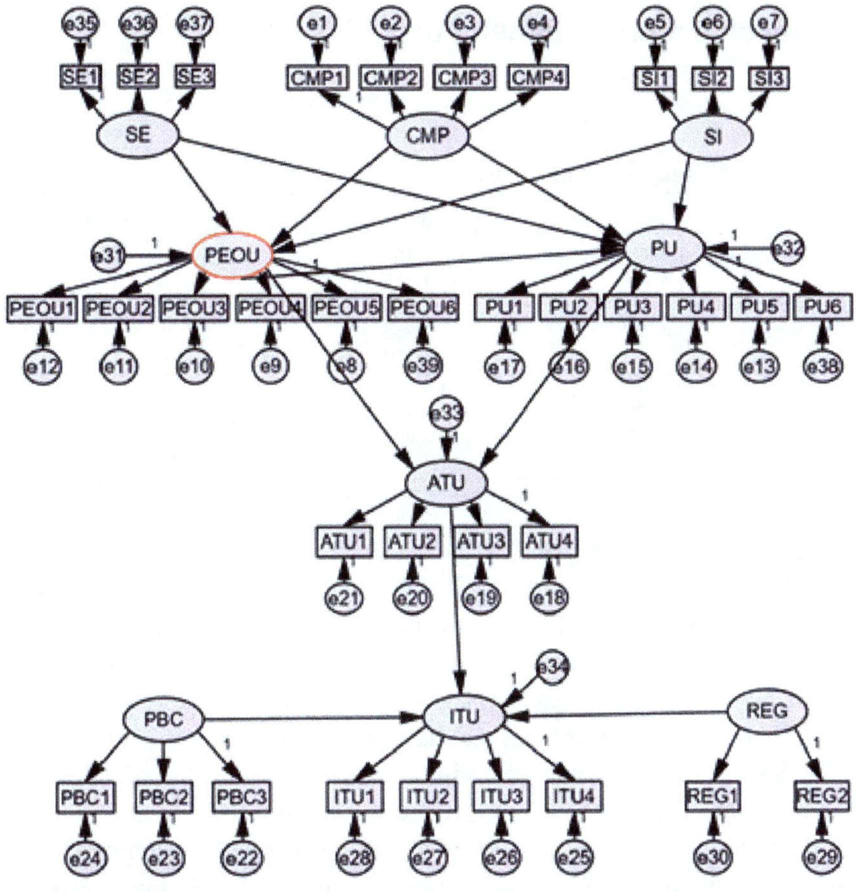

A visual diagram depicting the measurement model is shown in Figure 2 as a proposed model using Structural Equation Modeling (SEM). The measurement model contains the constructs: PU, PEOU, SI, CMP, SE, Attitude, ITU, REG and PBC. The descriptive statistics of the data and reliability mentioned in Table 4 and Table 5 are done using SPSS. The data has been tested for reliability using Cronbach's alpha reliability test. Confirmatory Factor Analysis (CFA) is used to check the data for reliability and validity. Factor loading indicates the correlation between the item and the latent variable. Composite reliability estimates the internal consistency of the measurement model and Average Variance Extracted test also is mentioned in the Table 10. The goodness of fit measures include Chi-square, P-value, Goodness-of-Fit (GFI), Adjusted Goodness-of-Fit (AGFI), Normed Fit Index (NFI), Comparative Fit Index (CFI), Root Mean Square Error of Approximation (RMSEA).

5.4 Cronbach's alpha reliability test

Table 4. Cronbach's alpha reliability test

Construct	Mean	Variance	Cronbach α
PU	4.997	1.181	0.673
PEOU	4.785	1.215	0.654
ATU	4.951	1.160	0.627
ITU	4.998	1.255	0.705
SE	4.780	1.129	0.611
SI	4.950	1.334	0.702
CMP	5.063	1.213	0.675
PBC	4.850	1.609	0.737
REG	4.666	1.154	0.623

Table 4 shows the test result of Cronbach's alpha. All nine constructs were above the acceptable limit of 0.5(*George, D., & Mallery, P. (2003*) and ranged from 0.611 to 0.737. PU (Perceive Usefulness) consists of 6 items and has a Cronbach α of 0.673. PEOU (Presided Ease of Use) also consists of 6 items and holds Cronbach α of 0.654. ATU (Attitude towards Usage) has 4 items in it and has Cronbach α of 0.627. ITU (Intention towards Usage) has 4 items in it and has a Cronbach α of 0.705. SE (Social Efficacy) has 3 items in it and has a Cronbach α of 0.611. Social Influence (SI) has 3 items with a Cronbach α of 0.702. CMP (Compatibility) has 4 items in it and holds a Cronbach α of 0.675. PBC (Perceived Behavioural control) has 3 items in it and holds a Cronbach α of 0.737. REG (Regulation) consists of 2 items and has a Cronbach α of 0.623.

5.5 Descriptive Statistics

Table 5. Descriptive Statistics

	Factor Loading	Mean	Std. Deviation	Variance
PU1	0.500	4.95	1.119	1.251
PU2	0.734	4.82	1.084	1.174
PU3	0.636	4.97	1.036	1.074
PU4	0.609	5.06	1.174	1.379
PU5	0.605	5.35	1.058	1.120
PU6	0.532	4.83	1.042	1.086
PEOU1	0.619	4.88	1.033	1.067
PEOU2	0.789	4.68	1.237	1.529
PEOU3	0.636	4.67	1.169	1.365
PEOU4	0.754	4.88	0.991	0.983
PEOU5	0.517	4.68	1.258	1.583
PEOU6	0.620	4.93	0.873	0.762
ATU1	0.829	5.09	0.939	0.883
ATU2	0.754	5.02	0.855	0.732
ATU3	0.798	5.00	1.001	1.002
ATU4	0.792	4.70	1.423	2.025
ITU1	0.629	5.10	0.997	0.994
ITU2	0.478	5.10	0.880	0.775
ITU3	0.885	4.73	1.540	2.372
ITU4	0.911	5.01	0.937	0.879
SE1	0.834	4.96	1.161	1.348
SE2	0.720	4.72	0.985	0.970
SE3	0.592	4.66	1.034	1.069
SI1	0.592	5.11	0.997	0.995
SI2	0.844	5.03	1.038	1.078
SI3	0.878	4.70	1.389	1.929
CMP1	0.904	4.98	1.031	1.063
CMP2	0.818	5.14	1.159	1.343
CMP3	0.737	4.90	1.199	1.438
CMP4	0.492	5.22	1.004	1.008

continued on following page

Table 5. Continued

	Factor Loading	Mean	Std. Deviation	Variance
PBC1	0.874	4.79	1.321	1.745
PBC2	0.784	4.57	1.420	2.018
PBC3	0.699	5.19	1.031	1.064
REG1	0.810	4.60	0.937	0.879
REG2	0.733	4.73	1.195	1.428

Table 5 shows the descriptive statistics of the data. This shows that the mean of the constructs varies from 4.57 to 5.22, have a standard deviation that ranges from 0.855 to 1.540, the variance of the constructs of the vary from 0.732 to 2.372 and all the factor loading are above 0.5.

5.6 Model Evaluation overall fit measurements

Table 6. Model Evaluation overall fit measurements

Indices	Value	Suggested value
Chi-square value	5100	-
DF	692	-
P value	0.00	-
Chi-square value/DF	7.369	< 5.00 (Hair et al., 1998)
GFI	0.696	> 0.90 (Hu and Bentler, 1999)
AGFI	0.661	> 0.90 (Hair et al. 2006)
PFI	0.637	> 0.90 (Hu and Bentler, 1999)
CFI	0.634	>0.9 (Daire et al., 2008)
RMSEA	0.130	< 0.08 (Hair et al. 2006)

Table 6 shows the Overall fit measurement of SEM. In this the chi-square value is 5100, DF is 692. The p-value is 0.00. The Chi-square value/DF is 7.369, the GFI (Goodness Fit Index) is 0.696, the AGFI (Adjusted Goodness of Fit Index) of the SEM is 0.661, the CFI (Comparative Fit Index) of the SEM is 0.634 and RMSEA (Root Mean Square Error of Approximation) is 0.130. All the values are below the accepted limit, so the model is not an excellent fit but a good fit.

Table 7. Calculation of AVE and CR

Statements		Statements	Factor Loading (FL)	Item Reliability	Delta	AVE	sum of(FL)	sum of delta	CR
SE1	<---	SE	0.594	0.353	0.647				
SE2	<---	SE	0.701	0.491	0.509				
SE3	<---	SE	0.268	0.072	0.928	0.305	1.563	2.084	0.540
CMP1	<---	CMP	0.892	0.796	0.204				
CMP2	<---	CMP	0.333	0.111	0.889				
CMP3	<---	CMP	0.619	0.383	0.617				
CMP4	<---	CMP	0.454	0.206	0.794	0.575	2.298	2.504	0.678
SI1	<---	SI	0.426	0.181	0.819				
SI2	<---	SI	0.703	0.494	0.506				
SI3	<---	SI	0.918	0.843	0.157	0.506	2.047	1.482	0.739
PEOU6	<---	PEOU	0.207	0.043	0.957				
PEOU5	<---	PEOU	0.552	0.305	0.695				
PEOU4	<---	PEOU	0.537	0.288	0.712				
PEOU3	<---	PEOU	0.404	0.163	0.837				
PEOU2	<---	PEOU	0.537	0.288	0.712				
PEOU1	<---	PEOU	0.387	0.150	0.850	0.206	2.624	4.763	0.691
PU6	<---	PU	0.405	0.164	0.836				
PU5	<---	PU	0.551	0.304	0.696				
PU4	<---	PU	0.667	0.445	0.555				
PU3	<---	PU	0.429	0.184	0.816				
PU2	<---	PU	0.381	0.145	0.855				
PU1	<---	PU	0.601	0.361	0.639	0.267	3.034	4.397	0.677
ATU4	<---	ATU	0.407	0.166	0.834				
ATU3	<---	ATU	0.808	0.653	0.347				
ATU2	<---	ATU	0.737	0.543	0.457				
ATU1	<---	ATU	0.421	0.177	0.823	0.385	2.373	2.461	0.696
ITU4	<---	ITU	0.962	0.925	0.075				
ITU3	<---	ITU	0.813	0.661	0.339				
ITU2	<---	ITU	0.409	0.167	0.833				
ITU1	<---	ITU	0.373	0.139	0.861	0.473	2.557	2.107	0.756
PBC3	<---	PBC	0.454	0.206	0.794				
PBC2	<---	PBC	0.92	0.846	0.154				
PBC1	<---	PBC	0.724	0.524	0.476	0.526	2.098	1.423	0.756
REG2	<---	REG	0.453	0.205	0.795				
REG1	<---	REG	1.028	1.057	-0.057	0.631	1.481	0.738	0.748

The average Variance Extracted (AVE) of all the constraints is greater than ƛ 2 between the variables, which will fulfil the criteria. In Table 7, the rule of thumb of CR (Construct Reliability) states that any value more than 0.6 is acceptable. The higher the CR (Construct Reliability), the higher the internal consistency.

5.7 Structural Model

Figure 3. Structural Equation Model (SEM)

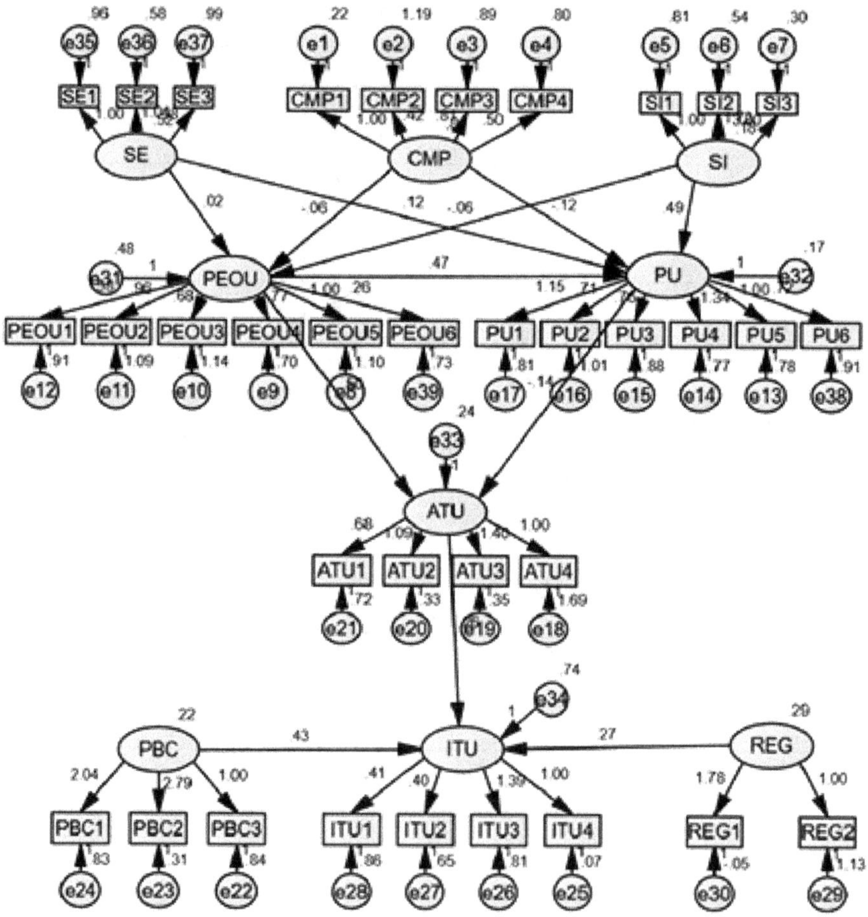

The set of multiple regressions were analysed using structural equation modelling (SEM) of AMOS. The structural model concerns direct and indirect relationships between the latent variables. The SEM technique is well suited to the research analysis. In the analysis, the hypotheses have a direct and significant relationship among variables. The research hypotheses are tested due to the consistency of the proposed measurement model with data. The hypothetical relationship between the different constructs is presented in Figure 3. From the hypothetical relationship to the construct, 8 hypotheses have been supported with path significant at p=0.05 level. Consequently, the study attempted to prove that there is a direct and positive association between all the other hypotheses. Generally, in SEM, the fit of the model evaluated by the chi-square measurement is not always easy because it is extremely sensitive to the sample size of the data. Owing to these constraints, different types of fit indices have been computed, and they denote independent sample size. The goodness of the fit and exhaustive link between the various hypotheses are presented in Tables 6 and 9.

5.8 Hypothesis Summary

Table 8. Variables in the Structural Equation Model Analysis

Relationship			Unstandardized co- efficient (B)	S.E	Standardised co-efficient (Beta)
PU	**<---**	**CMP**	**0.023**	**0.072**	**0.024**
PU	<---	SI	-0.056	0.049	-0.075
PU	<---	SE	-0.06	0.105	-0.037
PEOU	<---	SE	-0.118	0.038	-0.185
PEOU	<---	CMP	0.495	0.101	0.358
PEOU	<---	SI	0.119	0.055	0.147
PEOU	<---	PU	0.471	0.079	0.558
ATU	<---	PEOU	0.496	0.107	0.595
ATU	<---	PU	-0.141	0.088	-0.142
ITU	<---	ATU	0.257	0.09	0.164
ITU	<---	PBC	0.428	0.108	0.22
ITU	<---	REG	0.271	0.082	0.162

- From the above table 8, the unstandardized coefficient of social efficacy to perceived ease of use is -0.118. The estimate is negative; thus, effect will also be negative. This means the perceived ease of use will decrease by 0.118 for every one-unit increase in social efficacy.

- The unstandardised coefficient of compatibility to perceived ease of use is 0.495. The estimate is positive; thus, effect will also be positive. This means the perceived usefulness will increase by 0.495 for every one-unit increase in compatibility.
- The unstandardised coefficient of compatibility to perceived usefulness is 0.023. The estimate is positive; thus, effect will also be positive. This means the perceived usefulness will increase by 0.023 for every one-unit increase in compatibility.
- The unstandardised coefficient of social influence to perceived usefulness is -0.056 the estimate is negative thus effect will also be negative. This means the perceived usefulness will decrease by 0.056 for every one-unit increase in social influence.
- The unstandardised coefficient of social efficacy to perceived usefulness is -0.06. The estimate is negative; thus, effect will also be negative. This means the perceived usefulness will decrease by 0.06 for every one-unit increase in social efficacy.
- The unstandardised coefficient of social influence to perceived ease of use is 0.199. The estimate is positive; thus, effect will also be positive. This means the perceived ease of use will increase by 0.199 for every one-unit increase in social efficacy.
- The unstandardised coefficient of perceived usefulness to perceived ease of use is 0.471. The estimate is positive; thus, effect will also be positive. This means the attitude towards usage will increase by 0.471 for every one-unit increase in perceived ease of use.
- The unstandardised coefficient of attitude towards usage to intention to use is 0.257. The estimate is positive; thus, effect will also be positive. This means the intention of use will increase by 0.257 for every one-unit increase in the perceived behavioural control.
- The unstandardised coefficient of perceived usefulness to attitude towards usage is -0.141. The estimate is negative; thus, effect will also be negative. This means the attitude towards usage will decrease by 0.141 for every one-unit increase in perceived usefulness.
- The unstandardised coefficient of perceived ease of use to attitude towards usage is 0.496. The estimate is positive; thus, effect will also be positive. This means the attitude towards usage will increase by 0.496 for every one-unit increase in perceived ease of use.
- The unstandardised coefficient of perceived behavioural control to intention to use is 0.428. The estimate is positive; thus, effect will also be positive. This means the intention of use will increase by 0.428 for every one-unit increase in the perceived behavioural control.

- The unstandardised coefficient of regulation towards intention to use is 0.271. The estimate is positive; thus, effect will also be negative. This means the intention of use will increase by 0.271 for every one-unit increase in the regulation.
- The unstandardised coefficient of attitude towards usage to intention to use is 0.257. The estimate is positive; thus, effect will also be positive. This means the intention of use will increase by 0.257 for every one-unit increase in the regulation.

Table 9. Hypothesis Summary

Hypothesis	**Relationship**			**Unstandardized co-efficient (B)**	**P value**	**Results**
H1	PU	<---	SI	0.495	<0.001	supported
	PU	<---	CMP	-0.118	0.002	
	PU	<---	SE	0.13	0.031	
H2	PEOU	<---	SE	0.023	0.053	unsupported
	PEOU	<---	SI	-0.06	0.007	
	PEOU	<---	CMP	-0.056	0.003	
H3	PU	<---	PEOU	0.471	<0.001	supported
H4	ATU	<---	PU	-0.141	0.109	unsupported
H5	ATU	<---	PEOU	0.496	<0.001	supported
H6	ITU	<---	ATU	0.257	0.004	supported
H7	ITU	<---	REG	0.271	<0.001	supported
H8	ITU	<---	PBC	0.428	<0.001	supported

Hypothesis Summary shows the relationship among variables and hypothesis framed, unstandardised coefficient, based on the p-value, the results are estimated whether the hypothesis is supported or not unsupported. If the p-value is less than 0.05, the results are supported. If the p-value is more than 0.05, then the results are unsupported. Social influence, Compatibility and Self-efficacy are supported and have a significant relationship with perceived usefulness. Social influence, Compatibility and Self - Efficacy are not supported by perceived ease of use. Perceived Ease of Use has a significant relationship with Perceived Usefulness and is supportive. Perceived usefulness is not supportive of the Attitude to use. Attitude towards usage is supported and has a significant relationship with intention to Use. Regulation is the most important variable in the study; here, Regulation and Perceived Behavioural Control have a significant relationship, and the results of the hypothesis summary support it.

5.9 CFA Result of Goodness of Fit

Table 10. CFA A and CFA B

Indices	CFA A	CFA B	Suggested Value
Number of statements before CFA	22	13	
Chi-square/DF value	7.409	7.238	< 5.00 (Hair et al., 1998
P-value	0.000	0.00	> 0.05 (Hair et al., 1998)
GFI	0.769	0.834	> 0.90 (Hu and Bentler, 1999
AGFI	0.708	0.776	> 0.90 (Hair et al. 2006)
NFI	0.420	0.721	> 0.90 (Hu and Bentler, 1999)
CFI	0.443	0.742	> 0.90 (Daire et al., 2008)
RMSEA	0.136	0.135	< 0.08 (Hair et al. 2006)
Number of Statements after CFA	22	13	
Cronbach Alpha	0.622	0.697	>0.6

Table 10 shows the Overall fit measurement of CFA A and CFA B. In this the number of statements before CFA is 22 for CFA A and 13 for CFA B. The chi-square/DF value is 7.409 for CFA A and 7.238 for CFA B, GFI (Goodness Fit Index) of CFA A is 0.769 and 0.834 for CFA B, AGFI (Adjusted Goodness of Fit Index) of CFA A is 0.708 and 0.776 for CFA B, NFI (Normed Fit Index) for CFA A is 0.420 and 0.721 for CFA B, CFI (Comparative Fit Index) of CFA A is 0.443 and 0.742 for CFA B and RMSEA (Root Mean Square Error of Approximation) is 0.136 for CFA A and 0.135 For CFA B. No statements were deleted after the CFA test the number of statements remains 22 for the CFA A and 13 for CFA B. The Cronbach Alpha for CFA A is 0.622 and 0.697 for CFA B. Most of the values in the test do not meet the acceptable values, so the model does not have an excellent fit.

Figure 4. CFA A

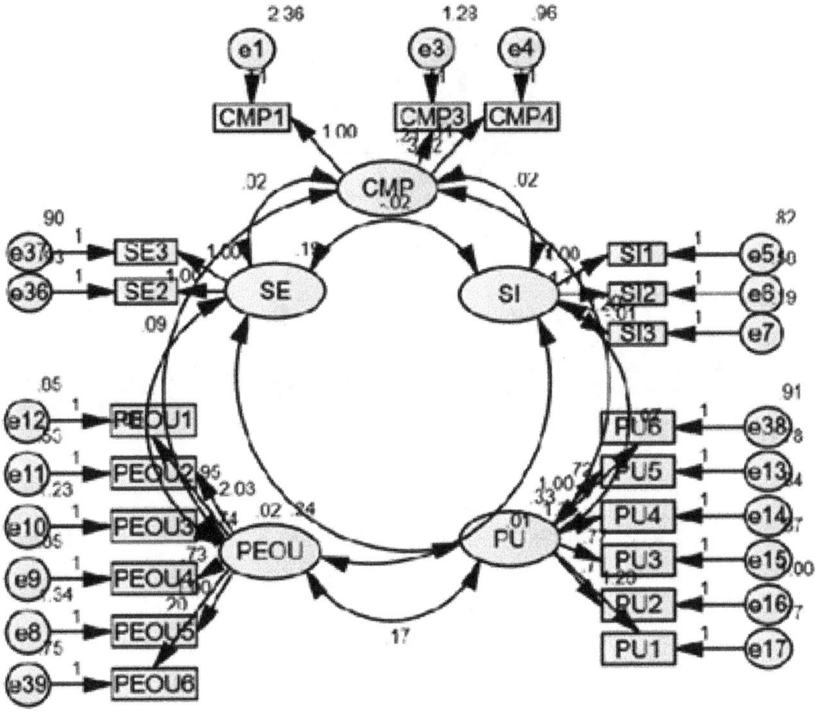

Figure 5. CFA B

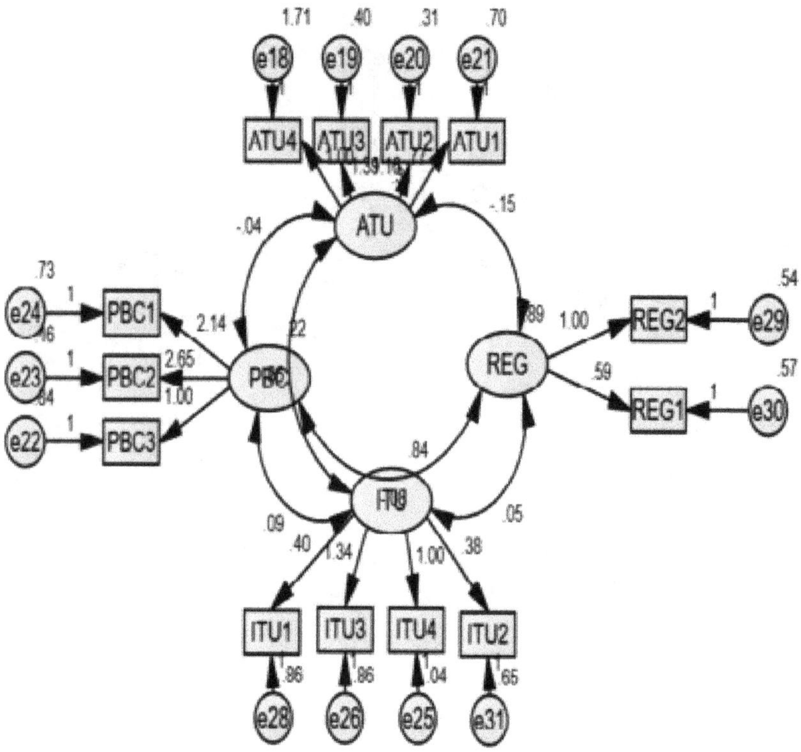

6. DISCUSSION AND CONCLUSION

As per our results and data analysis, we could tell that farmer feel that drone technology is useful and easy to operate, which could increase agricultural activities' efficiency, productivity and revenue. The study indicates the positive impact on the Attitude towards the Usage of Drone technology in agriculture, and Attitude directly impacts Intention to use. The farmer's intention towards accepting drone technology is affected by Regulation and Perceived Behavioural Control as they are found to be supported in the hypothesis summary. Farmers found Regulations pertaining to the flying of drones, usage of drones, safety measures, policy, licensing procedures, and prior government permissions as rigid and complicated. Farmers are not well aware of regulations framed by government policymakers, and this has made farmers reluctant to adopt drone-based technology. Cost, resources and opportunities play a major role in Perceived Behavioural Control. Farmers are highly concerned about the costs that are quite high in Drones despite many advantages. The Cost of drones is creating a burden for farmers to arrange loans, borrow or take any other further risks. Thus, Farmers are taking a step backwards in the phase of intention of Use because of Regulation and Perceived Behavioural Control, i.e., Cost.

As per the study we found that our demographic results also affect our results. As per analysis, the landholding of farmers with less than 4 acres is 36%, between 4-10 acres is 22.6%, and More than 10 acres is 41.4%. Logically, for farmers with more landholding, there is a high chance of adopting the drone technology considering the regulation and the cost of the drones, as it covers huge needs and improves efficiency with less time and effort. But for small landholders, drone technology is still useful and easy, improving productivity and much more advantageous. Still, the farmers with less land holding would think of the rigidness of regulation and policy procedures along with the consideration of huge cost factors.

Thus, our analysis provides results such as average and good model fit; thus, improvisation in the data is necessary, and further data collection is essential for the study to have more information. Hence, based on the study, we can conclude that Regulation, being the important variable, must be considered by agricultural specialists, government policymakers, and Agri-based start-ups. Hence, from the research, the farmers must benefit by accepting drone technology with government support by reducing the rigidness and cost of drones and providing farmers with much awareness related to different technologies that can be used in agricultural activities.

7. RESEARCH IMPLICATIONS

Firstly, government policymakers must consider the rigidity farmers face regarding drone technology. The policies must be framed in such a way that it must be feasible and reduce the rigidness of policy, procedures, permission and regulation of Drones by considering the usefulness of using drones in agriculture by farmers. Also, to reduce the heavy cost burden and make it cost-effective. Drone Startups must develop ties with the government to be cost-effective for farmers in the adoption of drones. Collaboration between Agri-based start-ups and the government helps farmers by providing many opportunities and scope for using drone technology in Agriculture. Thus, Government initiative towards villages to ensure the usage of different technologies among farmers results in an increase in economic growth and GDP in the agricultural sector in the country and also leads to overall development.

REFERENCES

Ajzen, I. (1991). The theory of planned behavior. *Organizational Behavior and Human Decision Processes, 50*(2), 179–211. DOI: 10.1016/0749-5978(91)90020-T

Ajzen, I. (2002). Perceived behavioral control, self-efficacy, locus of control, and the theory of planned behavior. *Journal of Applied Social Psychology, 32*(4), 665–683. DOI: 10.1111/j.1559-1816.2002.tb00236.x

Ajzen, I., & Fishbein, M. (1980). *Understanding attitudes and predicting social behaviour*. Prentice-Hall.

Akudugu, M., Nkegbe, P., Wongnaa, C., & Millar, K. (2023). Technology adoption behaviors of farmers during crises: What are the key factors to consider? *Journal of Agriculture and Food Research, 14*, 100694. DOI: 10.1016/j.jafr.2023.100694

Arhin, I., Yeboah, E., Liu, X., Liu, A., Chen, X., & Li, X. (2024). Integrating farmers' perception of sustainable agricultural technologies towards the development of sustainable tea production in China. *International Journal of Agricultural Sustainability, 22*(1), 2303886. Advance online publication. DOI: 10.1080/14735903.2024.2303886

Ayamga, M., Tekinerdogan, B., Kassahun, A., & Rambaldi, G. (2020). Developing a policy framework for adoption and management of drones for agriculture in Africa. *Technology Analysis and Strategic Management, 33*(8), 970–987. DOI: 10.1080/09537325.2020.1858047

Bandura, A. (1977). Self-efficacy: Toward a unifying theory of behavioral change. *Psychological Review, 84*(2), 191–215. DOI: 10.1037/0033-295X.84.2.191 PMID: 847061

Barathkumar, R., Selvanayaki, S., Deepa, N., Kannan, P., & Prahadeeswaran, M. (2024). Impact of drone technology on agriculture - farmers' perception analysis. *Plant Science Today, 11*(sp4). Advance online publication. DOI: 10.14719/pst.5934

Canicattì, M., & Vallone, M. (2024). Drones in vegetable crops: A systematic literature review. *Smart Agricultural Technology, 7*, 100396. DOI: 10.1016/j.atech.2024.100396

Cao, A., Guo, L., & Li, H. (2025). Understanding farmer cooperatives' intention to adopt digital technology. *International Journal of Agricultural Sustainability, 23*(1), 2464523. Advance online publication. DOI: 10.1080/14735903.2025.2464523

Cheruku, J. K., & Katekar, V. (2023). Digitalisation of Agriculture in India: The case for doubling farmers' income. Indian Institute of Public Administration, 194-205.

Daponte, P., De Vito, L., Glielmo, L., Iannelli, L., Liuzza, D., Picariello, F., & Silano, G. (2019). A review on the use of drones for precision agriculture. *IOP Conference Series. Earth and Environmental Science, 275*(1), 012022. DOI: 10.1088/1755-1315/275/1/012022

Davis, F. D. (1989). Perceived usefulness, perceived ease of use, and user acceptance of information technology. *Management Information Systems Quarterly, 13*(3), 319–340. DOI: 10.2307/249008

Fishbein, M., & Ajzen, I. (1975). *Belief, Attitude, Intention, and Behavior: An Introduction to Theory and Research.* Addison-Wesley., DOI: 10.1017/CBO9781107415324.004

Francis, A. J. J., Eccles, M. P. M., Johnston, M., Walker, A., Grimshaw, J., Foy, R., Kaner, E. F. S., Smith, L., Bonetti, D., & Francis, J. (2004). CONSTRUCTING QUESTIONNAIRES BASED ON THE THEORY OF PLANNED BEHAVIOUR A MANUAL for HEALTH SERVICES RESEARCHERS. In Direct. https://doi.org/DOI: 0-9540161-5-7

. G, P. M., C, D., & M, S. (2024). Farmers' awareness and perception of drone technology. *Journal of Experimental Agriculture International, 46*(10), 426–435. https://doi.org/DOI: 10.9734/jeai/2024/v46i102965

George, D., & Mallery, P. (2003). *SPSS for Windows step by step*. Pearson A and B.

Hair, J. F., Tatham, R. L., & Black, W. (1998). *Multivariate data analysis*. Prentice-Hall.

Hair, J. F., Black, W. C., Babin, B. J., Anderson, R. E., & Tatham, R. L. (2006). *'Multivariate data analysis*. Pearson Prentice Hall, Upper Saddle River.

Hua, L., & Wang, S. (2019). *Antecedents of consumers' intention to purchase energy-efficient appliances: An empirical study based on the technology acceptance model and theory of planned Behavior*. Sustainability., DOI: 10.3390/su11102994

Joshi, N., Patel, N., Sharma, P., & Singh, A. (2020). Revolution of Indian agriculture through drone technology. *Research Today*, *2*(5), 174–176.

Kamble, P. N., Ajotikar, M., Suryawanshi, P., & Tarde, V. (2024). Awareness of drone technology application in agriculture amongst progressive farmers. *International Journal of Agriculture Extension and Social Development*, *7*(12), 307–310. DOI: 10.33545/26180723.2024.v7.i12e.1418

Klerkx, L., Jakku, E., & Labarthe, P. (2019). A review of social science on digital agriculture. *NJAS Wageningen Journal of Life Sciences, 90–91*(1), 1–16. DOI: 10.1016/j.njas.2019.100315

Levi-Faur, D. (2011). Regulation and regulatory governance. *In Handbook on the Politics of Regulation.*

Liu, L., & Liu, K. (2023). Can digital technology promote sustainable agriculture? *Cogent Food & Agriculture, 9*(2), 2282234. Advance online publication. DOI: 10.1080/23311932.2023.2282234

Mack, G., Ritzel, C., Ammann, J., & Benni, N. E. (2024). Improving the understanding of farmers' non-compliance with agricultural policy regulations. *Journal of Rural Studies, 106*, 103190. DOI: 10.1016/j.jrurstud.2023.103190

Mahasneh, H. (2024). Drones in agriculture: Real-world applications and impactful case studies. *Journal of Natural Science Review, 2*(Special Issue), 643–656. DOI: 10.62810/jnsr.v2iSpecial.Issue.164

Mariappan, K., & Zhou, D. (2019). *A threat of farmers' suicide and the opportunity in organic farming for sustainable agricultural development in India.* Sustainability., DOI: 10.3390/su11082400

Mukherjee, A., Chakraborty, D., Maity, A., Ray, M., & Kumari, S. (2017). How is drone going to impact Indian agriculture? *Biotech Articles.* Retrieved from http://www.biotecharticles.com

Moore, G. C., & Benbasat, I. (1991). Development of an instrument to measure the perceptions of adopting an information technology innovation. *Information Systems Research, 2*(3), 192–222.

Noor, F., & Noel, A. S. (2023). Perception of farmers regarding drones for pesticide spraying. *Asian Journal of Advances in Agricultural Research, 22*(3), 9–1. DOI: 10.9734/ajaar/2023/v22i3440

Ouko, K. O., Mukhebi, A. W., Obiero, K. O., & Opondo, F. A. (2022). Using the technology acceptance model to understand fish farmers' intention to use black soldier fly larvae meal in Nile tilapia production in Kenya. *All Life, 15*(1), 884–900. DOI: 10.1080/26895293.2022.2112765

. P, M. P., C, D., & M, S. (2024). Farmers' awareness and perception of drone technology. *Journal of Experimental Agriculture International, 46*(10), 426–435. https://doi.org/DOI: 10.9734/jeai/2024/v46i102965

Phillips, L. A., Calantone, R., & Lee, M. T. (1994). International technology adoption: Behavior structure, demand certainty, and culture. *Journal of Business and Industrial Marketing*, *9*(2), 12–19. DOI: 10.1108/08858629410059762

Pingali, P. (2007). Agricultural growth and economic development: A view through the globalization lens. *Agricultural Economics*, *37*(s1), 1–12. DOI: 10.1111/j.1574-0862.2007.00231.x

Rajagopalan, R. P., & Krishna, R. (2018). Drones: Guidelines, regulations, and policy gaps in India.

Rani, A. L. K. A., Sharma, R., Verma, P., & Kumar, S. (2019). Drone: The green technology for future agriculture. *Harit Dhara*, *2*(1), 3–6.

Rezaei, R., Safa, L., & Ganjkhanloo, M. M. (2020). Understanding farmers' ecological conservation behavior regarding the use of integrated pest management—An application of the technology acceptance model. *Global Ecology and Conservation*, *22*, e00941. DOI: 10.1016/j.gecco.2020.e00941

Rogers, E. M. (1995). *Diffusion of innovations* (4th ed.). Free Press.

Selvan, S. S., Wahid, A., Patel, A., Kumar, V., & Sahu, P. (2021). Challenges in Indian agriculture. *Agricultural Reviews (Karnal)*, *42*(1), 1–12. DOI: 10.18805/ag.R-2103

Shang, L., Heckelei, T., Gerullis, M. K., Börner, J., & Rasch, S. (2021). Adoption and diffusion of digital farming technologies—Integrating farm-level evidence and system interaction. *Agricultural Systems*, *190*, 103074. DOI: 10.1016/j.agsy.2021.103074

Sharifzadeh, M. S., Damalas, C. A., Abdollahzadeh, G., & Ahmadi-Gorgi, H. (2017). Predicting adoption of biological control among Iranian rice farmers: An application of the extended technology acceptance model (TAM2). *Crop Protection (Guildford, Surrey)*, *96*, 88–96.

Sood, A., Bhardwaj, A. K., & Sharma, R. K. (2022). Towards sustainable agriculture: Key determinants of adopting artificial intelligence in agriculture. *Journal of Decision Systems*, *33*(4), 833–877. DOI: 10.1080/12460125.2022.2154419

Subramanian, K. S., Pazhanivelan, S., Srinivasan, G., Santhi, R., & Sathiah, N. (2021). Drones in insect pest management. *Frontiers in Agronomy*, *3*, 640885. DOI: 10.3389/fagro.2021.640885

Sundar, C. B. V., Asokhan, M., & Karthikeyan, C. (2023). Adoption of drones in agriculture: Social, economic, and personal factors. *International Journal of Environment and Climate Change*, *13*(8), 587–597. DOI: 10.9734/ijecc/2023/v13i81987

Suvittawat, A. (2024). Investigating farmers' perceptions of drone technology in Thailand: Exploring expectations, product quality, perceived value, and adoption in agriculture. *Agriculture*, *14*(12), 2183. DOI: 10.3390/agriculture14122183

Takagi, C., Purnomo, S. H., & Kim, M. (2020). Adopting smart agriculture among organic farmers in Taiwan. *Asian Journal of Technology Innovation*, *1*(1), 1–16. DOI: 10.1080/19761597.2020.1797514

Venkatesh, V. (2000). Determinants of perceived ease of use: Integrating control, intrinsic motivation, and emotion into the technology acceptance model. *Information Systems Research*, *11*(4), 342–365. DOI: 10.1287/isre.11.4.342.11872

Venkatesh, V., Morris, M. G., Davis, G. B., & Davis, F. D. (2003). User acceptance of information technology: Toward a unified view. *Management Information Systems Quarterly*, *27*(3), 425–478. DOI: 10.2307/30036540

Verma, P., & Sinha, N. (2018). Integrating perceived economic well-being into the technology acceptance model: The case of mobile-based agricultural extension service. *Technological Forecasting and Social Change*, *126*, 207–216. DOI: 10.1016/j.techfore.2017.08.013

Vincent, D. R., Deepa, N., Elavarasan, D., Srinivasan, K., Chauhdary, S. H., & Iwendi, C. (2019). Sensors-driven AI-based agriculture recommendation model for assessing land suitability. *Sensors (Basel)*, *19*(17), 3667. DOI: 10.3390/s19173667 PMID: 31450772

Zhang, Y., Luo, Y., Zhang, X., & Zhao, J. (2019). How green human resource management can promote green employee behavior in China: A technology acceptance model perspective. *Sustainability (Basel)*, *11*(19), 5408. DOI: 10.3390/su11195408

APPENDIX

Table 11. Final Measurement Scale for factors of Drone Technology

FACTORS	RATING SCALE						
Perceived Usefulness	**VSD (1)**	**SD (2)**	**D (3)**	**N (4)**	**A (5)**	**SA (6)**	**VSA (7)**
1: Using Drones in Agriculture would accomplish tasks more quickly							
2: Using Drones would enhance Agricultural performance							
3: Using Drones in Agriculture would increase productivity							
4: Using Drones would enhance effectiveness of the agricultural activities							
5: Using Drones would make it easier to perform the agricultural activities							
6: Drones found to be useful in Agriculture							
Perceived Ease of Use							
1: Learning to operate Drones in Agriculture would be easy							
2: Drones would make work easy as per requirements in Agriculture							
3: Operating with drones would be clear and understandable							
4: Drones in Agriculture would be flexible to operate with							
5: It is easy to become skilful in Agriculture by using Drones							
6: Drones are simple to use in Agriculture							
Attitude Towards Usage							
1: Using Drones in Agriculture would be beneficial							
2: Using Drones would be pleasant in Agricultural practices							
3: Using Drones in Agriculture would be good							
4: Using Drones in the Agricultural practices are valuable							
Intension to Use							
1: There is high possibility to use Drones in Agriculture within the foreseeable future							
2: Intend to use Drones in the Agricultural activities within the foreseeable future							
3: Complete willingness in using Drones in Agriculture within the foreseeable future							
4: Using Drones in Agriculture would be important in the foreseeable future							
Social Influence							
1: People who are important to you would think that you should use Drones in agriculture							
2: People would influence the behaviour of yours to use Drones in agriculture							
3: People's whose opinions are valued by you would prefer usage of Drones in agriculture							
Compatibility							

continued on following page

Table 11. Continued

FACTORS	RATING SCALE						
Perceived Usefulness	**VSD (1)**	**SD (2)**	**D (3)**	**N (4)**	**A (5)**	**SA (6)**	**VSA (7)**
1: Using Drones would be compatible with most conditions of the Agriculture							
2: Drones would fit well with required production practices and agricultural activities							
3: Using Drones in Agriculture would be consistent with the financial situations							
4: Using Drones in agriculture would be consistent with information and knowledge acquired along with prior experience							
Self-Efficacy							
1: You can be confident in using Drones for different agricultural practices							
2: You can be confident in using Drones for different agricultural practices							
3: By using Drones in Agriculture you can give technical advice to others							
Regulation							
1: You are well aware of all rules and regulation pertaining to the flying of Drones for agricultural activities							
2: The Effective Policy Registration process pertaining to use of Drones in agriculture are simple and user-friendly							
Perceived Behavioural Control							
1: Whether you would purchase the Drone or not is completely up to your decision making							
2: The resources, time and opportunities are the requirements to purchase Drones in Agriculture							
3: You can be confident, to perform the required activities and purchase Drones at your convenience							

VSD (1) = Very Strongly Disagree; SD (2) = Strongly Disagree; D (3) = Disagree; N (4) = Neither Disagree nor agree;

A (5) = Agree; SA (6) = Strongly Agree; VSA (7) = Very Strongly Agree

Chapter 5
Investigating Educators' Behavioural Intentions Towards Smart Tools in Education

G. S. Prakasha
https://orcid.org/0000-0002-1287-7606
Christ University, Bangalore, India

Sanskriti Rawat
https://orcid.org/0000-0002-6042-9074
Christ University, India

Robert Linton Tavis Ashton-Bell
https://orcid.org/0009-0000-2544-6138
Christ University, India

Maria Lapina
https://orcid.org/0000-0001-8117-9142
North-Caucasus Federal University, Russia

ABSTRACT

This study explores pre-service and in-service teachers' behavioral intention to adopt smart tools by examining their perceptions of usefulness, ease of use, enjoyment, and contextual readiness. Using a quantitative survey design, data was collected from 42 participants across the Delhi-NCR region through purposive sampling. A composite score of 15 Likert-scale items was used to measure behavioural intention, followed by quartile-based analysis, independent samples t-tests (for gender), and one-way ANOVA (for age groups). While results revealed no statistically significant

DOI: 10.4018/979-8-3373-7922-7.ch005

differences across gender and age, the quartile distribution indicated a varied spectrum of intention, with participants broadly categorised into low, moderate, and high-intention groups. The study highlights the need for teacher training programs to go beyond technical instruction. It advocates for a more nuanced, context-aware approach to technology readiness in teacher education that bridges the gap between vision and practice in the evolving ecosystem of Education 5.0.

INTRODUCTION

The world has undergone enormous changes and has seen various accounts of the industrial revolution in response to societal needs. Industry 4.0 envisages a super-smart society aiming at juxtaposing technological advancements with sustainability (Zengin et al., 2021) by digitising and automating life. The post-industry 4.0 societal structure combines technology and human values. One of the most heavily areas influenced by it is the education sector. Education 5.0 is the up-and-coming paradigm wherein the current generation of learners delves into technological innovations (Ahmad et al., 2023). It amalgamates technology and human values to deliver students more personalised learning (Prakash et al., 2023). Smart education is an extrapolated concept, considering the inevitable use of technology in education with a learner-centric approach (Seeling et al., 2022). While smart technologies enable time-saving and efficiency for teachers, challenges exist to their implementation, such as lack of training, limited access to technology, and infrastructural issues (Dmitrenko et al., 2023). Nevertheless, as the barriers between humans and machines have dissolved due to technology, redefining education for the next generation of educators is necessary by keeping human values intact (Tandon et al., 2020). A smart education system comprises integrated information and communication (ICT) frameworks to make teaching-learning more efficient and engaging. It links modern-day technologies such as smart boards, projectors, learning management systems, artificial intelligence (AI) powered tools, etc., to improve teaching and learning (Demir, 2021). An empirical study by Dai et al. (2023) found that teachers and students prefer smart classrooms in the physical mode equipped with spaces like smart infrastructure, teaching equipment and systems and sensing equipment such as facial recognition cameras, microphones, sensors, etc. Smart education is the foundation of society 5.0, whose future is built by teachers and new-generation learners. Smart education platforms enable teachers to enhance their pedagogical practices and equip them with necessary skills like collaboration, research, innovation, critical thinking and exchange of ideas online (Xing, 2023). Additionally, the new cohort of teachers is tech-savvy and responds best to technology-based education (Ziatdinov & Cilliers, 2021). Although in-service teachers find it hard to adapt to new technologies as they belong to an era where traditional teaching methods were prevalent, new-age teachers and learners, being tech-savvy, still lack the proper knowledge and skills to take advantage of smart technologies (Jan et al., 2020). Smart education is the need of the hour for a better and sustainable future (Andres et al., 2022). Moreover, educators must envision a 21st-century classroom through innovative and technology-integrated strategies (Prakasha, 2022). For such an educational leap, teachers of today must be ready to teach the next-generation learners effectively and possess the necessary competencies. Teacher training courses and teacher continual professional courses equip teachers with the necessary classroom skills and knowledge. Therefore, teachers must be digitally literate and competent to utilize technology and ensure smart education (Iqbal et al., 2020). Cutting-edge technological advancements like AI, mixed reality, the Internet of Things (IoT), etc., are integral to catering to learners' needs (Chirume, 2020). Moreover, cloud computing plays a significant role in enabling teachers to fasten the administrative processes like assessment, admission, communication with parents, etc. (Asadi et al., 2019). Yang et al. (2023) highlighted a composition of teachers' personalised learning space, wherein smart technologies like AI, big data and 5G played a crucial role in creating a learning community. Qarkaxhija et al. (2023) found that integrating smart technology aids in improving the creative potential of teachers, thus making room for more professional creativity. Moreover, Tao et al. (2021) proposed an ecosystem consisting of opportunities for self-regulation and personalised teaching and learning experiences for teachers and learners in the higher education setting. Similarly, Lu & Liang (2024) advocated digital literacy, awareness and skills among college teachers to make up for the unprecedented changes brought about by education 5.0. Hence, the fusion of smart education's innovative potential is reshaping the educational landscape, demanding comprehensive pedagogical and technological integration on the part of educators. While prospective teachers must cultivate digital literacy to engage their students (Nuryadi & Widiatmaka, 2023), veteran teachers must undergo training on smart teaching practices to enhance their pedagogy and to balance teacher agency, curriculum innovation and technology (Morgado et al., 2021). Additionally, new media literacy through training and development enables teachers to create educationally innovative content amidst the advent of smart education (Gao, 2024). Moreover, educators must possess relevant knowledge of smart technologies to better incorporate them in the classrooms, as offline teaching is also being converted to digital mode as per the market demand (Yang, 2022). Teachers should adopt pedagogies like e-learning, hybrid models, and flipped classrooms to cater to the learners' preferences. Personalised and automated learning methods, including AI-driven chatbots and tutors, align with learners' tech-savvy nature (Rakovac Bekeš & V. Galzina, 2023). When it comes to discipline-specific pedagogies, smart classrooms are instrumental in enhancing the subject-specific knowledge and skills of the learners. According to a study by Vakaliuk et al. (2020), proper training of future foreign language teachers in an exclusively developed innovative educational setting contributes to their readiness for innovative and creative activities in smart classrooms. Additionally, STEM (Science, Technology, Engineering and Maths) teachers require a more personalised and learner-oriented approach, which might not be possible due to various other factors such as increased workload, administrative tasks, etc. Hence, smart education and smart services enable STEM teachers to fulfil the learning needs of their students in a more streamlined manner (Antonova & Dankov, 2023). For the present study, the researchers seek to fulfil and answer the following objectives and questions:

RESEARCH OBJECTIVES

1. To understand teachers' behavioural intentions to use smart tools in teaching and learning.
2. To explore the factors contributing to teachers' decision to use smart tools.
3. To explore the influence of gender and age on teachers' decision to use smart tools.

RESEARCH QUESTIONS

1. How do teachers intend to use smart tools in teaching and learning?
2. What factors influence teachers' decision to use smart tools?
3. Do gender and age influence teachers' decision to use smart tools?

CONTEXT OF THE STUDY

Integrating smart tools and technology in education is gaining momentum globally, and India is no exception. With the increasing availability and accessibility of AI-driven applications, digital learning platforms, and interactive teaching aids, educators are gradually incorporating technology into their classrooms. However, teachers' adoption and effective use of these smart tools depend on multiple factors, including their perceptions, confidence, accessibility, and pedagogical standpoints (Davis, 1989; Teo, 2008). In the Indian context, teachers operate within diverse educational settings, ranging from well-equipped, affluent urban schools to resource-constrained and neglected rural classrooms. While government initiatives like the National Education Policy (NEP) 2020 emphasise digital transformation and technology-driven pedagogy, the extent to which teachers adopt smart tools varies significantly. Digital literacy, infrastructural facilities, and institutional support are crucial in shaping teachers' engagement with technology (Kumar & Daniel, 2016). Understanding teachers' perceptions of the usefulness, ease of use, and enjoyment of smart tools is essential for assessing their motivation to integrate technology into their teaching practices. Additionally, their confidence and willingness to experiment with digital tools influence how effectively they adopt and utilise these resources in real classroom scenarios. In the Indian context, where teachers often rely on peer networks and institutional feedback, examining the role of peer influence, feedback mechanisms, and accessibility can provide valuable insights into technology adoption patterns. Furthermore, teachers' core teaching beliefs—whether they lean toward traditional, teacher-centred approaches or more learner-centred, constructivist methods—affect how they perceive and integrate smart tools. Given the emphasis on personalised and experiential learning in modern education, aligning teachers' pedagogical beliefs with technological advancements is crucial for fostering meaningful engagement with smart tools. By exploring these aspects, this study aims to provide a nuanced understanding of the factors influencing teachers' adoption of smart tools in India.

THEORETICAL FRAMEWORK

The theoretical framework for the present study integrates the Technology Acceptance Model (TAM) (Davis, 1989), social cognitive theory (Bandura, 1986), social influence theory (Kelman, 1958), and cognitivist learning theories (Anderson et al., 2001) to explore teachers' adoption of smart tools in India. TAM concepts such as perceived usefulness, perceived ease of use, attitude towards use, and behavioural intention explain teachers' motivation and intention to use smart tools (Davis, 1989). Next, self-efficacy (Venkatesh & Bala, 2008), personal innovativeness (Jackson et al., 2013), and perceived enjoyment (Toral et al., 2007) contribute to teachers' confidence and willingness to experiment with technology. Furthermore, social factors, including peer influence (Lai & Chen, 2011) and perceived ubiquity value (Nikou & Economides, 2017), impact adoption decisions, while cognitive feedback (Nikou & Economides, 2017) enhances teaching effectiveness. Additionally, cognitivist teaching beliefs (Anderson et al., 2001) shape teachers' integration of smart tools into pedagogy. This framework comprehensively explains the key factors influencing the adoption of smart tools by teachers in India.

METHODOLOGY

The study employed a quantitative research design with a population of pre-service and in-service teachers in India. The sample for the present study consisted of 42 pre-service and in-service teachers in Delhi, NCR, selected through a purposive sampling technique. Khlaisang et al. (2021) developed a survey where the researchers used various standardised scales to explore the relationship among eleven variables regarding the use of smart tools by teachers: Perceived Usefulness (PU), Perceived Ease of Use (PEU), Attitude towards use (ATU), Computer self-efficacy (CSE), Perceived Enjoyment (ENJ), Personal Innovativeness (INNO), Peer Influence (PI), Perceived Ubiquity Value (PUV), Cognitivist Teaching Belief (COGT), Cognitive Feedback (CFB) and Behavioural Intention (BI). The researchers have modified the survey questionnaire according to the objectives of this study in relation to smart tools and both obtained face and content validity from the subject experts. After obtaining informed consent, the researchers approached the participants through LinkedIn, WhatsApp groups, etc. and circulated the survey. The study included male and female pre-service and in-service teachers from Delhi, NCR region. Of the approached participants, 42 consented to participate in the study. **Table 1** presents the item number, along with the item statement, corresponding to the eleven variables required for the study.

Table 1. Items included in the survey questionnaire

Item No.	Item Statement	Variable
1	Smart tools are beneficial to my teaching.	Perceived usefulness (PU)
2	Using smart tools would improve my teaching effectiveness.	
3	I find teaching with smart tools is easy to use.	Perceived ease of use (PEU)
4	Smart tools make teaching more interesting.	Attitude towards use (ATU)
5	I look forward to those aspects of my teaching that require me to use smart tools.	
6	I can learn how to use smart tools in teaching easily.	Computer self-efficacy (CSE)
7	I can use smart tools in teaching with minimal assistance.	
8	I enjoy using smart tools in teaching.	Perceived enjoyment (ENJ)
9	I like to experiment with new technology in my classroom.	Personal innovativeness (INNO)

continued on following page

Table 1. Continued

Item No.	Item Statement	Variable
10	If most of my colleagues started using smart tools for teaching, it would encourage me to do the same.	Peer Influence (PI)
11	Teaching with smart tools can provide personalised and adaptive information.	Perceived ubiquity value (PUV)
12	In teaching, examples should be given to students to enhance their ability to remember and understand in relation to prior knowledge to knowledge building.	Cognitivist teaching belief (COGT)
13	When using smart tools, my feedback to students' learning was clear and understandable.	Cognitive Feedback (CFB)
14	I intend to use smart tools in teaching in the future.	Behavioural intention (BI)
15	I plan to teach with smart tools in the future.	

RESULTS

This survey aimed to explore teachers' behavioural intentions toward using smart tools in education. A total of 42 pre-service and in-service teachers participated in the study, comprising 33 females and 9 males. The participants represented three age categories: 22–25, 26–30, and more than 30. The survey consisted of a 7-point Likert scale of 15 items. The scores from all 15 items were summed to generate a composite score out of 105, which was operationally defined as the "Behavioral Intention Score" (BIS) for each respondent. The descriptive statistics of the BIS were then calculated. To further illustrate the spread of behavioural intention among participants, the total scores were categorised into three quartiles based on the quartile deviation method:

- Q1 (Low Behavioral Intention): Scores less than or equal to the first quartile (≤ Q1)
- Q2 (Moderate Behavioral Intention): Scores between Q1 and Q3
- Q3 (High Behavioral Intention): Scores greater than the third quartile (> Q3)

A quartile deviation analysis assessed the spread and variability of the BIS. **Table 2** shows the descriptive statistics for all constructs used within the BIS calculation.

Table 2. Descriptive statistics of constructs within Behavioral Intention Score

S. No.	Constructs	No. of items	Mean	SD	Q1	Q3
1.	PU	2	5.71	1.225	4.883	6.537
2.	PEU	1	5.45	1.253	4.604	6.296
3.	ATU	2	5.69	1.318	4.800	6.580
4.	CSE	2	5.68	1.234	4.847	6.513
5.	ENJ	1	5.69	1.278	4.827	6.553
6.	INNO	1	5.83	1.188	5.028	6.632
7.	PI	1	5.50	1.518	4.475	6.525
8.	PUV	1	5.45	1.485	4.448	6.452
9.	COGT	1	5.79	1.220	4.967	6.614
10.	CFB	1	5.43	1.192	4.625	6.235
11.	BI	2	5.80	1.365	4.879	6.721
BIS		**15**	**5.63**	**1.297**	**4.762**	**6.514**

The mean of all results is above 5, with standard deviations ranging from 1.192 to 1.365, a similar finding to the results in Khlaisang et al. (2021). The results show that teachers' behavioral intention to adopt smart tools varied moderately. This implies a generally positive attitude toward adoption, with some participants expressing stronger intent than others. This spread could be further examined across demographic factors such as gender and age, which may influence teachers' openness to educational technology. **Figure 1** presents the quartile distribution of behavioural intention among the participants (n=42) regarding smart tools in education.

Figure 1. Quartile distribution of behavioural intention among the teachers

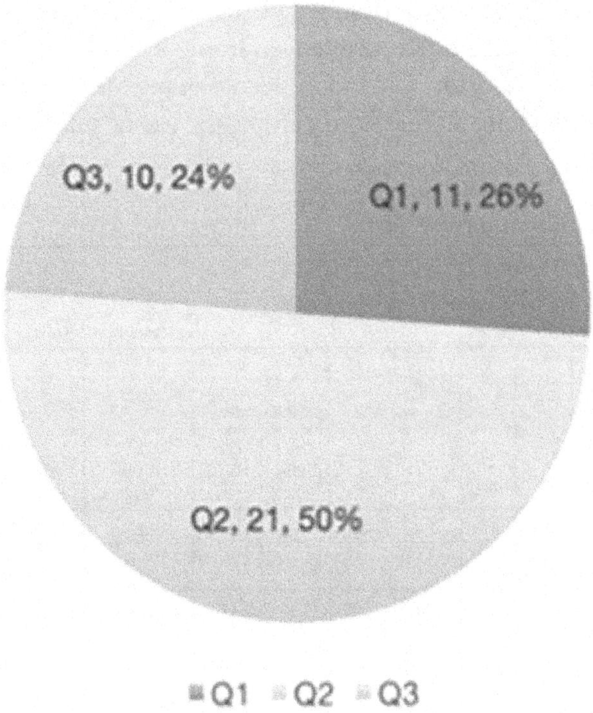

According to Figure 1, 11 participants (26.2%) fell into the Low Behavioral Intention (Q1) category. Furthermore, 21 participants (50%) belonged to the Moderate Behavioral Intention (Q2) category. Lastly, 10 participants (23.8%) were placed in the High Behavioral Intention (Q3) category. Thus, the distribution indicates that half of the participants demonstrated a moderate behavioural intention toward using smart tools. A smaller yet significant portion exhibited high intention, while about one-fourth of the respondents reported low intention. This pattern suggests a generally favourable attitude, with room to improve behavioural intention among the lower-scoring group through targeted interventions or support through training and workshops.

To further understand whether gender and age play a role in influencing the behavioural intention of the participants regarding the use of smart tools in education, an independent samples t-test was conducted to compare the behavioural intention scores between male and female participants. Levene's Test for Equality of Variances was found not significant ($F(1, 40) = 0.008$, $p = .930$), indicating that the assumption of equal variances was met. Therefore, results were interpreted using the row for equal variances assumed. There was no significant difference in Behavioral Intention scores between males and females: t (40) = 0.529, p = .599 (two-tailed). The mean difference between the groups was 3.23 with a standard error of 6.11, and the 95% confidence interval for the difference ranged from –9.11 to 15.57.

Table 3 presents the result of an independent sample t-test for behavioural intention among male and female teachers regarding the use of smart tools in education.

Table 3. Independent Sample t Test

		Levene's Test for Equality of Variances		t-test for Equality of Means							
						Significance				95% Confidence Interval of the Difference	
		F	Sig.	t	df	One-Sided p	Two-Sided p	Mean Difference	Std. Error Difference	Lower	Upper
Behavioral intention	Equal variances assumed	.008	.930	.529	40	.300	.599	3.232	6.106	-9.108	15.573
	Equal variances not assumed			.571	14.207	.289	.577	3.232	5.663	-8.897	15.361

The results from **Table 3** suggest that gender does not have a statistically significant effect on teachers' behavioural intention to use smart tools. Both male and female teachers reported similar levels of intention, with any observed difference likely due to random variation rather than a true effect.

Furthermore, researchers conducted a one-way ANOVA to determine whether there were significant differences in behavioural intention scores across three age groups (22-25 years, 26-30 years and more than 30 years). The analysis revealed no statistically significant difference in Behavioral Intention among the different age groups: F(2, 39) = 0.489, p = .617. The between-group variability (Sum of Squares = 259.75) was small relative to the within-group variability (Sum of Squares = 10,359.87), indicating that most of the variation in Behavioral Intention was within the groups rather than between them. **Table 4** presents the result of the one-way ANOVA w.r.t teachers' age regarding behavioural intention towards smart tools in education.

Table 4. One-way ANOVA w.r.t age

Behavioural Intention	Sum of Squares	df	Mean Square	F	Sig.
Between Groups	259.752	2	129.876	.489	.617
Within Groups	10359.867	39	265.638		
Total	10619.619	41			

The results from Table 4 indicate that the age group does not have a significant impact on teachers' behavioral intention. All three age groups showed similar levels of behavioural intention toward using smart tools, indicating that common assumptions of more experienced teachers, perhaps being close-minded to smart tools, may be unsubstantiated.

Figure 2 shows the distribution of the mean of the BIS among the respondents corresponding to their age groups.

Figure 2. Mean of Behavioural Intention w.r.t age

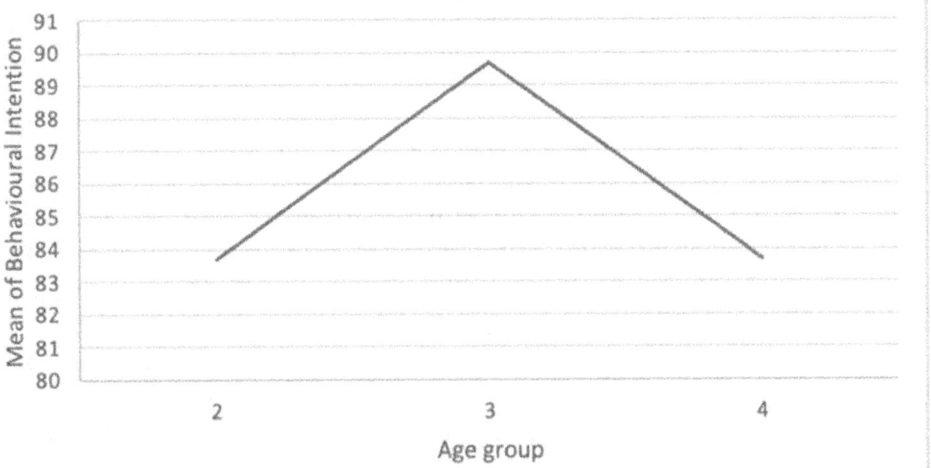

From **Figure 2**, it is evident that there is a non-linear relationship between respondents' age and their behavioural intention to towards smart tools. The mean BIS was found to be lowest among the youngest group (22-25 years) and highest among the mid age group (26-30 years), yet showed a decline again in the oldest group (more than 30 years). Although the ANOVA test revealed no significant difference as $p>0.05$, the findings from **Figure 2** suggest that mid-career teachers may possess higher behavioural intention towards dmart tools. On one hand, this might be due to their experience, practical exposure, in-field experiences etc. On the other hand, the youngest group, being still in the process of learning, might not have much experience in using smart tools in their pedagogical practices. Lastly, the older group (30+ years) may have experienced a lower behavioural intention due to lower exposure to technology during their training years or difficulty navigating technological changes. This might indicate the need for training to enhance teachers' exposure with technology.

DISCUSSION

The present study investigated pre-service and in-service teachers' behavioral intention to use smart tools in education, drawing on a sample of 42 participants across varying age groups and genders. Behavioural intention was measured through a 15-item Likert-scale instrument, and a composite score was generated by summing the items. This total score was used to analyse trends through quartile distribution, ANOVA, and t-tests, allowing us to explore how demographic factors and individual perceptions may shape teachers' willingness to adopt smart technologies.

- **Quartile Deviation Findings**

The quartile deviation analysis revealed that for all eleven constructs, the mean was 5.63, which meant that most respondents' intention related to the use of smart tools was found to be agreeable. Furthermore, for Q1, the mean was found to be 4.762, which meant that around 26.2% of the respondents were inclined away from using smart tools or had a neutral attitude towards them. Lastly, for Q2, the mean was found to be 6.514, which meant that around 23.8% of the respondents had a positive attitude towards the use of smart tools for teaching and learning. This finding offers insights into Research Question 1, which sought to understand the intent of teachers towards smart tools. The moderate to high behavioural intention among the majority suggests that many teachers recognise the potential of smart tools. However, the presence of a low-intent group signals the need for interventions (training, hands-on workshops, etc.) that address usability and perceived value in practice.

- **Influence of Gender: Independent Samples t-Test**

An independent sample t-test was conducted to determine whether gender influenced behavioral intention. This result helps address research question 3, which explored whether gender had an influence on teachers' intent to use smart tools. The result was not statistically significant ($t(40) = 0.529$, $p = .599$), indicating that male and female teachers did not differ significantly in their behavioural intention to use smart tools. The lack of a gender gap suggests that both male and female participants demonstrate similar confidence levels or readiness to adopt smart tools, challenging assumptions that gender plays a decisive role in tech adoption in teaching.

- **Influence of Age: One-Way ANOVA**

A one-way ANOVA was conducted to examine whether behavioral intention varied by age group. It partially informs the rest of the research questions 3, which sought to explore the influence of age on teachers' intent to use smart tools in education. The results were also not statistically significant ($F(2, 39) = 0.489$, $p = .617$), indicating that age does not significantly influence pre-service teachers' intention to use smart tools. This finding suggests that teachers share relatively similar attitudes toward smart tool adoption regardless of age. The lack of significant difference implies that beliefs around teaching and willingness to integrate new tools are more evenly distributed across age groups than expected. However, in terms of differences among the age groups w.r.t mean of behavioural intention scores, the findings revealed a meaningful trend. Teachers in the 26-30 age group were found to possess a higher behvaioural intention than teachers belonging to age groups 22-25 years and more than 30 years. This pattern may suggest that teachers in the age group of 26-30 years may experience an optimal blend of training, classroom experiences and technological competence, influencing their intent towards smart tools. Whereas, the youngest age group might still be in the learning phase and hence might not have experiences of using or assessing smart tools for teaching and learning. Lastly, the oldest group of teachers, (30+ years), might have specific beliefs or inclinations when it comes to integration of technology in education. This may have led to their intent being lesser.

- **Broader Implications for Influencing Factors**

While demographic variables like age and gender did not significantly affect behavioural intention, the variability revealed through quartile analysis hints at other influential factors such as peer support, feedback, and accessibility (Research Question 2). These contextual and environmental aspects (CSE, ENJ, INNO, PI, PUV and COGT), which were identified as extended variables in Khlaisang et al. (2021), may better explain the differences in behavioural intention and should be explored further in future research through qualitative methods or targeted survey items.

CONCLUSION

The current study explored teachers' behavioral intention to adopt smart tools by examining their perceptions, confidence, contextual influences, and pedagogical standpoints. The findings offer a layered understanding of how these factors play out in the teacher preparation landscape. Although no significant differences were found across gender or age groups, the quartile-based distribution revealed a generally favourable intention towards smart tools with small levels of variation in the behavioral intention among the teachers. While half of the respondents fell into the moderate intention category, 26.2% of the respondents fell in the first quartile, and 23.8% of them fell in the third quartile. This divergence suggests that behavioural intention is not solely shaped by demographic factors but likely influenced by more nuanced, experiential elements within the teacher training environment, such as computer self-efficacy, perceived enjoyment, personal innovativeness, peer influence, perceived ubiquity value and cognitivist teaching beliefs. Furthermore, the absence of significant gender or age effects challenges common assumptions about generational digital divides or gender-based differences in technology use. Instead, it indicates a growing normalisation of digital tool use among teachers, irrespective of background — possibly reflecting their exposure to technology-rich academic environments. However, the presence of a low-intention group underscores the need for targeted support. These participants may face internal (e.g., lack of confidence) or external (e.g., lack of peer encouragement or inadequate feedback systems) barriers that limit their intention to use smart tools. As such, future training programs should emphasise skill development and tool familiarity, as well as confidence-building, peer collaboration and reflective practice that aligns tool use with teachers' evolving instructional beliefs. Therefore, the present study highlights the complexity of technology adoption in education, specifically in the Indian context. Teachers' intentions are shaped not just by individual attitudes but by the interplay of beliefs, contextual supports, and the perceived alignment of tools with pedagogical goals. For smart tools to be meaningfully integrated into classrooms, teacher education must address both the technical and human aspects of change.

Conflicts of interest

Authors have no competing interests.

REFERENCES

Ahmad, S., Umirzakova, S., Mujtaba, G., Amin, M. S., & Whangbo, T. (2023). Education 5.0: requirements, enabling technologies, and future directions (arXiv:2307.15846). *arXiv*. https://doi.org//arXiv.2307.15846DOI: 10.48550

Anderson, L. W., Krathwohl, D. R., & Bloom, B. S. (2001). *A taxonomy for learning, teaching, and assessing: A revision of Bloom's taxonomy of educational objectives*. Longman.

Andres, B., Sempere-Ripoll, F., Esteso, A., & Alemany, M. M. E. (2022). Mapping between industry 5.0 and education 5.0. *EDULEARN22 Proceedings,* 2921–2926. https://doi.org/DOI: 10.21125/edulearn.2022.0739

Antonova, A., & Dankov, Y. (2023). Smart services in education: Facilitating teachers to deliver personalized learning experiences. In *Data Science and Algorithms in Systems* (pp. 108–117). Springer International Publishing. DOI: 10.1007/978-3-031-21438-7_9

Asadi, Z., Abdekhoda, M., & Nadrian, H. (2019). Understanding and predicting teachers' intention to use cloud computing in smart education. *Interactive Technology and Smart Education, 17*(1), 14–27. DOI: 10.1108/ITSE-05-2019-0019

Bandura, A. (1986). *Social foundations of thought and action: A social cognitive theory*. Prentice-Hall.

Chirume, S. (2020). Sustainable professional development of primary school mathematics teachers in zimbabwe through philosophy of education 5.0: challenges and prospects. *Sumerianz Journal of Social Science*, 150–161. https://doi.org/DOI: 10.47752/sjss.312.150.161

Dai, Z., Xiong, J., Zhao, L., & Zhu, X. (2023). Smart classroom learning environment preferences of higher education teachers and students in China: An ecological perspective. *Heliyon, 9*(6), e16769. DOI: 10.1016/j.heliyon.2023.e16769 PMID: 37303554

Davis, F. D. (1989). Perceived usefulness, perceived ease of use, and user acceptance of information technology. *Management Information Systems Quarterly, 13*(3), 319–340. DOI: 10.2307/249008

Demir, K. A. (2021). Smart education framework. *Smart Learning Environments, 8*(1), 1–36. DOI: 10.1186/s40561-021-00170-x PMID: 40477293

Dmitrenko, N. Y., Voloshyna, O. V., Kizim, S. S., Mnyshenko, K. V., & Nahorniak, S. V. (2023). Smart education in the prospective teachers' training. *CTE Workshop Proceedings*, 10, 414–429. DOI: 10.55056/cte.568

Gao, J. (2024). *Research on the cultivation of new media literacy among college teachers from the perspective of smart education*. Media and Communication Research., DOI: 10.23977/mediacr.2024.050325

Iqbal, H. M. N., Parra-Saldivar, R., Zavala-Yoe, R., & Ramirez-Mendoza, R. A. (2020). Smart educational tools and learning management systems: Supportive framework. [IJIDeM]. *International Journal on Interactive Design and Manufacturing*, *14*(4), 1179–1193. DOI: 10.1007/s12008-020-00695-4

Jackson, C. M., Mun, Y. Y., & Park, J. S. (2013). An empirical test of three mediation models for the relationship between personal innovativeness and user acceptance of technology. *Information & Management*, *50*(4), 154–161. DOI: 10.1016/j.im.2013.02.006

Jan, A., Sultana, I., & Adnan, M. (2020). Digital media and smart education in Pakistan: Challenges and prospects for the teachers in the age of E-learning. *Global Educational Studies Review.* https://www.academia.edu/download/79117620/4fgmQJvxwr.pdf

Kelman, H. C. (1958). Compliance, identification, and internalization: Three processes of attitude change. *The Journal of Conflict Resolution*, *2*(1), 51–60. DOI: 10.1177/002200275800200106

Khlaisang, J., Songkram, N., Huang, F., & Teo, T. (2021). Teachers' perception of the use of mobile technologies with smart applications to enhance students' thinking skills: A study among primary school teachers in Thailand. *Interactive Learning Environments*, *31*(8), 5037–5058. DOI: 10.1080/10494820.2021.1993933

Kumar, P., & Daniel, B. K. (2016). Integration of digital technologies in the classroom: A case study from India. *Education and Information Technologies*, *21*(5), 1297–1316.

Lai, J. Y., & Chen, C. C. (2011). Self-efficacy, effort, job performance, and job satisfaction: An exploratory study of the internet-enabled work environment. *Journal of International Management*, *18*(1), 1–15.

Lu, W., & Liang, H. (2024). Research on strategies and measures for enhancing digital literacy of college teachers in the era of smart education. *2024 IEEE First International Conference on Data Intelligence and Innovative Application (DIIA)*, 1–3. DOI: 10.1109/DIIA62678.2024.10871896

Nikou, S. A., & Economides, A. A. (2017). Mobile-based assessment: Investigating the factors that influence behavioral intention to use. *Computers & Education, 109*, 56–73. DOI: 10.1016/j.compedu.2017.02.005

Nuryadi, M. H., & Widiatmaka, P. (2023). Strengthening civic literacy among students through digital literacy in society 5.0. [EduLearn]. *Journal of Education and Learning*, *17*(2), 215–220. DOI: 10.11591/edulearn.v17i2.20746

Prakash, S., Muniammal, M. A., Maruthavanan, M., Natarajan, S., Thangavel, K., & Kumar, S. (2023). Education 5.0 revolutionizing learning for the future (Vol.1).

Prakasha, G. S. (2022). *Next-generation teacher competencies: competency balance scorecard, competency development cycle and strategies. in handbook of research on promoting global citizenship education*. IGI Global., DOI: 10.4018/978-1-7998-9542-8.ch014

Qarkaxhija, J., Assanova, Z., Igenbayeva, R., Toxanova, S., Yensebayeva, G., Myrzanova, A., & Turlugulova, N. (2023). Development of the creative potential of future teachers based on smart education. [iJET]. *International Journal of Emerging Technologies in Learning*, *18*(24), 19–33. DOI: 10.3991/ijet.v18i24.43205

Rakovac Bekeš, E., & Galzina, V. (2023). Exploring the pedagogical use of ai-powered chatbots educational perceptions and practices. *International Convention on Information and Communication Technology, Electronics and Microelectronics.*

Seeling, P., Roberts, S., & Weible, J. (2022). Towards education 5.0: instruction with learners in the loop. *Proceedings of the 23rd Annual Conference on Information Technology Education*, 92–93. https://doi.org/DOI: 10.1145/3537674.3555793

Tandon, R., & Tandon, S. (2020). Education 4.0: A new paradigm in transforming the future of education in India. *International Journal of Innovative Science. Engineering & Technology*, *7*(2), 32–54.

Tao, S., Li, Y., Dong, X., Gunasekaran, N., & Aziz, A. (2021). Smart educational learning strategies for teachers and students in the higher education system. *J. Multiple Valued Log. Soft Comput.*, *36*, 99–115.

Teo, T., Lee, C. B., & Chai, C. S. (2008). Understanding pre-service teachers' computer attitudes: Applying and extending the Technology Acceptance Model. *Journal of Computer Assisted Learning*, *24*(2), 128–143. DOI: 10.1111/j.1365-2729.2007.00247.x

Toral, S. L., Martínez-Torres, M. R., & Barrero, F. (2007). Perceived usefulness and ease of use of e-collaboration tools in engineering education. *Computers & Education*, *49*(4), 1097–1113.

Vakaliuk, T., Osova, O., Antoniuk, D., & Novitska, I. (2020). Future Foreign Language Teachers' Readiness for Innovative Activity in the Context of Smart Education: *The Results of the Pedagogical Experiment. Universal Journal of Educational Research, 8*(12B), 8223–8232. DOI: 10.13189/ujer.2020.082626

Venkatesh, V., & Bala, H. (2008). Technology Acceptance Model 3 and a research agenda on interventions. *Decision Sciences, 39*(2), 273–315. DOI: 10.1111/j.1540-5915.2008.00192.x

Xing, P. (2023). A practical study on the application of smart education platform for teaching and research by primary and secondary school teachers. *International Journal of New Developments in Education, 5*(8). Advance online publication. DOI: 10.25236/IJNDE.2023.050814

Yang, R. (2022). A study on the model of online learning support services for teachers in the context of smart education. *Proceedings of the 14th International Conference on Education Technology and Computers*, 337–343. DOI: 10.1145/3572549.3572603

Yang, R., Zhao, B., Liu, M., & Wu, Z. (2023). Teachers' professional development trends and reflections from the perspective of smart education. In Atlantis Highlights in Computer Sciences (pp. 1130–1135). *Atlantis Press International BV.*

Zengin, Y., Naktiyok, S., Kaygın, E., Kavak, O., & Topçuoğlu, E. (2021). An investigation upon industry 4.0 and society 5.0 within the context of sustainable development goals. *Sustainability (Basel), 13*(5), 5. Advance online publication. DOI: 10.3390/su13052682

Ziatdinov, R., & Cilliers, J. (2021). Generation Alpha: Understanding the next cohort of university students. *European Journal of Contemporary Education, 10*(3). Advance online publication. DOI: 10.13187/ejced.2021.3.783

Chapter 6
AI in Education: Designing Contextualized Learning to Boost Motivation and Ensure Equity

Carmen Rodríguez-Jiménez
https://orcid.org/0000-0001-8623-8316
University of Granada, Spain

Adrián Segura-Robles
https://orcid.org/0000-0003-0753-7129
University of Granada, Spain

Manuel Enrique Lorenzo-Martín
University of Granada, Spain

Carmen Rocío Fernández-Fernández
University of Granada, Spain

ABSTRACT

The text analyses how Artificial Intelligence (AI) can be integrated into education to design contextualized learning situations that increase student motivation and promote equity. Through tools like ChatGPT, content can be personalized, and the curriculum can be connected to students' local realities. The proposed methodology includes context analysis, definition of competencies, the use of prompts, and continuous evaluation with feedback. Additionally, the value of Universal Design for Learning (UDL) is highlighted as an inclusive framework. However, the implementation of AI entails ethical challenges such as data privacy, algorithmic bias,

DOI: 10.4018/979-8-3373-7922-7.ch006

and the need for teacher training. The text concludes that AI should not replace teachers but complement them, and it advocates for inclusive policies and a culture of critical thinking to foster more motivating, personalized, and equitable education.

1. INTRODUCTION

This research paper analyses how artificial intelligence (AI) can be used as a key tool to design contextualized learning situations that connect the curriculum with students' local realities, promoting their motivation, the development of transversal competencies, and educational equity. It also presents a series of proposals for integrating AI into curriculum design, based on the analysis of the educational context, the definition of competencies aligned with local realities, and the validation of these designs through feedback from experts, students, and teachers. This proposal seeks to foster a more equitable and relevant education that prepares students to face the challenges of a constantly changing world.

The rise of AI has transformed multiple sectors, including education. In this context, AI emerges as a key tool to personalize learning experiences and connect curricular content with students' realities. This ability to contextualize has the potential to overcome barriers that hinder motivation and student engagement, especially in environments marked by social, economic, or cultural inequalities (Bearman et al., 2023).

Contemporary education faces complex challenges in a world defined by VUCA environments (volatility, uncertainty, complexity, and ambiguity) and BANI conditions (brittleness, anxiety, non-linearity, and incomprehensibility). These conditions of constant change and uncertainty demand that educational systems and professionals adopt strategies that respond to classroom diversity and adapt quickly to global transformations. Within this framework, the integration of AI into curriculum design appears as a promising response, promoting contextualized learning situations that enhance student motivation, the development of transversal competencies, and educational equity (Bearman et al., 2023; Zhang & Aslan, 2021).

In the last decade, AI has significantly advanced, enabling the development of technologies capable of analyzing large volumes of data, identifying patterns, and generating solutions tailored to individual needs. In the field of education, these capabilities could translate into more personalized and effective learning experiences. Tools like ChatGPT and NotebookLM make it possible to generate educational content adapted to students' contexts and characteristics through the use of specific instructions or prompts (Firat, 2023).

Additionally, the Universal Design for Learning (UDL) stands out as a key theoretical framework for guiding the integration of AI in education. This approach seeks to remove learning barriers by ensuring that all students can participate equitably in educational activities. AI, by facilitating the creation of adaptive and personalized materials, aligns with UDL principles and promotes inclusive education that respects student diversity (Biggs & Tang, 2007).

Moreover, AI enables the automation of repetitive tasks such as the creation of assessment rubrics or didactic activities, thereby reducing the administrative burden on teaching staff. This frees up time for tasks with greater pedagogical value, such as direct interaction with students and methodological innovation (Cisternas-San Martín et al., 2025).

One of the greatest challenges in education is making the curriculum relevant to students. Often, students perceive school content as disconnected from their lives and contexts, which negatively affects their motivation and engagement. In this regard, AI offers tools to design contextualized learning situations by integrating curricular content with students' local and cultural realities. For example, through well-designed prompts, teachers can generate materials that include references to the community's history, geography, or social issues, making learning more meaningful (Firat, 2023).

Despite its benefits, the implementation of AI in education raises important ethical and pedagogical challenges. Ensuring student data privacy is essential, as AI technologies collect large volumes of personal information. Furthermore, AI algorithms may reproduce biases present in training data, thereby perpetuating inequalities rather than mitigating them (Zhang & Aslan, 2021).

It is also crucial to ensure human oversight in the use of these technologies. Teachers must play an active role as critical mediators, reviewing and adapting AI-generated content to ensure its pedagogical and cultural appropriateness. This requires teacher training not only in the technical use of these tools but also in understanding their ethical and pedagogical implications (Cisternas-San Martín et al., 2025).

Finally, the integration of AI must be accompanied by inclusive policies that ensure equitable access to these technologies, particularly in economically or technologically disadvantaged contexts. This includes providing adequate infrastructure, training teachers, and promoting a digital ethic centered on the needs of students and their environments (Bearman et al., 2023).

2. SCIENTIFIC LITERATURE REVIEW

2.1. Artificial Intelligence and Education

The integration of AI into the educational field has experienced remarkable progress in recent decades, positioning itself as a key tool for transforming teaching and learning. These advances have not only enabled personalized learning but also optimized educational processes through the analysis of large volumes of data and the creation of solutions tailored to students' individual needs (Bearman et al., 2023; Zhang & Aslan, 2021).

In their critical review, Bearman et al. (2023) highlight that AI has enabled the creation of personalized learning environments in which platforms can adapt content and activities to students' learning styles and levels. This is particularly useful in heterogeneous classrooms, where differences in cognitive abilities, interests, and cultural experiences can pose significant challenges for educators. According to the authors, AI's ability to analyze data on academic performance and student preferences facilitates the identification of learning gaps and allows for early intervention.

In their 2023 work, the authors present several contributions that we consider key:

- They propose three major discourses surrounding AI: first, as a tool for improving efficiency in administrative and educational processes; second, as a means of personalizing learning; and third, as a technology that presents ethical and critical challenges for higher education.
- They emphasize the need to consider not only the technological benefits of AI but also the social, ethical, and cultural implications it brings to educational institutions.
- They highlight gaps in the existing literature regarding the long-term impact of AI and advocate for more critical and contextual approaches. In this regard, we hope our contribution adds value to this direction.

On the other hand, Zhang and Aslan (2021) have explored how AI can enhance the adaptability of educational systems in the face of global challenges, such as school closures during the COVID-19 pandemic. Their research highlights the role of AI technologies in offering flexible solutions such as automated online tutoring, the creation of adaptive learning resources, and continuous assessment platforms. These innovations have not only ensured learning continuity during emergencies but have also opened up new possibilities for teaching in remote or underserved contexts.

Among their main contributions, we highlight the following:

- They identify three key areas of impact: personalized learning, automation of administrative tasks, and the improvement of the learning experience through simulations and augmented reality.
- They propose future research directions, including the development of more ethical, transparent, and culturally adapted AI systems—an approach that directly connects to the concept of contextualized learning.

- They emphasize the need for interdisciplinary research to evaluate the educational impact of AI from pedagogical, ethical, and technological perspectives.

Among the most prominent tools in this field are ChatGPT and NotebookLM, which can generate personalized educational content based on user-provided instructions. These platforms enable teachers to design instructional materials tailored to specific contexts, incorporating local and cultural elements that enhance the relevance of learning. According to Firat (2023), these tools are particularly useful in self-directed learning and open education, where students can access personalized content without relying on a face-to-face instructor. The main contributions highlighted in this reference are:

- The study explores how tools like ChatGPT can revolutionize self-directed and open education by facilitating access to knowledge and promoting independent learning.
- It identifies several benefits of ChatGPT: its ability to respond to questions instantly, generate personalized educational materials, and provide interactive tutoring.
- It also discusses limitations such as potential bias in the model's responses and the need to teach students how to use these tools critically.
- It concludes that ChatGPT can democratize learning by lowering access barriers, but it requires supervision and digital literacy to avoid misunderstandings.

In their article *"Artificial Intelligence as an Assistant for Developing a Competency-Based Continuing Education Curriculum,"* Cisternas-San Martín, Guzmán Muñoz, and Rivas Poblete (2025) address the potential of AI as a key tool to transform curriculum design in continuing education. The authors examine how AI can support the creation of more relevant and up-to-date educational programs by analyzing labor market demands and trends. In addition, they propose a methodological model that enables the automation of essential processes such as identifying required skills and personalizing content, thus contributing to greater efficiency and relevance in curriculum development. However, the study also critically reflects on the ethical and technological limitations inherent in the use of AI in this context, emphasizing the need for a responsible and strategic approach to its implementation. Key contributions from this study include:

- It examines the use of AI tools to assist in designing continuing education curricula focused on professional competencies.

- It proposes a methodological model that integrates AI to analyze labor demands and trends, facilitating the creation of more relevant and updated educational programs.
- It highlights how AI can automate processes such as identifying necessary skills and adapting content, contributing to greater personalization and efficiency in curriculum development.
- The study includes a critical analysis of the ethical and technological limitations of using AI in this context.

Despite all these advancements, and as reflected in the aforementioned contributions, it is essential to address the challenges related to the implementation of AI in the classroom. Among these, we highlight the lack of teacher training in the use of such technologies—particularly generative AI—and the inequalities in access to technological infrastructure, which limit the full reach of the potential benefits that AI can offer.

2.2. Universal Design for Learning as an Inclusive Framework

Universal Design for Learning (UDL), initially proposed by the Center for Applied Special Technology (CAST), has gained international recognition as a key theoretical framework for promoting educational inclusion. Its implementation focuses on three core principles: providing multiple means of representation, offering multiple means of action and expression, and ensuring multiple means of engagement to meet the diverse learning styles and needs of students. This approach ensures that all learners, regardless of their abilities or backgrounds, can participate equitably in educational processes (Biggs & Tang, 2007).

In Spain, UDL has been formally integrated into the educational system through the Organic Law Amending the Organic Law of Education (LOMLOE, 2020). This legislation establishes UDL as a guiding principle for curriculum design, aiming to ensure inclusion and equity at all levels of education. According to the LOMLOE, teachers must design learning situations based on UDL principles in order to address student diversity and foster their holistic development. The inclusion of UDL in Spanish educational policy represents a significant step toward building a more inclusive and equitable system (Ministry of Education and Vocational Training, 2020).

UDL has emerged as a fundamental framework for ensuring educational inclusion by providing a flexible and adaptive curriculum that responds to student diversity. According to Lorenzo Lledó (2023), UDL promotes the creation of educational environments that not only ensure equitable access to learning but also transform individual differences into pedagogical opportunities. The integration of technologies such as AI can strengthen the implementation of these proposals by optimizing inclusive educational processes.

The relationship between UDL and AI has become a topic of growing interest in educational research. AI can facilitate UDL implementation by generating educational materials adapted to the individual characteristics of students. For instance, tools like ChatGPT can create personalized explanations, examples, and activities aligned with UDL principles. Additionally, AI can provide real-time feedback and data analysis on student progress, enabling teachers to adjust their pedagogical strategies accordingly.

The implementation of UDL combined with AI has recently been highlighted as an effective solution for personalizing learning and promoting inclusion in the classroom. According to Yépez Álvarez et al. (2024), AI offers tools to adapt educational content to students' individual needs, aligning with UDL principles.

The authors emphasize that AI platforms such as ChatGPT and other emerging technologies can facilitate access to personalized materials that respect the diversity of learning styles, which is especially valuable in contexts where cultural, linguistic, or cognitive differences pose challenges.

Moreover, Yépez Álvarez et al. (2024) stress that although AI automates numerous educational processes, it is crucial to maintain human oversight to ensure pedagogical and ethical quality in the implementation of these technologies.

However, implementing UDL through AI is not without challenges. One of the main concerns is ensuring that the technologies used comply with accessibility standards and do not perpetuate existing biases. As noted by Biggs and Tang (2007), the flexibility inherent to UDL requires ongoing monitoring to ensure that educational materials are appropriate for all students.

Effective UDL implementation requires continuous teacher training. AI can support this by offering personalized learning platforms for educators, including interactive modules adapted to their specific needs and levels of knowledge. In conclusion, these are two closely related elements that, in order to be successfully integrated and coexist, will require substantial teacher training.

2.3. Ethical and Pedagogical Challenges

The incorporation of AI into the educational field has generated a profound transformation in pedagogical methodologies and in teaching-learning dynamics. However, this advancement not only opens up new possibilities but also presents significant ethical and didactic challenges. These issues arise in a context where AI is redefining the way knowledge is accessed and constructed, making it essential to ensure quality education aligned with the Sustainable Development Goals (SDGs) (Cruz et al., 2024; Cedeño et al., 2024).

From a pedagogical perspective, AI has demonstrated its capacity to adapt educational content to students' individual needs, promoting greater engagement in the learning process (Torre et al., 2024; Gómez, 2023). However, this technological personalization demands deep ethical reflection, as its implementation may widen access gaps if technological equity is not ensured and student data privacy is not adequately protected (Alcocer et al., 2024). In this sense, cooperation between teachers, technology experts, and policymakers is essential to establish regulatory frameworks that govern its use in a fair and inclusive manner (Torre et al., 2024; Alcocer et al., 2024).

One of the main challenges for teachers is the effective integration of AI into their educational practices. Without proper training in the use of these tools, it becomes difficult to harness their transformative potential (Mujica-Sequera, 2024; Pérez & González, 2024). Therefore, ongoing training in digital competencies emerges as an urgent need for educators to adapt their methodologies to the current environment without compromising the quality of teaching (Vela et al., 2024; Pintado et al., 2023). As AI becomes an essential component in higher education, existing pedagogical strategies must be reviewed and redesigned to reflect this technological evolution (Cedeño et al., 2024; Marcano et al., 2024).

On the ethical front, there is concern that the use of AI may contribute to a more superficial educational experience, reducing human interactions that are key to students' holistic development (Alcocer et al., 2024; Muñoz et al., 2024). For this reason, it is crucial that AI is applied as a complement to face-to-face teaching, not as a substitute, thereby promoting a balanced education that values the human dimension of learning (Loayza, 2024; Rodríguez et al., 2023).

Despite the challenges, teachers' perceptions of AI tend to be favorable, as its potential to optimize processes and improve educational quality is widely recognized (Padilla, 2019; Lema et al., 2024). However, its implementation must be strategic and must take into account both its benefits and ethical implications. To this end, the establishment of a regulatory framework is recommended to ensure equitable access, protection of personal data, and the promotion of inclusive education (Alcocer et al., 2024; Cotrina-Aliaga et al., 2021).

2.3.1. Privacy and Data Management

AI relies on large volumes of data to personalize instruction, including academic, behavioral, and emotional information about students. While this data enables the design of individualized educational experiences, its misuse can compromise student privacy and violate data protection regulations. According to Williamson and

Eynon (2020), it is essential to establish clear legal and ethical frameworks to ensure the security of information. Furthermore, the implementation of anonymization and encryption technologies can help minimize the risks associated with handling sensitive data.

2.3.2. Algorithmic Bias and Educational Justice

Algorithmic bias is a critical ethical challenge in the use of AI in education. Algorithms are trained on historical data that may reflect preexisting structural inequalities, potentially perpetuating or even exacerbating educational disparities. For example, automated assessment systems have shown a tendency to favor certain social groups over others due to biases present in the training data (Holmes et al., 2021). According to Cisternas-San Martín et al. (2025), those responsible for the design and implementation of such tools must adopt auditing strategies and ethical evaluations of algorithms to ensure fair and inclusive outcomes.

2.3.3. Teacher Training and Pedagogical Mediation

Although AI can automate various teaching tasks, the role of educators remains essential as critical mediators in the teaching and learning process. This requires teachers to be trained not only in the technical use of AI-based tools but also in understanding their pedagogical and ethical implications (Firat, 2023). For example, educators must be able to assess the quality of AI-generated recommendations and adapt them to the specific needs of their students. Ongoing teacher training is crucial to ensure that these technologies are used both effectively and ethically. Luckin (2018) emphasizes that teachers must understand how AI can complement their work, fostering collaboration between humans and machines to maximize educational benefits.

2.3.4. Sustainability and Accessibility

Another important challenge is access to AI tools in diverse educational contexts. According to Zhang and Aslan (2021), many educational institutions in low- and middle-income countries lack the technological infrastructure necessary to implement these technologies, which could widen educational inequalities. It is crucial to develop inclusive policies that ensure equitable access to AI, alongside investments in infrastructure and resources. Additionally, the environmental impact of AI use—particularly in terms of energy consumption—must be addressed through the design of more sustainable systems. Williamson and Eynon (2020) argue that investing in technological infrastructure is essential to ensure that AI can be implemented equitably.

2.3.5. Ethics in the Design and Use of AI

The ethical design of AI technologies must be at the core of their development and implementation in education. This includes ensuring algorithmic transparency, clearly explaining how decisions are made and how they affect students. Williamson and Eynon (2020) emphasize that educational institutions should collaborate with experts in ethics and technology to ensure that AI applications align with the principles of justice, equity, and student well-being. To achieve this, they propose involving multiple stakeholders in the design of these technologies, including educators, software developers, and education policymakers.

While AI offers significant benefits for the development of both transversal and curricular skills, it also raises ethical and legal concerns. The implementation of AI in sectors such as banking or business has transformed processes and improved efficiency, but it also requires addressing ethical issues and ensuring that AI applications align with social values. This calls for a comprehensive understanding of AI's implications and the development of curricula that integrate these considerations.

- **Equity and Bias**: AI systems in education can perpetuate existing socio-cultural biases and inequalities. The risk of bias in AI-generated content can lead to unfair treatment and the dissemination of inaccurate information, requiring specific skills and rigorous data monitoring to ensure equitable educational experiences (Holmes et al., 2021).
- **Privacy Concerns**: The use of AI often involves the collection and use of student data, raising significant privacy concerns—especially when minors are involved. It is essential to establish clear policies and guidelines to regulate data collection, use, and protection, ensuring that data is used solely for educational purposes and that robust security measures are in place (Klimova, 2023).

3. CONTEXTUALIZED LEARNING THROUGH AI

Contextualized learning is a methodology that seeks to connect academic content with meaningful situations that are close to the students' environment. This approach contributes to increased motivation, deeper comprehension, and better transfer of acquired knowledge (Johnson, 2020). It is based on the premise that learning becomes more meaningful when students can establish connections between what they learn and the issues or scenarios they observe in their daily lives (Herrington & Oliver, 2022).

Artificial Intelligence (AI), with its ability to process large volumes of information and adapt to specific contexts, presents itself as a powerful tool for implementing more dynamic and effective contextualized learning. AI-based tools such as conversational systems and virtual assistants enable the design of personalized and realistic learning situations, facilitating the acquisition of both cognitive and socio-emotional skills (Luckin et al., 2023). Below are several examples of how AI can be used in the classroom:

- **Role simulation**: The use of AI to simulate professional or historical roles allows students to adopt different perspectives and deepen their understanding of complex phenomena through a critical and situated lens. For example, by asking an AI: "Act as a historian expert in World War II. Explain the causes and consequences of the Hiroshima bombing, considering the political and social context of the time," the student receives a response that not only provides factual information but also includes nuanced insights into the sociopolitical circumstances of that era. This technique, framed within gamification and immersive storytelling, supports the development of historical thinking and critical empathy (Gee, 2021).
- **Real-world scenarios**: Another powerful resource is the creation of realistic scenarios in which AI acts as an expert analyzing contemporary issues. For instance: "You are an environmental scientist. Describe the impact of climate change on the forests and seas of [City], as well as viable solutions for their conservation, considering economic and social challenges." With these types of prompts, students are confronted with complex problems that require the integration of knowledge from various domains (environment, economics, sociology, etc.), thereby fostering the development of transversal competencies and real-world problem-solving skills (UNESCO, 2022).
- **Local case analysis**: AI also enables the generation of analyses from a local or historical perspective. Prompts such as: "Act as a relevant politician from [City]. Examine the impact of migratory movements and their effects on the local population," or "Act as [Relevant Historical Figure] and explain how you experienced the passage of different civilizations through [City]," enhance students' ability to reflect on their immediate environment. These activities encourage active citizenship and social engagement by fostering critical analysis of issues that affect their local community (Salinas et al., 2023).

Furthermore, the localized contexts generated by AI enhance meaningful learning, as students recognize the direct relevance of the content studied and how it relates to their own realities (Coll & Rochera, 2022).

3.1. AI as a Tool to Foster Student Motivation

Motivation is an essential factor in the learning process, as it directly influences student engagement, perseverance, and academic performance (Ryan & Deci, 2020). In this regard, AI emerges as a key tool for personalizing instruction and promoting both intrinsic and extrinsic motivation among students.

Personalization is one of the most valued aspects of AI in educational settings. Intelligent systems can adapt content, activities, and teaching strategies to the individual needs, interests, and learning paces of each student (Holmes et al., 2019). This adaptation creates a more inclusive and meaningful environment, enhancing the perception of competence and autonomy—two core components of motivation according to Self-Determination Theory (Deci & Ryan, 2017).

For instance, educational platforms with AI algorithms can offer individualized learning paths, personalized recommendations, and immediate, specific feedback. These features help reduce frustration caused by tasks that are too difficult or irrelevant, while increasing students' satisfaction and engagement with the activity (Zawacki-Richter et al., 2019).

In addition, AI promotes extrinsic motivation through gamification elements such as badges, challenges, or personalized leaderboards, as well as reward systems tailored to students' progress and performance (Huang et al., 2022). These environments not only capture attention but also allow students to perceive a sense of achievement and advancement, thereby increasing their involvement in the task.

Furthermore, AI supports social motivation, as many AI-based platforms integrate collaborative spaces and adaptive learning networks that allow students to interact and share learning experiences with peers who have similar learning profiles (Luckin et al., 2023).

In this way, AI not only contributes to the personalization of learning but also strengthens the emotional and social components of the educational process—key aspects for sustaining long-term motivation.

4. METHODOLOGY: PROPOSAL FOR THE IMPLEMENTATION OF AI IN CURRICULUM DESIGN

The incorporation of AI into curriculum design implies a profound transformation in the traditional way of understanding and developing educational processes. This methodological proposal is presented as a structured framework aimed at achieving the effective and efficient inclusion of AI, ensuring that it aligns with current educational demands and the training objectives of educational institutions.

The proposal begins with a thorough analysis of the specific educational context in which it is to be implemented. This includes collecting data on student needs, clearly defining competencies aligned with those needs, precisely developing instructions or "prompts" for the efficient use of advanced technological tools and establishing a rigorous process of evaluation and feedback to continuously optimize the curriculum design.

4.1. Context Analysis

Context analysis is the first and fundamental step toward the effective implementation of AI within the curriculum. This process is not limited to a superficial description of the environment but requires a detailed study of the internal and external variables that directly influence the educational process.

This analysis involves identifying key elements such as students' socioeconomic profiles, local cultural factors, and the technological infrastructure of the educational institution. The educational reality must be examined from a comprehensive and holistic perspective, identifying not only technological shortcomings but also potential strengths that may support the implementation of educational innovations. This information-gathering process allows for the development of a complete and realistic diagnosis. Such contextual diagnosis serves as the essential foundation for designing a curriculum that is adapted to the specific needs of students, ensuring its relevance and practical utility.

4.2. Definition of Competencies

The precise and clear definition of the competencies to be developed is a crucial stage in AI-driven curriculum design. These competencies must directly emerge from the contextual analysis, ensuring complete alignment between the identified real needs and the proposed educational objectives.

It is essential to identify advanced digital technical skills that enable the conscious and critical use of AI, along with competencies related to solving real-world problems through technology. Likewise, it is vital to include transversal competencies such as critical thinking, creativity, and collaboration. The resulting curriculum should equip students to interact critically with emerging technologies, understand their ethical and social implications, and develop the ability to design and produce creative technological solutions to concrete problems.

4.3. Use of Prompts in AI Tools

The use of prompts in AI tools has significantly transformed the contemporary educational landscape. AI tools are widely recognized for their ability to personalize the educational experience, allowing content and methodologies to be adapted to the individual needs of students.

This approach is achieved through the analysis of data on students' performance and learning styles, facilitating the delivery of appropriate educational resources and promoting more effective and participatory learning (Vera-Flores et al., 2024; Torre et al., 2024).

Moreover, these tools enable students to receive individualized guidance tailored to their learning methods and academic requirements, which is fundamental for maintaining motivation and engagement in the classroom (Vera et al., 2024). Recent studies have also noted that AI can facilitate the creation of educational simulations—scenarios in which students apply their knowledge in realistic and complex situations (Rodríguez et al., 2024).

This not only fosters active learning but also prepares students to face real-world challenges by effectively combining theory and practice (Alcocer et al., 2024). However, it is crucial to balance the integration of these tools with human intervention, such as teacher mediation, to prevent students from becoming overly dependent on technology, which could hinder their autonomy (Estrada-Araoz et al., 2024).

The use of AI tools and prompts in the educational field offers significant opportunities for advancing personalized learning and creating meaningful and interactive learning contexts. Furthermore, their use in education is transforming the way students learn and engage with educational content. AI tools like ChatGPT and other large language models (LLMs) are being integrated into educational environments to enhance learning experiences through personalized and interactive interaction. Managing prompts effectively is especially important to support the appropriate use of AI in education, particularly in specific subject areas.

- **STEM education enhancement**: AI prompts are used to create educational content, quizzes, and simplified explanations, making complex STEM topics more accessible and appealing to students. This approach promotes interactive learning rather than rote memorization (Li, 2024).
- **Health sciences education**: In health sciences education, the proper use of prompts is crucial to aligning AI outputs with educational objectives, improving teaching effectiveness, and enhancing student understanding of complex topics (Sun, 2024).
- **Mathematics instruction**: AI tools like ChatGPT and MathGPT are being used in calculus education to facilitate problem-solving and conceptual un-

derstanding through well-designed prompts, leading to better student engagement and learning outcomes (Torres-Peña, 2024).

4.4. Prompt Examples for Tools like ChatGPT and NotebookLM

As discussed in the previous section, the use of specific prompts is essential for certain subject areas, but proper use can—and should—be applied across all educational disciplines. Below are a series of practical examples that can be used by teachers from any field:

- **Lesson plan design**: *"Act as a [subject] teacher for [educational level] students. Design an interactive and engaging lesson plan on [specific topic]. Include learning objectives, activities, materials, and assessment methods."*
- **Creation of educational materials**: *"Act as an art education teacher and design an activity for analyzing works of art. Provide students with a selection of paintings or sculptures from different artistic periods and create a series of questions to guide their interpretation."*
- **Assessment and feedback**: *"Act as a [subject] teacher and evaluate the following student essay on [topic]. Provide constructive feedback on structure, content, grammar, and style. Suggest areas for improvement and offer additional resources."*
- **Adaptation to diverse needs**: *"Act as an inclusive education teacher. A student with [specific need] is struggling to understand [concept]. Develop teaching strategies and adapted materials to help them achieve the learning objectives."*
- **Fostering critical thinking**: *"Act as a [subject] teacher and design a critical thinking activity for the class on [specific topic]. Include questions and scenarios that challenge students to analyze and evaluate information."*

These prompts are designed to encourage AI to emulate the skills and reasoning of a teacher in a variety of educational situations.

4.5. Assessment and Feedback

Ongoing assessment and feedback are crucial elements to ensure the success and effectiveness of a curriculum designed with the integration of AI. This process allows for the identification of strengths and weaknesses in the curriculum design, as well as the implementation of evidence-based adjustments for continuous optimization. Evaluation should be carried out in three clearly defined phases:

- **Validation of the curriculum design through expert judgment**: The first phase involves validating the curriculum design with input from teachers and experts in education, pedagogy, and educational technology, as well as specialists in AI applied to the school context. These professionals assess aspects such as the coherence of the curriculum, the methodological suitability, the technological and pedagogical relevance of the proposed AI tools, and the clarity in defining and aligning specific and transversal competencies with the identified needs.
- **Feedback from students and teachers after implementation**: Once the curriculum has been validated and implemented in a real educational setting, direct feedback will be collected from both students and teachers. Instruments such as surveys, individual interviews, focus groups, and analysis of academic results following the practical application of the curriculum will be used. In this phase, it is essential to gather specific information about students' general perceptions regarding the use of AI-based technological tools, the clarity and practical usefulness of the prompts, and the identification of any obstacles encountered during the learning process. Likewise, feedback will be gathered from teachers regarding the usability of the curriculum, the impact observed on student learning, and their professional assessment of the proposal's strengths and areas for improvement.
- **Adjustments to optimize the proposal**: Based on the information gathered during the feedback phase, the final stage involves making specific adjustments necessary to optimize and refine the curriculum proposal. These adjustments may include modifying AI-generated educational materials, revising and improving the evaluation indicators used, adapting or redefining the competencies to be developed, or implementing methodological changes in classroom application.

These adjustments should not be viewed as failures of the initial design but rather as an integral part of an iterative process of continuous improvement aimed at ensuring the curriculum's relevance, effectiveness, and educational quality. Once these modifications are applied, the evaluation cycle begins again, allowing for the maintenance of a dynamic, up-to-date curriculum permanently adapted to students' emerging needs and the evolving educational context.

5. THE IMPACT OF AI ON THE LEARNING PROCESS

Numerous studies have demonstrated the benefits of implementing AI in the educational field. The following is a synthesis of some of the positive effects that this type of resource can have in educational contexts.

5.1. Impact of AI on Student Motivation

The study of the relationship between motivation and AI is an emerging area of research that explores how AI can influence human development. Below is a synthesis of studies that investigate the role of this technology in enhancing motivation within educational settings.

The integration of AI into educational environments can stimulate intrinsic motivation by emphasizing personal meaning and relevance (Artemova, 2024). Other studies have shown that such tools can also positively influence motivation in foreign language learning or native language development (Yang, 2024), improving learning outcomes and promoting self-regulated learning (Moybeka et al., 2023). Furthermore, curricula that incorporate knowledge related to the use and understanding of AI in primary education have been associated with higher levels of intrinsic and professional motivation among students (Lin et al., 2021), highlighting the importance of contextualizing instruction around the needs of contemporary society.

5.2. Development of Transversal Competencies

Transversal competencies are essential in any educational context. Among the most relevant are teamwork, creativity, communication, entrepreneurial spirit, and critical thinking. The integration of AI into educational and professional environments offers new opportunities and challenges for the development of these skills.

Resources provided by various AI models are being used to enhance the acquisition of transversal competencies. Among the most significant findings, it has been shown that AI can accelerate teamwork by providing innovative educational methods (Graczyk-Kucharska, 2024). Positive correlations have also been found between AI-driven educational methods and the development of other transversal skills such as creativity and communication (Graczyk-Kucharska, 2019).

Paradoxically, critical thinking stands out as one of the transversal competencies most inherently tied to these technologies, as it should never be pushed into the background. The very act of using AI in educational settings exemplifies the importance of not blindly trusting the information generated (Michalon & Camacho-Zuñiga, 2023).

5.3. Curriculum Alignment

We have examined the impact of AI on various aspects of education, but its influence on the learning process itself goes even further, offering significant benefits across the entire educational system. Among these benefits, the following stand out:

- **Improved learning outcomes**: It has been demonstrated that AI technologies significantly enhance learning achievements. Studies indicate that AI can lead to better understanding and retention of content, as well as increased student engagement and motivation (Sutrisman et al., 2024; Zheng et al., 2021).
- **Personalized learning**: AI enables the customization of educational content to meet individual student needs, fostering personalized learning paths. This includes curriculum adaptation and real-time feedback, enhancing the learning experience and supporting the development of professional competencies (Tapalova, 2022).
- **Administrative efficiency**: AI assists educators by automating tasks such as grading and feedback, allowing teachers to focus more on the quality of the teaching-learning process. This administrative efficiency contributes to higher teaching quality and improved educational outcomes (Chen et al., 2020).
- **Intelligent tutoring systems**: AI-powered tutoring systems provide personalized support and guidance, helping students learn at their own pace and according to their unique learning styles (Kamalov, 2023).

5.4. Technical Challenges

The integration of AI in education also presents several technical challenges that must be addressed in order to maximize its potential. These challenges affect both the effectiveness and equitable implementation of AI in educational environments.

- **Equitable access**: There is a risk of reinforcing existing inequalities if AI technologies are not accessible to all students. Ensuring equitable access to AI-driven educational resources is crucial to preventing the widening of the digital divide (Sytnyk & Podlinyayeva, 2024).
- **Technical barriers**: The development and implementation of AI applications in education can be technically complex. This includes the need for sophisticated infrastructure, the difficulty of creating effective AI models, and the integration of AI systems with existing educational technologies (Familoni & Onyebuchi, 2024).

- **Teacher training and support**: Educators require adequate training to effectively use AI tools. Without proper support, teachers may struggle to integrate AI into their teaching practices, which limits the potential benefits of AI in education (Chan & Zary, 2019).
- **Evaluation and validation**: Rigorous evaluation and validation of AI applications are necessary to ensure their effectiveness and reliability. This includes assessing the impact of AI on learning outcomes and ensuring that AI systems are aligned with educational goals (Luan et al., 2020).

6. CONCLUSIONS AND RECOMMENDATIONS

6.1. Conclusions

The integration of AI into education represents a significant opportunity to substantially enhance the quality and relevance of educational processes. This study has demonstrated that AI enables the creation of contextualized learning situations adapted to the local and cultural realities of students, thereby fostering greater motivation and engagement in the educational process. Contextualized learning, facilitated by technologies such as ChatGPT and NotebookLM, has proven to be an essential factor in connecting the curriculum to students' everyday lives, increasing intrinsic motivation and contributing to the holistic development of both specific and transversal competencies.

One of the most relevant contributions of this research is the methodological proposal for integrating AI into curriculum design. This methodology is grounded in a rigorous and in-depth analysis of the educational context, identifying both specific needs and potential strengths of the environment in which it is implemented. Furthermore, it clearly emphasizes the importance of defining specific and transversal competencies aligned with local realities and current educational demands. This definition not only ensures the educational relevance of the curriculum but also promotes a more comprehensive education, equipping students with key tools to face the challenges of today's world, which is characterized by highly complex and rapidly changing environments.

The effective implementation of AI in education also depends heavily on the proper use of pedagogical prompts in technological platforms. The design and application of specific prompts make it possible to fully harness the adaptive and personalized capabilities of AI tools. The examples presented in this work illustrate how well-crafted prompts encourage active, critical, and contextually grounded learning in various curricular areas, including STEM, social sciences, mathematics, health sciences, and inclusive education. However, the implementation of these technologies must always be mediated by the professional judgment of educators, who must guide their use in a critical and ethically responsible manner.

This research also underscores the relevance of Universal Design for Learning (UDL) as an inclusive framework guiding the integration of AI. This approach ensures not only educational equity but also curriculum adaptability to the diversity of students and contexts. The combination of UDL and AI represents a powerful tool for removing barriers to learning, promoting equitable, flexible, and personalized education that responds to individual student needs.

Despite the numerous advantages identified, the incorporation of AI into education is not without significant ethical and pedagogical challenges. Particularly concerning are issues related to student data privacy, the risk of algorithmic bias that may reproduce and perpetuate educational inequalities, and the urgent need to properly train teachers in the technical, ethical, and pedagogical use of these technological tools. Additionally, accessibility and technological sustainability must be ensured, especially in educational contexts with socioeconomic limitations.

Finally, it is essential to highlight that AI does not aim to replace the fundamental role of the teacher, but rather to complement and enrich it. AI provides educators with the ability to automate repetitive administrative tasks and focus on more impactful pedagogical activities, such as direct student interaction, critical learning mediation, and ongoing methodological innovation.

The effective integration of AI in education requires a careful balance between technological innovation, pedagogical relevance, ethical responsibility, and continuous teacher training. When these factors are properly aligned, AI becomes a powerful tool for comprehensive educational improvement, promoting more motivating, inclusive, and relevant learning experiences that can meet the educational challenges of both the present and the future.

6.2. Recommendations

Finally, the following set of recommendations is proposed for the effective implementation of AI models in teaching practice:

- **Ongoing and specialized teacher training**: It is recommended to implement continuous professional development programs for educators, focused not only on the technical use of AI but also on its pedagogical and ethical implications. This training should aim to enhance teachers' critical thinking regarding AI-generated content and its didactic integration into various educational contexts.
- **Contextualized curriculum design**: It is essential to align AI with a curriculum design that considers local realities and the specific needs of students.

The use of prompts tailored to students' cultural and contextual backgrounds should be encouraged to promote more meaningful and motivating learning.

- **Development of inclusive and ethical policies**: Educational authorities must develop policies that ensure equitable access to AI, with special attention to vulnerable settings or those with limited technological infrastructure. These policies should include principles of technological sustainability, data privacy protection, and algorithmic bias auditing.
- **Teacher supervision and mediation**: Despite the advances in AI, it is crucial to ensure that the teacher's role is not diminished but enhanced. It is recommended to establish mechanisms for human oversight in the creation and validation of AI-generated materials to ensure their pedagogical and cultural appropriateness.
- **Interdisciplinary research and ongoing evaluation**: It is advised to promote educational research on the use of AI, particularly from pedagogical, technological, and ethical perspectives. Moreover, it is essential to establish systems for the continuous evaluation of AI's impact on student motivation, academic performance, and inclusion.
- **Promotion of a culture of critical thinking**: Finally, it is necessary to educate both students and teachers in digital literacy and critical thinking so that they can use AI responsibly and reflectively, avoiding uncritical dependence on automated responses.

REFERENCES

Alcocer, A., Cabrera, A., & García, E. (2024). La inteligencia artificial en la educación: desafíos éticos y perspectivas hacia una nueva enseñanza. *Latam Revista Latinoamericana De Ciencias Sociales Y Humanidades, 5*(6). DOI: 10.56712/latam.v5i6.3019

Alcocer, A., Cabrera, A., & García, E. (2024). La inteligencia artificial en la educación: desafíos éticos y perspectivas hacia una nueva enseñanza. *Latam Revista Latinoamericana De Ciencias Sociales Y Humanidades, 5*(6). DOI: 10.56712/latam.v5i6.3019

Artemova, I. (2024). Bridging Motivation and AI in Education: An Activity Theory Perspective. Digital Education Review. .DOI: 10.1344/der.2024.45.59-67

Bearman, M., Ryan, J., & Ajjawi, R. (2023). Discourses of artificial intelligence in higher education: A critical literature review. *Higher Education, 86*(2), 369–385. DOI: 10.1007/s10734-022-00937-2

Biggs, J., & Tang, C. (2007). *Teaching for quality learning at university: What the student does* (3rd ed.). McGraw-Hill Education.

Cedeño, E., Quintero, A., Quiñónez, O., Zamora, M., & Prado, N. (2024). Análisis de tendencias y futuro de la inteligencia artificial en la educación superior: Perspectivas y desafíos. *Ciencia Latina Revista Científica Multidisciplinar, 8*(1), 3061–3076. DOI: 10.37811/cl_rcm.v8i1.9637

Chan, K., & Zary, N. (2019). Applications and Challenges of Implementing Artificial Intelligence in Medical Education: Integrative Review. *JMIR Medical Education, 5*(1), e13930. Advance online publication. DOI: 10.2196/13930 PMID: 31199295

Chen, L., Chen, P., & Lin, Z. (2020). Artificial Intelligence in Education: A Review. *IEEE Access : Practical Innovations, Open Solutions, 8*, 75264–75278. DOI: 10.1109/ACCESS.2020.2988510

Cisternas-San Martín, N., Guzmán Muñoz, E., & Rivas Poblete, A. (2025). Inteligencia artificial como asistente para desarrollar un currículum de educación continua basado en competencias. *Revista de Investigación Educacional, 43*(43). Advance online publication. DOI: 10.6018/rie.619441

Coll, C., & Rochera, M. J. (2022). *Aprendizaje significativo y contextos reales en la educación del siglo XXI*. Editorial UOC.

Cotrina-Aliaga, J., Vera-Flores, M., Ortiz-Cotrina, W., & Sosa-Celi, P. (2021). *Uso de la inteligencia artificial (ia) como estrategia en la educación superior*. Revista Iberoamericana De La Educación., DOI: 10.31876/ie.vi.81

Cruz, C., Bejarano, J., Santana, X., & Villamarín, J. (2024). El impacto de la inteligencia artificial en la enseñanza de la economía y la administración: Tendencias, desafíos y oportunidades. *Revista Social Fronteriza*, *4*(2), e42239. DOI: 10.59814/resofro.2024.4(2)239

Deci, E. L., & Ryan, R. M. (2017). *Self-determination theory: Basic psychological needs in motivation, development, and wellness*. Guilford Press.

Estrada-Araoz, E., Quispe-Aquise, J., Malaga-Yllpa, Y., Larico-Uchamaco, G., Pizarro-Osorio, G., Mendoza-Zuñiga, M., Velasquez-Bernal, A. C., Roque-Guizada, C. E., & Huamaní-Pérez, M. (2024). Role of artificial intelligence in education: Perspectives of peruvian basic education teachers. *Data & Metadata*, *3*, 325. DOI: 10.56294/dm2024325

Familoni, B., & Onyebuchi, N.Babajide Tolulope FamiloniNneamaka Chisom Onyebuchi. (2024). ADVANCEMENTS AND CHALLENGES IN AI INTEGRATION FOR TECHNICAL LITERACY: A SYSTEMATIC REVIEW. *Engineering Science & Technology Journal*, *5*(4), 1415–1430. DOI: 10.51594/estj.v5i4.1042

Firat, M. (2023). How ChatGPT can transform autodidactic experiences and open education. *Open Education Faculty, Anadolu University*. http://doi.org/DOI: 10.31219/osf.io/9ge8m

Gee, J. P. (2021). *What Video Games Have to Teach Us About Learning and Literacy* (2nd ed.). St. Martin's Press.

Gómez, W. (2023). La inteligencia artificial y su incidencia en la educación: transformando el aprendizaje para el siglo xxi. *Revista Internacional De Pedagogía E Innovación Educativa, 3*(2), 217-230. DOI: 10.51660/ripie.v3i2.133

Graczyk-Kucharska, M., Olszewski, R., Przybyła, J., Łuszkiewicz, J., Hojka, K., & Spychała, M. (2024). *Modelling Accelerating Acquisition of Teamwork Competences with Transversal Competences and Artificial Intelligence*. Krakow Review of Economics and Management/Zeszyty Naukowe Uniwersytetu Ekonomicznego w Krakowie., DOI: 10.15678/krem.13131

Graczyk-Kucharska, M., Özmen, A., Szafrański, M., Weber, G., Goliński, M., & Spychała, M. (2019). Knowledge accelerator by transversal competences and multivariate adaptive regression splines. *Central European Journal of Operations Research*, *28*(2), 645–669. DOI: 10.1007/s10100-019-00636-x

Herrington, J., & Oliver, R. (2022). Authentic learning environments and tasks. *Educational Technology Research and Development, 70*(1), 1–15. DOI: 10.1007/s11423-021-10041-9

Holmes, W., Bialik, M., & Fadel, C. (2021). *Artificial Intelligence in Education: Promises and Implications for Teaching and Learning.* https://curriculumredesign.org/wp-content/uploads/AI-in-Education.pdf

Holmes, W., Porayska-Pomsta, K., Holstein, K., Sutherland, E., Baker, T., Shum, S., Santos, O., Rodrigo, M., Cukurova, M., Bittencourt, I., & Koedinger, K. (2021). Ethics of AI in Education: Towards a Community-Wide Framework. *International Journal of Artificial Intelligence in Education, 32*(3), 504–526. DOI: 10.1007/s40593-021-00239-1

Huang, R., Tlili, A., Chang, T. W., Nascimbeni, F., & Burgos, D. (2022). Roles of AI in fostering self-directed learning and enhancing learner motivation. *Educational Technology Research and Development, 70*(4), 1925–1943. DOI: 10.1007/s11423-022-10116-4

Johnson, A. P. (2020). *Contextual Teaching and Learning: An Introduction and an Overview*. SAGE Publications.

Kamalov, F., Calonge, D., & Gurrib, I. (2023). New Era of Artificial Intelligence in Education: Towards a Sustainable Multifaceted Revolution. *Sustainability (Basel), 15*(16), 12451. DOI: 10.3390/su151612451

Klimova, B., Pikhart, M., & Kacetl, J. (2023). Ethical issues of the use of AI-driven mobile apps for education. *Frontiers in Public Health, 10*, 1118116. DOI: 10.3389/fpubh.2022.1118116 PMID: 36711343

Lema, R., Ruano, A., Rivas, N., Erazo, C., & Arias, R. (2024). Perspectivas de la educación con la inteligencia artificial a un cercano plazo. *Ciencia Latina Revista Científica Multidisciplinar, 8*(2), 5522–5536. DOI: 10.37811/cl_rcm.v8i2.10966

Ley Orgánica (2020). de 29 de diciembre, por la que se modifica la Ley Orgánica 2/2006, de 3 de mayo, de Educación (LOMLOE). *Boletín Oficial del Estado*, nº 340, de 30 de diciembre de 2020, pp. 122868-122953. https://www.boe.es/boe/dias/2020/12/30/pdfs/BOE-A-2020-17264.pdf

Li, M. (2024). Using Prompt Engineering to Enhance STEM Education. *2024 IEEE Integrated STEM Education Conference (ISEC)*, 1-2. DOI: 10.1109/ISEC61299.2024.10664775

Lin, P., Chai, C., Jong, M., Dai, Y., Guo, Y., & Qin, J. (2021). Modeling the structural relationship among primary students' motivation to learn artificial intelligence. *omputers and Education. Artificial Intelligence, 2*, 100006. DOI: 10.1016/j.caeai.2020.100006

Loayza, K. (2024). Transformando la educación básica: retos y perspectivas de la inteligencia artificial. *Vitalia, 5*(2), 01-17. DOI: 10.61368/r.s.d.h.v5i2.113

Lorenzo Lledó, A. (2023). El modelo educativo inclusivo desde el diseño universal para el aprendizaje (DUA). En *Atención a la Diversidad en los Centros Educativos*. Universidad de Alicante.

Luan, H., Géczy, P., Lai, H., Gobert, J., Yang, S., Ogata, H., Baltes, J., Guerra, R., Li, P., & Tsai, C. (2020). Challenges and Future Directions of Big Data and Artificial Intelligence in Education. *Frontiers in Psychology, 11*, 580820. Advance online publication. DOI: 10.3389/fpsyg.2020.580820 PMID: 33192896

Luckin, R. (2018). *Machine Learning and Human Intelligence: The Future of Education for the 21st Century*. UCL Institute of Education Press.

Luckin, R., Holmes, W., Griffiths, M., & Forcier, L. B. (2023). *Artificial Intelligence and Education: Promises and Implications for Teaching and Learning*. UNESCO Publishing.

Marcano, B., Rodríguez, H., & Lopez, L. (2024). Impacto de la inteligencia artificial en la mejora de la práctica docente. Actas IDI-UNICyT, 461-466. DOI: 10.47300/actasidi-unicyt-2023-71

Michalon, B., & Camacho-Zuñiga, C. (2023). ChatGPT, a brand-new tool to strengthen timeless competencies. *Frontiers in Education, 8*, 1251163. DOI: 10.3389/feduc.2023.1251163

Moybeka, A., Syariatin, N., Tatipang, D., Mushthoza, D., Dewi, N., & Tineh, S. (2023). Artificial Intelligence and English Classroom: The Implications of AI Toward EFL Students' Motivation. *Edumaspul: Jurnal Pendidikan, 7*(2), 2444–2454. DOI: 10.33487/edumaspul.v7i2.6669

Mujica-Sequera, R. (2024). Clasificación de las herramientas de la inteligencia artificial en la educación. *Revista Docentes 2 0, 17*(1), 31-40. DOI: 10.37843/rted.v17i1.513

Muñoz, G., Delgado, J., & Dávalos, J. (2024). Inteligencia artificial y gobernanza en la gestión académica y administrativa de la educación superior. *Revista Social Fronteriza, 4*(6), e46508. DOI: 10.59814/resofro.2024.4(6)508

Padilla, R. (2019). La llegada de la inteligencia artificial a la educación. *Revista De Investigación en Tecnologías De La Información, 7*(14), 260–270. DOI: 10.36825/RITI.07.14.022

Pérez, O., & González, N. (2024). Formación docente para el uso de la inteligencia artificial. *Ciencia Latina Revista Científica Multidisciplinar, 8*(5), 11772–11788. DOI: 10.37811/cl_rcm.v8i5.14594

Pintado, A., Prado, R., Peláez, C., & Aguilar, W. (2023). Inteligencia artificial y sostenibilidad: el compromiso de una institución de educación superior. *Magazine De Las Ciencias Revista De Investigación E Innovación, 8*(4), 12-28. DOI: 10.33262/rmc.v8i4.2954

Rodríguez, A., Romero, M., Agreda, C., Saldarriaga, J., & Saldarriaga, M. (2024). Impacto de la inteligencia artificial en las prácticas educativas: percepciones y actitudes del profesorado. *Latam Revista Latinoamericana De Ciencias Sociales Y Humanidades, 5*(2). DOI: 10.56712/latam.v5i2.1933

Rodríguez, M., Rubio, A., Lingán, A., Rubio, D., Bocanegra, J., & Carrillo-Flores, J. (2023). Inteligencia artificial en la educación digital y los resultados de la valoración del aprendizaje. DOI: 10.31219/osf.io/c3pmd

Ryan, R. M., & Deci, E. L. (2020). Intrinsic and extrinsic motivations: Classic definitions and new directions. *Contemporary Educational Psychology, 61*, 101860. DOI: 10.1016/j.cedpsych.2020.101860 PMID: 10620381

Salinas, J., De Benito, B., & Pérez-Mateo, M. (2023). *Innovación educativa y desarrollo profesional docente en entornos digitales*. Narcea Ediciones.

Sun, G. (2024). Prompt Engineering for Nurse Educators. *Nurse Educator, 49*(6), 293–299. Advance online publication. DOI: 10.1097/NNE.0000000000001705 PMID: 38968442

Sutrisman, H., Simanjuntak, R., Prihartanto, A., & Kusumo, B. (2024). The Impact of Using AI in Learning on Understanding of Material by Young Students. *International Journal of Educational Research, 1*(3), 24–32. Advance online publication. DOI: 10.62951/ijer.v1i3.43

Sytnyk, L., & Podlinyayeva, O. (2024). AI in education: main possibilities and challenges. In Proceedings of the 8th International Scientific and Practical Conference" International Scientific Discussion: Problems, Tasks and Prospects (pp. 569-579). DOI: 10.51582/interconf.19-20.05.2024.058

Tapalova, E., Vorobyev, A., & Smirnov, S. (2022). The impact of contextual learning on students' motivation and performance: An empirical study. *International Journal of Educational Technology in Higher Education, 19*(1), 1–18. DOI: 10.1186/s41239-022-00336-4 PMID: 35013716

Tapalova, O., Zhiyenbayeva, N., & Gura, D. (2022). Artificial Intelligence in Education: AIEd for Personalised Learning Pathways. *Electronic Journal of e-Learning, 20*(5), 639–653. DOI: 10.34190/ejel.20.5.2597

Torre, C., López, M., Basantes, D., & Chango, E. (2024). El papel de la inteligencia artificial en la adaptación curricular: Perspectivas para la mejora continua de la educación digital. *Revista Imaginario Social, 7*(2). Advance online publication. DOI: 10.59155/is.v7i2.186

Torres-Peña, R., Peña-González, D., Chacuto-López, E., Ariza, E., & Vergara, D. (2024). Updating Calculus Teaching with AI: A Classroom Experience. *Education Sciences, 14*(9), 1019. Advance online publication. DOI: 10.3390/educsci14091019

UNESCO. (2022). *Reimagining our futures together: A new social contract for education*. UNESCO Publishing.

Vela, G., Naranjo, B., Quinte, R., Villafuerte, V., & Velasco, J. (2024). Inclusión de la inteligencia artificial en la docencia universitaria. *Latam Revista Latinoamericana De Ciencias Sociales Y Humanidades, 5*(1). DOI: 10.56712/latam.v5i1.1642

Vera, D., Gutiérrez, M., Chuquillan, D., & Carrión, D. (2024). Avances tecnológicos para la inclusión: El rol de la inteligencia artificial en la diversificación del aprendizaje escolar. *Revista Social Fronteriza, 4*(3), e43265. DOI: 10.59814/resofro.2024.4(3)265

Vera-Flores, K., Cerna-Vinueza, A., Cornejo, M., & Vergel-Parejo, E. (2024). Inteligencia artificial en el campo de la educación de lengua y literatura. *Mqrinvestigar, 8*(4), 5013–5037. DOI: 10.56048/MQR20225.8.4.2024.5013-5037

Williamson, B., & Eynon, R. (2020). Education data futures: Data-intensive technologies and the aims of education. *Learning, Media and Technology, 45*(1), 36–51.

Yang, T. (2024). Impact of Artificial Intelligence Software on English Learning Motivation and Achievement. *SHS Web of Conferences*. DOI: 10.1051/shsconf/202419302011

Yépez Álvarez, L. D., Jirón Febre, T. I., Rumbaut Rangel, D., & Jurado Martínez, G. (2024). El papel de la inteligencia artificial en la personalización de la educación. *Magazine de las Ciencias, 9*(3), 31–40. DOI: 10.33262/rmc.v9i3.3160

Zawacki-Richter, O., Marín, V. I., Bond, M., & Gouverneur, F. (2019). Systematic review of research on artificial intelligence applications in higher education: A framework for future research. *International Journal of Educational Technology in Higher Education*, *16*(1), 1–27. DOI: 10.1186/s41239-019-0171-0

Zhang, K., & Aslan, A. B. (2021). AI technologies for education: Recent research & future directions. *Educational Technology Journal*, *2*, 100025. DOI: 10.1016/j.caeai.2021.100025

Zheng, L., Niu, J., Zhong, L., & Gyasi, J. (2021). The effectiveness of artificial intelligence on learning achievement and learning perception: A meta-analysis. *Interactive Learning Environments*, *31*(9), 5650–5664. DOI: 10.1080/10494820.2021.2015693

Chapter 7

AI Technologies and Multicultural Education Insights Into Teachers' Attitudes and Challenges in Higher Education

Amr Mahmoud Mohamed
https://orcid.org/0000-0003-3891-963X
North Private College of Nursing, Saudi Arabia

Samia Mouas
https://orcid.org/0000-0002-5259-5582
Batna 2 University, Algeria

Yara Abdelaty
Port Said University, Egypt

Abdullahi Yusuf
https://orcid.org/0000-0003-2487-0564
Sokoto State University, Nigeria

ABSTRACT

This study analyzes the role of AI technology, particularly Artificial Intelligence (AI) tools, in fostering multiculturalism and improving English instruction across diverse educational settings.This study employs a 24-item Likert scale survey to examine the perspectives of 115 English as a Foreign Language (EFL) instructors about artificial intelligence and multiculturalism. Results indicate a reasonable

DOI: 10.4018/979-8-3373-7922-7.ch007

consensus on the advantages of artificial intelligence in promoting cultural diversity; yet its effectiveness is constrained by obstacles such as language barriers and cultural disparities.The research emphasizes the necessity of integrating artificial intelligence and offering opportunities for EFL teachers to improve their intercultural communication skills and cultural competence. It underscores the necessity of inclusion and representation in fostering supportive educational settings by advocating for instructional methods that promote active engagement, collaboration, and culturally pertinent content.

INTRODUCTION

The ability to understand and value cultural variety is crucial in a globalized context, particularly in English as a Foreign Language (EFL) education (Bennett, 2022; Omer et al., 2025; Al-khresheh et al., 2024; Byram, 2021). As English becomes a worldwide lingua franca, EFL educators can improve multicultural awareness and intercultural competency, thus preparing students for global citizenship (Baker, 2022; Sun, 2024). Educational regulations sometimes inadequately address the challenges of multicultural education, hindering teachers' efforts to include diversity into language instruction (Deardorff, 2020).

Advancements in artificial intelligence (AI), defined by robots replicating human cognitive functions like learning and reasoning, have substantial transformational promise in education. Artificial intelligence enhances educational experiences, personalizes instruction (Boudouaia et al., 2024; Mohamed, 2024), and promotes intercultural communication across diverse academic settings (Sarwari et al., 2024).

In multicultural EFL courses, AI facilitates cross-cultural discourse and boosts student participation (Hohenstein et al., 2023). Virtual field excursions and cultural exchange programs augment cultural awareness (Yi, 2021), requiring educators to integrate technology proficiency with cultural sensitivity to tailor materials for diverse backgrounds. The attitudes of educators toward AI play a crucial role in its effectiveness in promoting inclusivity (Lee, 2020). The increasing diversity in higher education, with English as the medium of instruction, highlights the necessity for inclusive pedagogies that respect linguistic and cultural diversity (Byram, 2020; Kubota, 2020).

Culturally relevant materials, critical engagement with social issues, and linguistic support reduce barriers, promote respect, and improve communication skills by exposing individuals to diverse linguistic forms (Belda-Medina, 2022).

Although the role of AI in education is extensively documented (Alenezi et al., 2023; Yusuf et al., 2024b), its use to enhance multiculturalism in EFL classrooms, particularly in higher education, has not been thoroughly investigated. Recent studies emphasize language acquisition and personalized learning (Boudouaia et al., 2024; Jmaiel et al.,2025), while intercultural competence remains underexplored. Research on teachers' perceptions of AI (Chounta et al., 2022; Mohamed & Shaaban, 2023; Ofosu-Ampong, 2024) seldom focuses on EFL contexts, resulting in a lack of insight into attitudes toward AI as a multicultural resource.

This research examines the influence of artificial intelligence on the advancement of multiculturalism within EFL classrooms, emphasizing the viewpoints of educators and the challenges encountered in higher education. This study analyzes the impact of attitudes on teaching practices, investigates gender-based differences, and identifies effective strategies to promote inclusivity, providing insights for equitable language education.

The Study Questions

1. How do EFL teachers perceive the use of AI technologies to foster multiculturalism in the EFL classroom?
2. Do male and female teachers differ statistically in their attitudes towards multiculturalism and teaching English in higher education?
3. How do EFL teachers' attitudes toward multiculturalism and diversity influence their teaching practices?
4. What are the challenges EFL teachers face when teaching English to students from diverse cultural backgrounds?
5. What is the impact of multiculturalism and diversity on the learning experience in higher education?

LITERATURE REVIEW

Culturalism is a theoretical perspective that emphasizes the significance of culture in shaping individuals and societies, highlighting the importance of acknowledging cultural differences to foster social justice and equality (Hall, 2006). This approach considers culture and society as distinct and independent realms, separate from political and economic systems and processes (Mitchell, 1999). As a critical method of investigation, culturalism examines the intricate relationship between culture, power, and politics, as well as how cultural practices and representations contribute to both legitimizing and challenging prevailing power structures (Appadurai, 1996).

Multiculturalism in education is a concept that recognizes and values the diverse cultural backgrounds of students in educational settings. The aim is to create a learning environment that acknowledges and values the diverse ethnic, linguistic, and cultural backgrounds of students. Multiculturalism plays a crucial role in English language instruction within higher education, enhancing language acquisition and fostering intercultural communication (Setyono & Widodo, 2019; Tsang et al., 2023).

Recent research underscores the importance of multiculturalism in English language education at higher education institutions. Zhang and Zhou (2019) conducted a study indicating that the incorporation of multiculturalism in English language instruction significantly enhances students' linguistic performance and intercultural competency. Kim (2020) and Keengwe (2010) conducted studies illustrating the beneficial effects of integrating multiculturalism into English language training on student motivation and engagement with the language.

Diverse ways can be utilized to incorporate multiculturalism into English language training in higher education settings. An effective technique is employing culturally pertinent information and resources that reflect the experiences and viewpoints of students from varied ethnic backgrounds. This approach prompts students to express their cultural experiences and viewpoints in the classroom, therefore promoting intercultural discussion and comprehension (Alaca, 2017; Kahn & Agnew, 2017).

AI TECHNOLOGIES AND MULTICULTURALISM

Studies suggest that AI technology can either promote or hinder cultural diversity and inclusivity in language education. The application of AI technology in language teaching correlates with improved intercultural competence and heightened student involvement (Klimova & Chen, 2024; Long & Lin, 2022; Xia et al., 2024). The deployment of AI technologies entails drawbacks, including the perpetuation of power imbalances and inequality (Mac Fadden et al., 2024). Cultural sensitivity, relevance, support, equitable access, instructional adaptation, and integration are vital practices in multicultural education when utilizing educational AI technology to address educational inequality (Feng et al., 2024). Additional research is necessary to authenticate these methodologies and to understand the application of AI technology by educators in multicultural settings. The study promotes the integration of AI technology in educational environments to prepare administrators, educators, and students for professional competition.

Educators advocate for inclusivity and multiculturalism in the educational setting (Chisholm, 1998; Kayalar, 2016). It is imperative to recognize and appreciate the cultural backgrounds of the students. Multiculturalism promotes social equity and cultivates multicultural comprehension. The employment of culturally sensitive resources and the incorporation of varied viewpoints can enhance multiculturalism and inclusivity. Formulating ground principles, confronting biases, and encouraging cross-cultural collaboration are vital for cultivating a polite and inclusive classroom atmosphere. Educators can promote diversity and inclusion by engaging in community initiatives and contesting discriminatory policies. Promoting multiculturalism and inclusivity is essential for cultivating a supportive school atmosphere. Culturally responsive practices, social justice advocacy, and inclusive educational environments promote intercultural understanding (Hoffman et al., 2022; Karataş et al., 2023; Lau & Shea, 2022; Szelei et al., 2020; Gorski & Dalton, 2020).

Obstacles related to pedagogy, culture, and language impede the instruction of English to varied student demographics. Linguistic problems stem from disparities in language proficiency, vocabulary, and grammatical comprehension, adversely affecting students' academic achievement and their confidence in English communication. Educators must modify their instructional approaches to accommodate diverse student proficiency levels and provide supplementary assistance to those encountering difficulties. Cultural barriers emerge from disparities in customs, beliefs, and values, resulting in diverse learning methodologies and expectations among pupils. Educators must modify their pedagogical approaches to demonstrate inclusivity and cultural sensitivity (Kurbanova et al., 2022; Lau & Shea, 2022; Tsang et al., 2023).

From a pedagogical standpoint, it is challenging to fulfill the learning requirements and preferences of many different students. To accommodate varying learning styles and aptitudes, educators must use a variety of instructional methods. While struggling students should be encouraged, advanced students should be pushed. Systemic barriers could also make it more difficult to teach English to many kinds of students. Often, non-native English teachers lack resources and help. Teachers might also experience linguistic or cultural prejudice, which would make it more difficult for them to instruct various pupils (Rao & Chen, 2020; Tsang et al., 2023).

This study shows that diversity and multiculturalism have a moderate impact on the overall quality of higher education. There is broad agreement on the need for inclusion, representation, and cultural competency, as seen in item 19. Additionally, Item 22 suggests that resource accessibility and intercultural communication will have little impact. In order to meet the demands of culturally diverse students, higher education institutions must acknowledge the inherent difficulties and their consequences and provide an inclusive learning environment.

Establishing an inclusive educational environment that acknowledges diverse cultural backgrounds necessitates the promotion of cultural awareness and comprehension. The development of intercultural communication skills is essential, and education serves as a significant instrument in fostering awareness of cultural variety. Activities such as role-plays and group discussions facilitate teachers in promoting intercultural communication. Engaging in active listening and appreciating students' perspectives fosters effective intercultural communication. These tactics foster a classroom environment that values cultural diversity and reciprocal respect.

Culturally responsive education acknowledges and incorporates students' cultural origins in the classroom. This method entails integrating cultural viewpoints into dialogues, prompting students to convey their experiences, and employing culturally pertinent resources. These tactics improve engagement and academic achievement among students from various ethnic backgrounds. Cultural sensitivity is enhanced by enabling students to investigate their cultural identities, confront biases, and establish collaborations with families and communities. Educators can foster an inclusive learning environment that values the diverse cultures of their students by enhancing intercultural communication skills, utilizing adaptive teaching methods, and promoting critical thinking (Al-khresheh et al., 2022; Berlian & Huda, 2022; Debnam et al., 2023).

Language proficiency and intercultural competence are two domains where multiculturalism and inclusion significantly influence student learning results (Dimitrov & Deardorff, 2023). Liu et al. (2023) assert that students' enthusiasm and academic performance are enhanced when the classroom emphasises multiculturalism and inclusivity. Intercultural communication and language acquisition are enhanced by multiculturalism's application of cross-cultural dialogue and culturally pertinent resources. Moreover, promoting global thinking, empathy, and appreciation of diversity among pupils promotes their social and emotional growth (Dooly & Darvin, 2022). The self-esteem, resilience, confidence, and overall health of students are dramatically influenced, hence enhancing their academic performance.

Additionally, promoting inclusivity and multiculturalism can help students in ways other than their academic endeavors. Students will flourish in a world that is becoming more interconnected if they can develop intercultural competencies and respect for variety. Furthermore, they tend to become more responsible and involved members of society and prioritize justice and equal opportunity (Spellman et al., 2022). Students' academic performance is significantly impacted by inclusivity and multiculturalism. Establishing a welcoming and culturally sensitive learning environment can improve future achievements, social and emotional growth, language skills, and intercultural competency.

Instructing English to a diverse student body necessitates proficient cross-cultural communication. Cultural awareness, promotion of cross-cultural dialogue, and adaptation to cultural differences are essential. Promotion of cross-cultural communication is essential, and it is important to incorporate culturally appropriate information. Thus, they enhance intercultural communication skills and language acquisition (Aririguzoh, 2022; Lifintsev & Wellbrock, 2019).

Enhancing students' intercultural competency within cross-cultural communication is crucial for English instruction. Intercultural competence encompasses various skills, such as empathy, open-mindedness, and adaptability in diverse cultural contexts. Promoting intercultural competence in English language instruction equips students with essential skills for success in a globalized environment (Alenezi et al., 2023; Al-khresheh et al., 2023; Lee, 2020).

Finally, although cross-cultural communication is crucial for EFL classes, little is known about how to best help students develop intercultural competence, cultural awareness, and responsiveness. There should be additional research on how to use AI to improve EFL students' cross-cultural communication skills. Filling this knowledge gap is critical for creating equitable and inclusive classrooms that can educate children to thrive in today's globalized world.

METHODOLOGY

Context and Participants

This study used a convenience sample to enrol students from multicultural Saudi Arabian universities. It was convenience sampling, not a random sample, that was used to pick the participants (Farrokhi & Mahmoudi-Hamidabad, 2012). A variety of channels were used for recruitment, including email invitations, social media platforms, and professional networks. The sample consisted of 51 male participants and 64 female participants. A 24-item Likert scale questionnaire was required for participants. Qualified participants must currently teach English as a foreign language in a higher education institution. The questionnaire takes approximately 10-15 minutes to complete. Participants' responses are treated as confidential and are used only for research purposes. Without any repercussions, participants can withdraw from the study at any time. Participants provided informed consent. Research is conducted in accordance with all ethical considerations.

Procedures

This research will utilize a quantitative research design employing a cross-sectional survey methodology to investigate the perspectives of EFL teachers in higher education on embracing multiculturalism and teaching English in a diverse world. Data will be gathered through an online survey platform and analyzed using descriptive and inferential statistics. Ethical considerations will be upheld throughout the study.

Instruments

To gather information on EFL teachers' perceptions of multiculturalism in teaching EFL, a 24-item Likert-scale questionnaire was developed. The questionnaire comprised of six items for each of the four dimensions of multiculturalism:

1) Attitudes towards multiculturalism and diversity, 2) Use of AI technologies for teaching English in a multicultural classroom, 3) Challenges encountered while teaching English to culturally diverse students, and 4) The effect of multiculturalism and diversity on the overall learning experience.

After a literature review and pilot testing, a questionnaire was developed to determine participants' views and approaches to multiculturalism. It included 24 Likert-scale items. About 15 minutes were needed to complete the online survey. Inferential and descriptive statistics were used to identify patterns. Participants' perspectives and practices regarding multiculturalism in EFL teaching were thoroughly analyzed.

A jury of experts in EFL teaching and multiculturalism evaluated the questionnaire's content, relevance, and clarity to ensure validity and reliability. The questionnaire was refined based on their feedback, making it suitable for measuring multiculturalism in EFL teaching perceptions and practices. Additionally, four experts assessed the questionnaire's construct validity and content validity.

Moreover, statistical analyses were conducted to examine relationships between item scores, dimensions, and the total score (Table 1).

Table 1. Correlation Analysis of Item Score, Dimension, and Total Score

Questionnaire's Dimensions							
Attitudes towards multiculturalism and diversity		**Using AI technologies for teaching English in a multicultural classroom**		**Challenges in teaching English to culturally diverse students.**		**The impact of multiculturalism and diversity on the overall learning experience**	
Item No	**Correlation coefficient**	**Item No**	**Correlation coefficient**	**Item No**	**Correlation coefficient**	**Item No**	**Correlation coefficient**
1	.666**	7	.665**	13	.637**	19	.658**
2	.617**	8	.643**	14	.610**	20	.677**
3	.617**	9	.651**	15	.652**	21	.758**
4	.691**	10	.746**	16	.676**	22	.715**
5	.615**	11	.611**	17	.709**	23	.640**
6	.645**	12	.615**	18	.672**	24	.782**
Dimension correlation with the total score							
.703**		.844**		.693**		.770**	

**. significant at the 0.01 level.

The correlation coefficients between each dimension's total score and the scale's overall score, as well as between the items and each dimension's total score, are shown in Table 1. The scale appears to have excellent internal consistency, as indicated by the statistical significance of all correlations at the 0.01 level.

The questionnaire's reliability was evaluated using measures like Cronbach's alpha, Spearman-Brown, and Guttman Split-Half (Tables 2 and 3), which provided insights into internal consistency and stability over time. Translation and back-translation procedures were carried out, followed by a pilot study involving six teachers to assess the items' clarity and appropriateness, instructions, and format.

By employing the jury method and conducting statistical analyses, the questionnaire's validity and reliability were rigorously assessed, ensuring that it effectively captured the desired data and provided meaningful insights into participants' perspectives on multiculturalism in EFL teaching.

To assess the consistency of the scale statements, internal consistency was measured by examining the correlation between the dimension and the total score of the scale using a table.

Calculation of Scale Stability using Cronbach's Alpha

It is clear from the following table that the value of Cronbach's alpha coefficient for you after the dimensions of the scale.

Table 2. The values of the stability coefficient for each dimension of the scale

Dimensions	Number of Items	Cronbach's alpha
Attitudes towards multiculturalism and diversity	6	0.714
Using AI technologies for teaching English in a multicultural classroom	6	0.834
Challenges in teaching English to culturally diverse students.	6	0.792
The impact of multiculturalism and diversity on the overall learning experience	6	0.758

The stability coefficient values in Table 2 fall between 0.714 and 0.834, which are both high and statistically significant.

By splitting each sub-dimension into odd and even items and using the Spearman-Brown-Guttmann equation to get the correlation coefficients between the two halves, the split-half reliability was assessed. Table 3 presents the findings.

Table 3. The values of the stability coefficient for each dimension of the scale

Dimensions	Number of Items	Spearman-Brown	Guttman Split-Half
Attitudes towards multiculturalism and diversity	6	0.721	0.789
Using AI technologies for teaching English in a multicultural classroom	6	0.734	0.796
Challenges in teaching English to culturally diverse students.	6	0.695	0.764
The impact of multiculturalism and diversity on the overall learning experience	6	0.714	0.794

Furthermore, the reliability coefficient of the scale was computed using Test-Retest. The reliability coefficient for the test-retest technique was 0.764, meaning that the scale is appropriate for the application and has a satisfactory degree of stability when tested using the test-retest method.

Data Collection

An online questionnaire of 24 Likert-type items was designed to gather data about EFL teachers' perceptions of multiculturalism. The questionnaire was divided into four dimensions, with six items for each dimension: attitudes towards multiculturalism and diversity, using AI technologies for teaching English in a multicultural classroom, challenges in teaching English to culturally diverse students, and the impact of multiculturalism and diversity on the overall learning experience. Participants were contacted through social media groups, specifically WhatsApp, to ensure maximum participation. In a given timeframe, participants were asked to complete the online questionnaire.

Participants' confidentiality and anonymity were ensured. Participants' data was kept confidential and only accessible by the research team. All participants' identities and personal information were kept anonymous.

Data Analysis

This study analyzed questionnaire data using SPSS. Participants' responses were analyzed using descriptive statistics, including means, standard deviations, and frequencies. By utilizing statistics, we were able to examine data systematically and identify key trends and patterns. Using SPSS enhanced the rigor of the analysis and enabled the interpretation of questionnaire data, providing valuable quantitative findings.

RESULTS AND DISCUSSION

Question 1: How Does AI Technologies Use Influence EFL Teachers' Pedagogical Practices For Multiculturalism?

To answer this question, standard deviations, means, and the ranking of each item are calculated as follows:

*Table 4. The means and standard deviations for the field of (**Using AI technologies for teaching English in a multicultural classroom**)*

No.	Items	Mean	Std. Devation	Order	level
8	The use of AI technologies in the EFL classroom can help promote cultural diversity.	3.670	1.233	1	high
9	Incorporating AI technologies into my teaching practices can enhance students' understanding and appreciation of different cultures.	3.591	1.213	2	moderate
7	I feel confident in my ability to effectively incorporate AI technologies to promote multiculturalism in the EFL classroom.	3.539	1.216	3	moderate
10	The use of AI technologies in the EFL classroom can help students learn about cultures that are different from their own.	3.487	1.280	4	moderate
11	Using AI technologies to foster multiculturalism in the EFL classroom can lead to more inclusive and equitable teaching practices. .	3.452	1.251	5	moderate
12	AI technologies use has increased opportunities for students to share their cultural backgrounds and experiences in my teaching practices.	3.383	1.274	6	moderate
Total Degree		3.520	0.816	moderate	

In Table 4, the means and standard deviations for six Likert scale items related to using AI technologies in a multicultural classroom are presented. Six items were scored between 3.383 and 3.670, with a total mean score of 3.520, which indicates moderate agreement between participants.

Using AI technologies in the EFL classroom can help promote cultural diversity, according to item 8, which received the highest average score of 3.670 based on a standard deviation of 1.233. Among all items, item 12 received the lowest mean score, 3.383, with a standard deviation of 1.274, indicating a moderate level of agreement that AI technologies uses have increased student opportunities to share their cultural backgrounds and experiences.

Based on their mean scores, the "Order level" column in the table ranks the items from first to last, with item 8 ranking first and item 12 ranking last. All six items fall under the moderate level of agreement, according to the "Degree" column.

Study participants appear to be in agreement about the potential for AI technologies in the EFL classroom to foster cultural diversity. However, there seems to be some variation in participants' confidence in their abilities to use AI technologies for multiculturalism, as well as how AI technologies may have made sharing cultural backgrounds and experiences easier. Using AI technologies to enhance cultural diversity and inclusivity could be beneficial for EFL classrooms based on the results.

Question 2: Do Male And Female Teachers Differ Statistically In Their Perceived Attitudes, Strategies, Challenges, And Impact Of Multiculturalism In Teaching English In Higher Education?

To investigate the relationship between the levels of attitudes towards multiculturalism and teaching English in higher education according to gender, the Mann-Whitney U test was conducted to determine the mean differences between male and female students' perceived attitudes, strategies, challenges, and impact of multiculturalism in English teaching:

Table 5. T-test for gender-based differences in the attitudes towards multiculturalism and teaching English in higher education

Item	Gender	N	Mean Rank	Test Statistics
Attitudes towards multiculturalism and diversity	Male	51	67.85	1129.5***
	Female	64	50.15	
Strategies and approaches used for teaching English in a multicultural classroom	Male	51	59.66	1547.5
	Female	64	56.68	
Challenges in teaching English to culturally diverse students.	Male	51	56.00	1734.0
	Female	64	59.59	
The impact of multiculturalism and diversity on the overall learning experience	Male	51	62.91	1381.5
	Female	64	54.09	

*** < 0.01

According to the data presented in Table 5, the independent t-test results reveal significant differences between males and females in the dimension of "Attitudes towards multiculturalism and diversity" (Test statistics = 1129.5, p-value < 0.01), indicating that males have higher mean scores compared to females. However, no statistically significant differences were found between males and females in the other three dimensions.

The absence of statistically significant differences between males and females in the dimensions of "Strategies and approaches used for teaching English in a multicultural classroom" (Test statistics = 1547.5, p-value > 0.05), " Challenges in teaching English to culturally diverse students" (Test statistics = 1734.0, p-value > 0.05), and "The impact of multiculturalism and diversity on the overall learning experience" (Test statistics = 1381.5, p-value > 0.05) suggests that gender does not play a significant role in shaping attitudes and perceptions related to these aspects.

On the other hand, the statistically significant difference observed in the dimension of "Attitudes towards multiculturalism and diversity" (Test statistics = 1129.5, p-value < 0.01) indicates that there is a significant variation in attitudes between males and females. Moreover, the fact that male students obtained higher mean scores than their female counterparts suggests that males tend to have more positive attitudes towards multiculturalism and diversity in the context of higher education.

While there may be a statistically significant difference, the magnitude of the difference and its practical implications should be considered in further analysis and interpretation of the results. Additionally, the specific context and characteristics of the sample should be considered when generalizing these findings to a larger population.

Question 3: How do EFL Teachers' Attitudes Toward Multiculturalism And Diversity Influence Their Teaching Practices?

To answer this question, standard deviations, means, and the order of each item was calculated as follows:

Table 6. The means and standard deviations for the " Attitudes towards multiculturalism and diversity in the classroom" domain "N=115"

	Items	Mean	Std. Devation	Order	level
1	I believe that embracing multiculturalism is important in the classroom.	3.809	1.206	1	high
2	I am comfortable teaching students from diverse cultural backgrounds.	3.704	1.263	2	high
3	I believe that cultural sensitivity is important when teaching English to students from diverse cultural backgrounds.	3.591	1.263	3	moderate
4	I am willing to modify my teaching approach to meet the needs of students from diverse cultural backgrounds.	3.539	1.245	4	moderate
6	I feel confident in my ability to incorporate multiculturalism into my teaching practices.	3.522	1.273	5	moderate
5	I believe that diversity in the classroom enhances the learning experience for all students.	3.330	1.262	6	moderate
Total Degree (Attitudes towards multiculturalism and diversity in the classroom)		3.582	0.803	moderate	

The results, presented in Table 6, categorised the 6 items into high, moderate, and total degrees based on their mean score and standard deviation. The participants demonstrated moderate attitudes towards multiculturalism and diversity in the classroom, as indicated by a mean score of 3.582 and a standard deviation of 0.803. This suggests that the teachers had a moderately positive outlook on incorporating multiculturalism and diversity into their teaching practices.

Among the items, Item 1 had the highest level of agreement, with a mean score of 3.809 and a standard deviation of 1.206. This indicates that the EFL teachers expressed a relatively high level of agreement regarding this aspect related to multiculturalism and diversity in their teaching practices. Conversely, Item 5 exhibited the lowest level of agreement, reflected by a mean score of 3.33 and a standard deviation of 1.262. This indicates that the teachers exhibited reduced consensus on this item, highlighting a potential area of variability in their attitudes toward multiculturalism and diversity (Yang & Montgomery, 2013).

The findings indicate that the overall attitudes of EFL teachers towards multiculturalism and diversity in the classroom are moderate, though variations exist in their perceptions and agreement on specific aspects. Analysing these variations can assist in pinpointing areas that may need further support or specialised professional development to improve teachers' attitudes and practices concerning multiculturalism and diversity in the EFL classroom.

Question 4: What Are The Challenges EFL Teachers Face When Teaching English To Students From Diverse Cultural Backgrounds?

To answer this question, standard deviations, means, and the ranking of each item is calculated as follows:

Table 7. The means and standard deviations for the field of (Challenges in teaching English to culturally diverse students) N=199.

No.	Items	Mean	Std. Devation	Order	level
14	I find it difficult to address cultural differences and expectations in the classroom.	3.574	1.264	1	moderate
13	I find it challenging to overcome language barriers when teaching English to students from diverse cultural backgrounds.	3.539	1.259	2	moderate
15	I find it challenging to create an inclusive classroom environment for all students.	3.496	1.287	3	moderate
18	I find it difficult to incorporate cultural sensitivity into my teaching practices.	3.452	1.299	4	moderate

continued on following page

Table 7. Continued

No.	Items	Mean	Std. Devation	Order	level
17	I find it challenging to adapt my teaching approach to meet the needs of students from diverse cultural backgrounds.	3.426	1.250	5	moderate
16	I find it difficult to address varying levels of English language proficiency in the classroom.	3.330	1.262	6	moderate
Total Degree		3.470	0.838	moderate	

Table 7 provides insight into the difficulties educators encounter when teaching English to culturally diverse students. On average, participants scored 3.470, and their standard deviation was 0.838, indicating moderate challenges. It follows that while the challenges are not excessively severe in this setting, they still require attention and intervention to improve teaching and learning.

The most widespread and significant relevance of item 14 was acknowledged as the highest degree of agreement among participants. This item has a mean score of 3.575 and a standard deviation of 1.264, which indicates that a prominent challenge is experienced. High levels of agreement may be due to cultural differences and language barriers. Communication, language acquisition, and engagement can all be affected by teaching English to culturally diverse students. It is crucial to adapt instructional methods to address these challenges.

Item 16, on the other hand, had the lowest level of consensus among the participants, with an average score of 3.330 and a standard deviation of 1.262. In comparison to other items, this task was seen as having less importance. The lease agreement could be justified by factors related to student involvement and educational methodologies.

Instruction and acquisition of knowledge are most efficacious when students engage actively. Numerous impediments, including a deficiency of motivation and ambition, may hinder students from culturally diverse backgrounds from achieving full engagement. In light of these challenges, educators must adopt engaging and dynamic teaching methodologies that address the unique interests and needs of their students (Pham & Tran, 2015).

The study determined that good teaching strategies are essential for assisting students from diverse cultural origins in overcoming challenges in learning English. To be effective, teachers must employ pedagogical practices that accommodate their students' distinct learning styles and experiences. To surmount challenges and foster positive learning experiences, it is essential to employ tactics that encourage active engagement, cooperation, and culturally pertinent information.

Question 5: What Is The Impact Of Multiculturalism And Diversity On The Learning Experience In Higher Education?

To answer this question, standard deviations, means, and the ranking of each item is calculated as follows:

Table 8. The means and standard deviations for the field of (The impact of multiculturalism and diversity on the overall learning experience in higher education) N=199.

No.	Items	Mean	Std. Devation	Order	level
19	I believe that a diverse student body enhances the learning experience for all students.	3.748	1.248	1	high
24	I think it is important to celebrate cultural diversity in the classroom.	3.626	1.203	2	moderate
20	I think it is important to promote inclusivity and equity in the classroom.	3.617	1.254	3	moderate
21	I believe that exposure to different cultures and perspectives is valuable for students' personal and professional growth.	3.591	1.277	4	moderate
23	I believe that diversity in the classroom promotes critical thinking and problem-solving skills.	3.591	1.263	5	moderate
22	I think it is important to create a safe and welcoming environment for students from diverse cultural backgrounds.	3.443	1.258	6	moderate
Total Degree		3.620	0.841	moderate	

Table 8 reveals the influence of multiculturalism and diversity on the learning experience in higher education. The participants expressed moderate challenges, as evidenced by an average score of 3.470 and a standard deviation of 0.838. This implies that while the challenges are not excessively severe, they necessitate attention and intervention to improve the teaching and learning environment in this setting.

Item 19 received the highest agreement among the participants, with a mean score of 3.748 and a standard deviation of 1.248. This suggests that the challenge addressed by this item is perceived as particularly impactful and widely experienced. Possible reasons for this substantial consensus may encompass:

1- Inclusion and Representation: In higher education, inclusion and representation may be important factors. The curriculum, teaching materials, and classroom discussions may fail to represent students from diverse cultural backgrounds. It is important to ensure that all students feel included and represented in the learning process, as shown by the greater agreement on this item (Burke & Hardware, 2015).

2- Cultural Competence: As well as challenges associated with cultural competence, item 19 might also pertain to educators and students. It is possible to positively impact learning experiences by understanding and valuing different cultural perspectives. It appears that the high level of agreement on this item indicates that higher education needs to promote cultural competency (Washington, 2020).

In contrast, Item 22 exhibited the lowest level of agreement among participants, reflected by a mean score of 3.443 and a standard deviation of 1.258. This indicates that this specific item was regarded as having a lesser impact relative to other items. Potential rationales for the lease agreement may encompass:

1- ***Intercultural Communication:*** Item 22 may investigate matters related to intercultural communication throughout higher education. This attribute, while notable, may be considered to have a diminished effect relative to other obstacles. This may arise from ongoing endeavors to enhance intercultural communication or from students' prior encounters with this facet of the educational environment (Hoff, 2020).

2-***Access to resources:*** Item 22 may pose difficulties with resource accessibility for pupils from various cultural backgrounds. These impediments, however significant, may be seen as having a relatively negligible effect on the overall learning experience. This arises from continuous initiatives to foster fair access to resources within higher education institutions (Gay, 2013; O'Hara & Pritchard, 2008).

Diversity and multiculturalism moderately affect the overall quality of higher education, according to this study. As can be seen from item 19, there is a general consensus regarding the importance of cultural competency, representation, and inclusion. Additionally, Item 22 suggests that resource accessibility and intercultural communication will have little impact. In order to meet the demands of culturally diverse students, higher education institutions must acknowledge the inherent difficulties and their consequences and provide an inclusive learning environment.

CONCLUSION

The integration of AI technology and intercultural competence is fundamental to English language teaching in today's multicultural context. The results indicate that EFL educators have a favorable attitude towards the incorporation of AI technologies to enhance multiculturalism in educational environments. The survey discusses a lack of multicultural education training, alongside linguistic barriers and cultural misinterpretations. The study suggests that EFL teachers should be provided with opportunities to enhance their intercultural communication and cultural competency, alongside the integration of AI technologies into multicultural education within EFL courses. Implementing these suggestions can enhance cultural awareness in the classroom and better equip teachers to address the challenges of teaching in a multicultural environment.

Limitations

This study has limitations that must be acknowledged, despite providing valuable insights. A convenience sample may restrict the generalizability of the findings to the specific group of EFL teachers studied. Future research must include a more varied and representative sample to improve external validity. Furthermore, self-report instruments, such as Likert scales, may result in response biases and social desirability influences. Employing various data gathering methods, including interviews and classroom observations, may yield more thorough and impartial insights into instructors' opinions and practices.

Recommendations

The study's findings yield several insightful suggestions for future investigation and application. Longitudinal research is essential to investigate the long-term effects of incorporating multicultural education into EFL curricula and teacher professional development programs. This project will yield long-term consequences that will be elucidated through these investigations. Longitudinal research could facilitate the examination of teachers' attitudes, behaviors, and academic achievement over time. Furthermore, analyzing the perspectives and experiences of students from various cultural backgrounds in an EFL classroom enhances our comprehension of the dynamics of multicultural education. EFL teachers, teacher educators, and curriculum developers must collaborate effectively to create and apply inclusive, culturally sensitive instructional materials and techniques. Supporting EFL teachers in developing cultural competence and intercultural communication skills is essential for the implementation of inclusive learning environments.

Pedagogical Implications

There are significant pedagogical implications of the study's findings for teachers of English language instruction. In the first place, AI technologies integration has been shown to improve students' language learning experiences, especially when multiculturalism and favorable attitudes towards AI technologies are considered. Additionally, the study's highlighted issues, such as linguistic and cultural barriers, raise the importance of cultivating intercultural competency and communication skills among educators. Educators should be given the opportunity to pursue multicultural education training through ELT courses that incorporate AI technologies. As a result, educators can better prepare students for teaching in diverse classrooms and help them develop cultural sensitivity.

REFERENCES

Al-khresheh, M., Mohamed, A., & Bin Ali, R. (2023). A Study on Writing Apprehension among Saudi Undergraduate EFL Students, *Language Teaching. Research Quarterly*, *34*, 1–18. DOI: 10.32038/ltrq.2023.34.01

Al-khresheh, M., Mohamed, A. M., & Asif, M. (2022). Teachers' Perspectives towards Online Professional Development Programs during the Period of COVID-19 Pandemic in the Saudi EFL Context. *FWU Journal of Social Sciences*, *16*(2). Advance online publication. DOI: 10.51709/19951272/Summer2022/1

Al-khresheh, M. H., Mohamed, A. M., & Shaaban, T. S. (2024). A quasi-experimental study on the effectiveness of augmented reality technology on english vocabulary learning among early childhood pupils with learning disabilities. Innoeduca. *International Journal of Technology and Educational Innovation*, *10*(2), 5–26. DOI: 10.24310/ijtei.102.2024.17823

Alaca, B. (2017). *Kindergarten teachers' perspectives on culturally responsive education*. University of Toronto.

Alenezi, M., Mohamed, A., & Shaaban, T. (2023). Revolutionizing EFL special education: how ChatGPT is transforming the way teachers approach language learning. Innoeduca. *International Journal of AI technologies and Educational Innovation*, *9*(2), 5–23. https://doi.org/DOI: 10.24310/innoeduca.2023.v9i2.16774

Appadurai, A. (1996). *Modernity at large: Cultural dimensions of globalization* (Vol. 1). University of Minnesota Press.

Aririguzoh, S. (2022). Communication competencies, culture and SDGs: effective processes to cross-cultural communication. *Humanities and Social Sciences Communications*, *9*(1), 1-11. https://doi.org/DOI: 10.1057/s41599-022-01109-4

Baker, W. (2022). *Intercultural and transcultural awareness in language teaching*. Cambridge University Press. DOI: 10.1017/9781108874120

Belda-Medina, J. (2022). Promoting inclusiveness, creativity and critical thinking through digital storytelling among EFL teacher candidates. *International Journal of Inclusive Education*, *26*(2), 109-123. https://doi.org/DOI: 10.1080/13603116.2021.2011440

Bennett, M. J. (2022). Intercultural communication: A current perspective. In Basic Concepts of Intercultural Communication: Selected Readings (ed. Bennett, J.) (Intercultural Press, 1998).

Berlian, Z., & Huda, M. (2022). Reflecting culturally responsive and communicative teaching (CRCT) through partnership commitment. *Education Sciences, 12*(5), 295. https://doi.org/DOI: 10.3390/educsci12050295

Boudouaia, A., Mouas, S., & Kouider, B. (2024). A study on ChatGPT-4 as an innovative approach to enhancing English as a foreign language writing learning. Journal of Educational Computing Research. https://doi.org/ Borenstein, M., Hedges, L. V., Higgins, J. P. T., & Rothstein, H. R. (2009). *Introduction to Meta-Analysis.* John Wiley & Sons. https://doi.org/DOI: 10.1177/07356331241247 46

Burke, A., & Hardware, S. (2015). Honouring ESL students' lived experiences in school learning with multiliteracies pedagogy. Language, Culture and Curriculum, 28(2), 143-157. https://doi.org/DOI: 10.1080/07908318.2015.1027214

Byram, M. (2020). *Teaching and assessing intercultural communicative competence: Revisited.* Multilingual matters.

Byram, M. (2021). *Teaching and assessing intercultural communicative competence: Revisited.* Multilingual matters., DOI: 10.21832/9781800410251

Chisholm, I. M. (1998). Six elements for technology integration in multicultural classrooms. *Journal of Information Technology for Teacher Education, 7*(2), 247–268. DOI: 10.1080/14759399800200033

Chounta, I. A., Bardone, E., Raudsep, A., & Pedaste, M. (2022). Exploring teachers' perceptions of artificial intelligence as a tool to support their practice in Estonian K-12 education. *International Journal of Artificial Intelligence in Education, 32*(3), 725–755. DOI: 10.1007/s40593-021-00243-5

Deardorff, D. K. (2020). *Manual for developing intercultural competencies: Story circles.* UNESCO Publishing.

Debnam, K. J., Smith, L. H., Aguayo, D., Reinke, W. M., & Herman, K. C. (2023). Nominated exemplar teacher perceptions of culturally responsive practices in the classroom. *Teaching and Teacher Education, 125*, 104062. https://doi.org/DOI: 10.1016/j.tate.2023.104062

Dimitrov, N., & Deardorff, D. (2023). Intercultural competence as the core to developing globally engaged teachers. *At school in the world: Developing globally engaged teachers*, 3-27.

Dooly, M., & Darvin, R. (2022). Intercultural communicative competence in the digital age: critical digital literacy and inquiry-based pedagogy. *Language and Intercultural Communication, 22*(3), 354-366. https://doi.org/DOI: 10.1080/14708477.2022.2063304

Farrokhi, F., & Mahmoudi-Hamidabad, A. (2012). Rethinking convenience sampling: Defining quality criteria. *Theory & practice in language studies, 2*(4). https://doi.org/DOI: 10.4304/tpls.2.4.784-792

Feng, X., Zhang, N., Yang, D., Lin, W., & Maulana, R. (2024). From awareness to action: Multicultural attitudes and differentiated instruction of teachers in Chinese teacher education programmes. *Learning Environments Research*, •••, 1–42. DOI: 10.1007/s10984-024-09518-9

Gay, G. (2013). Teaching to and through cultural diversity. *Curriculum inquiry, 43*(1), 48-70. https://doi.org/DOI: 10.1111/curi.12002

Gorski, P. C., & Dalton, K. (2020). Striving for critical reflection in multicultural and social justice teacher education: Introducing a typology of reflection approaches. *Journal of Teacher Education, 71*(3), 357-368. https://doi.org/DOI: 10.1177/0022487119883545

Hall, S. (2006). Cultural studies and its theoretical legacies. In *Stuart Hall* (pp. 272–285). Routledge.

Hoff, H. E. (2020). The evolution of intercultural communicative competence: Conceptualisations, critiques and consequences for 21st century classroom practice. *Intercultural Communication Education, 3*(2), 55-74. https://doi.org/DOI: 10.29140/ice.v3n2.264

Hoffman, D. L., Leong, P., Ka'aloa, R. P. H., & Paek, S. (2022). Teachers' Perspectives on Culturally-Relevant Computing: Principles and Processes. *TechTrends, 66*(3), 423-435. https://doi.org/DOI: 10.1007/s11528-022-00733-w

Hohenstein, J., Kizilcec, R. F., DiFranzo, D., Aghajari, Z., Mieczkowski, H., Levy, K., Naaman, M., Hancock, J., & Jung, M. F. (2023). Artificial intelligence in communication impacts language and social relationships. *Scientific Reports, 13*(1), 5487. DOI: 10.1038/s41598-023-30938-9 PMID: 37015964

Jmaiel, H. A., Abukhait, R. O., Mohamed, A. M., Shaaban, T. S., Al-khresheh, M. H., & AL-Qadri, A. H. (2025). The role of ChatGPT in enhancing EFL students' ESP writing skills: an experimental study of gender and major differences. Discover Education, 4(1), 1-19. https://doi.org/DOI: 10.1007/s44217-025-00700-6

Kahn, H. E., & Agnew, M. (2017). Global learning through difference: Considerations for teaching, learning, and the internationalization of higher education. *Journal of Studies in International Education, 21*(1), 52-64. https://doi.org/DOI: 10.1177/102831531562202

Karataş, S., Eckstein, K., Noack, P., Rubini, M., & Crocetti, E. (2023). Meeting in school: Cultural diversity approaches of teachers and intergroup contact among ethnic minority and majority adolescents. *Child Development, 94*(1), 237-253. https://doi.org/DOI: 10.1111/cdev.13854

Kayalar, F. (2016). Cross-Cultural Comparison of Teachers' Views upon Integration and Use of technology in Classroom. *Turkish Online Journal of Educational AI technologies -TOJET, 15*(2), 11-19. https://eric.ed.gov/?id=EJ1096412

Keengwe, J. (2010). Fostering cross-cultural competence in preservice teachers through multicultural education experiences. *Early Childhood Education Journal, 38*(3), 197–204. DOI: 10.1007/s10643-010-0401-5

Kim, D. (2020). Learning language, learning culture: Teaching language to the whole student. *ECNU Review of Education, 3*(3), 519-541. https://doi.org/DOI: 10.1177/2096531120936693

Klimova, B., & Chen, J. H. (2024). The impact of AI on enhancing students' intercultural communication competence at the university level: A review study. *Language Teaching Research Quarterly*, *43*, 102–120. DOI: 10.32038/ltrq.2024.43.06

Kubota, R. (2020). Promoting and problematizing multi/plural approaches in language pedagogy. *Plurilingual pedagogies: Critical and creative endeavors for equitable language in education*, 303-321. https://doi.org/DOI: 10.1007/978-3-030-36983-5_14

Kurbanova, F. K., Botirova, M., & Abdurahmonova, S. (2022). Promoting learning and developing cultural competence. *Academic research in educational sciences, 3*(3), 914-917.

Lau, W. S., & Shea, M. (2022). Empowering English learners in the classroom through culturally responsive social-emotional teaching practices. *Journal of Multilingual and Multicultural Development*, 1-18. \DOI: 10.1080/01434632.2022.2078337

Lee, J. (2020). Informal digital learning of English and strategic competence for cross-cultural communication: Perception of varieties of English as a mediator. *ReCALL, 32*(1), 47–62. DOI: 10.1017/S0958344019000181

Lifintsev, D., & Wellbrock, W. (2019). Cross-cultural communication in the digital age. *Estudos em Comunicação, 1*(28), 93-104. https://doi.org/DOI: 10.25768/fal.ec.n28.a05

Liu, G., Ma, C., Bao, J., & Liu, Z. (2023). Toward a model of informal digital learning of English and intercultural competence: a large-scale structural equation modeling approach. *Computer Assisted Language Learning*, 1-25. https://doi.org/ DOI: 10.1080/09588221.2023.2191652

Long, J., & Lin, J. (2022). An empirical study on cultivating college students' cross-cultural communicative competence based on the artificial-intelligence English-teaching mode. *Frontiers in Psychology*, *13*, 976310. DOI: 10.3389/ fpsyg.2022.976310 PMID: 36562067

Mac Fadden, I., García-Alonso, E. M., & López Meneses, E. (2024). Science Mapping of AI as an Educational Tool Exploring Digital Inequalities: A Sociological Perspective. *Multimodal Technologies and Interaction*, *8*(12), 106. DOI: 10.3390/ mti8120106

Mitchell, T., & Steinmetz, G. (1999). Society, economy, and the state effect. *State/ culture: State-formation after the cultural turn, 76*, 90.

Mohamed, A. (2024). Artificial intelligence in language education: Opportunities and challenges. *TESOL Quarterly*, *58*(2), 210–230. DOI: 10.1002/tesq.5678

Mohamed, A., & Shaaban, T. (2023). Investigating College Students' Metacognitive Awareness in Enhancing ESP Writing Proficiency. *Journal of Teaching English for Specific and Academic Purposes*, *11*(2), 387–401. DOI: 10.22190/ JTESAP230420029M

O'Hara, S., & Pritchard, R. H. (2008). Meeting the challenge of diversity: Professional development for teacher educators. *Teacher Education Quarterly, 35*(1), 43-61. https://www.jstor.org/stable/23479030

Ofosu-Ampong, K. (2024). Beyond the hype: Exploring faculty perceptions and acceptability of AI in teaching practices. *Discover Education*, *3*(1), 38. DOI: 10.1007/ s44217-024-00128-4

Omer, W., Mohamed, A., & Al-khresheh, M. (2025). The Integration of English Language Skills in Business Education: Insights from Accounting in the Saudi Context. *Journal of Teaching in International Business* https://doi.org/DOI: 10.1080/08975930.2025.2452529

Pham, L., & Tran, L. (2015). Understanding the symbolic capital of intercultural interactions: A case study of international students in Australia. *International Studies in Sociology of Education, 25*(3), 204-224. https://doi.org/DOI: 10.1080/09620214.2015.1069720

Rao, Z., & Chen, H. (2020). Teachers' perceptions of difficulties in team teaching between local-and native-English-speaking teachers in EFL teaching. *Journal of Multilingual and Multicultural Development, 41*(4), 333-347. https://doi.org/DOI: 10.1080/01434632.2019.1620753

Sarwari, A. Q., Javed, M. N., Mohd Adnan, H., & Abdul Wahab, M. N. (2024). Assessment of the impacts of artificial intelligence (AI) on intercultural communication among postgraduate students in a multicultural university environment. *Scientific Reports*, *14*(1), 13849. DOI: 10.1038/s41598-024-63276-5 PMID: 38879546

Setyono, B., & Widodo, H. P. (2019). The representation of multicultural values in the Indonesian Ministry of Education and Culture-Endorsed EFL textbook: a critical discourse analysis. *Intercultural Education, 30*(4), 383-397. https://doi.org/DOI: 10.1080/14675986.2019.1548102

Spellman, K., Dillenbeck, J., Edwards, N. N., & Bohecker, L. (2022). Supporting marginalized students in counselor education and supervision programs. *Journal of Counselor Leadership and Advocacy, 9*(1), 32-44. https://doi.org/DOI: 10.1080/2326716X.2021.2007429

Sun, L. (2024). Enhancing intercultural competence of Chinese English majors through AI-enabled Collaborative Online International Learning (COIL) in the digital era. *Education and Information Technologies*, •••, 1–33. DOI: 10.1007/s10639-024-13143-7

Szelei, N., Tinoca, L., & Pinho, A. S. (2020). Professional development for cultural diversity: The challenges of teacher learning in context. *Professional development in education, 46*(5), 780-796. https://doi.org/DOI: 10.1080/19415257.2019.1642233

Tsang, A., Aubrey, S., & Yuan, R. (2023). Multiculturalism and multilingualism in higher education: intercultural activity participation and opportunities for language learning. *International Journal of Multilingualism*, 1-17. https://doi.org/DOI: 10.1080/14790718.2022.2164769

Washington, A. N. (2020, February). When twice as good isn't enough: The case for cultural competence in computing. In *Proceedings of the 51st ACM technical symposium on computer science education* (pp. 213-219). https://doi.org/DOI: 10.1145/3328778.3366792

Xia, Y., Shin, S.-Y., & Kim, J.-C. (2024). Cross-Cultural Intelligent Language Learning System (CILS): Leveraging AI to Facilitate Language Learning Strategies in Cross-Cultural Communication. *Applied Sciences (Basel, Switzerland)*, *14*(13), 5651. DOI: 10.3390/app14135651

Yang, Y., & Montgomery, D. (2013). Gaps or bridges in multicultural teacher education: AQ study of attitudes toward student diversity. *Teaching and Teacher Education, 30*, 27-37. https://doi.org/DOI: 10.1016/j.tate.2012.10.003

Yi, Y. (2021). Establishing the concept of AI literacy. *Jahr–European Journal of Bioethics*, *12*(2), 353–368.

Yusuf, A., Pervin, N., Román-González, M., & Noor, N. M. (2024b). Generative AI in education and research: A systematic mapping review. *Review of Education*, *12*(2), e3489. DOI: 10.1002/rev3.3489

Zhang, X., & Zhou, M. (2019). Interventions to promote learners' intercultural competence: A meta-analysis. *International journal of intercultural relations, 71*, 31-47. https://doi.org/DOI: 10.1016/j.ijintrel.2019.04.006

Chapter 8
Aligning Smart Education With Smart Economy:
Multi-Dimensional Educational Quality Assessment

Andreas Marinosyan
https://orcid.org/0000-0003-0577-2360
Moscow City University, Russia

ABSTRACT

This chapter proposes a mathematical framework for assessing educational quality that bridges the gap between educational processes and real-world impacts within a "smart economy," defined as a network of relationships mediated by knowledge. Current educational assessment practices rely on internal metrics that often fail to capture how graduates' capabilities align with evolving economic and societal needs. The proposed framework integrates diverse stakeholder perspectives through two interconnected "fields": the field of student qualities and the field of educational programs. By representing student qualities and programs as vectors in multi-dimensional spaces, the framework formalizes how education transforms capabilities. The chapter presents mathematical formulations from both upward (building from student-program interactions) and downward (deriving from economic system dynamics) perspectives. While acknowledging implementation challenges and limitations, the framework offers a path toward assessment that is holistic, externally aligned, and resistant to metric corruption.

DOI: 10.4018/979-8-3373-7922-7.ch008

INTRODUCTION

The accelerating pace of technological advancement – marked by breakthroughs in artificial intelligence, automation, biotechnology, and digital connectivity – is fundamentally reshaping our global landscape. This transformation has given rise to an environment characterized by volatility, uncertainty, complexity, and ambiguity (VUCA), and more recently described as exhibiting brittleness, anxiety, nonlinearity, and incomprehensibility (BANI). These profound shifts present challenges and opportunities for educational systems worldwide, demanding a fundamental reconsideration of how we prepare learners for their futures. The rapidly evolving nature of work requires educational institutions to move beyond the traditional paradigm of knowledge transfer toward cultivating capacities for continuous learning, critical thinking, and adaptive resilience. This reality demands educational approaches that foster not only domain expertise but also transformative learning capacities that enable ongoing adaptation.

Despite these imperatives, a significant gap persists in how we assess the effectiveness of educational programs. Current evaluation practices predominantly rely on internal metrics – such as grades, test scores, completion rates, and faculty qualifications – which, while valuable within the educational context, often fail to adequately capture the external alignment of graduates' capabilities with the evolving needs of the economy and broader society. Traditional assessment models typically focus on measuring inputs (resources allocated) and immediate outputs (knowledge acquisition), rather than longer-term outcomes and impacts. When external metrics are considered, they tend to be limited to employment rates or starting salaries, which represent an extremely narrow view of educational value and are vulnerable to economic fluctuations unrelated to educational quality.

This disconnect between educational assessment and real-world impact is particularly problematic in the context of what we term the "smart economy." We define the smart economy not merely as an economy driven by technological innovation, but more fundamentally as a dynamic network of relationships between people mediated by knowledge. This knowledge extends beyond technical expertise and factual information to encompass the crucial ability to transform oneself and one's actions to meet the evolving demands of others – employers, collaborators, clients, administrative institutions, and society as a whole. Even seemingly simple economic transactions are ultimately driven by the knowledge of how to transform resources into something that fulfills a specific need. This definition underscores the critical role of continuous learning, adaptation, and the proactive application of knowledge in creating value within modern economic and social systems.

To address these challenges, we propose a multi-field algorithmic framework designed to provide a quantitative assessment of educational programs that bridges the gap between educational processes and their real-world impacts. Unlike traditional input-output models, our framework integrates diverse stakeholder perspectives – including those of industries and society, individual learners, educational entities, and academic experts – to holistically evaluate the transformative impact of education. Drawing on transformative learning theories, the framework emphasizes not only the acquisition of domain-specific competencies but also the cultivation of meta-skills that empower individuals to navigate and contribute meaningfully to the dynamic smart economy.

LITERATURE REVIEW

This literature review situates our assessment framework within broader scholarly contexts spanning educational assessment, transformative learning, and economic theory. By examining the evolution of smart education and smart economics concepts, along with their theoretical underpinnings, we establish the foundation for our integrated approach to educational quality assessment.

The concept of ***smart education*** has emerged as a response to the profound technological and social transformations of the 21st century. While educational systems have continuously evolved throughout history, the current transition represents a particularly significant paradigm shift – one characterized by the integration of digital technologies, shifting pedagogical approaches, and a reimagining of the very purpose of education in an increasingly complex world (Zhu et al., 2016). The historical evolution of educational systems reveals a pattern of adapting to changing societal demands and technological affordances. From the standardized, industrial-era model designed to prepare workers for factory employment to the information-age emphasis on knowledge acquisition, each educational paradigm has reflected the dominant economic and social structures of its time. The emerging smart education paradigm represents a response to the demands of what has been termed the "smart economy" – an economic landscape characterized by knowledge-intensive activities, rapid technological change, and the primacy of innovation and adaptability.

The dominant discourse surrounding smart education has largely centered on technological integration, particularly the incorporation of digital tools, learning analytics, and adaptive systems into educational processes. As Zhu and He (2012) note, the essence of smart education lies in "creating intelligent environments by using smart technologies, so that smart pedagogies can be facilitated as to provide personalized learning services and empower learners" (p. 6). A systematic literature review by Martín et al. (2019) found that IoT technologies were predominant in smart education research (46 out of 56 papers), followed by data analysis approaches, with learning analytics and big data being the most commonly employed frameworks. Their analysis further categorized smart education systems into four main types based on their objectives: identification (e.g., student tracking), data management, information display, and virtual environments. Building on this technological foundation, Demir (2021) proposes a more structured approach through his Smart Education Framework, which organizes technologies into distinct layers: core layer – new or improved learning/teaching approaches (personalized learning, flipped learning, etc.); essential/transforming technologies (learning management systems, smart classrooms, virtual classrooms); enriching technologies (technologies that enhance the smart education experience); and supporting technologies (general-purpose technologies that support the infrastructure).

While technological infrastructure is crucial, smart education extends beyond mere technological implementation. Hwang (2014) describes smart learning as a technology-enhanced learning environment that adapts to learners' needs and preferences, enabling personalized and contextualized learning experiences (p. 2). Similarly, Bajaj & Sharma (2018) highlight this perspective by noting that smart education is "about delivering personalized learning, anywhere and anytime" (p. 840). This focus on personalization and flexibility acknowledges the diverse pathways through which learning occurs and the need for educational systems that can adapt to individual learners' needs, preferences, and contexts. As Demir (2021) emphasizes, "smart education is not just about technology. It is also about new teaching and learning approaches" (p. 3). This dual nature of smart education is further elaborated by Díaz-Parra et al. (2022), who characterize it as both "the influence of diverse technologies to enhance education quality" and "the process of optimally managing human, economic, and technology resources from educational institutions and research centres" (p. 65–66).

The evolution of smart education has not been without critical perspectives that highlight potential limitations and challenges. Fawns (2022) challenges both "technology first" and "pedagogy first" positions as oversimplifications that decontextualize technology and make educators vulnerable to different forms of determinism. Technological determinism, as Fawns notes, incorrectly views technology as driving social change with outcomes that can be predicted by design and functionality. This perspective often leads to unrealistic expectations about technological solutions to educational challenges. Adding nuance to this discussion, Díaz-Parra et al. (2022) distinguish between "education-hard problems" and "education-soft problems." While the former can be addressed through technological optimization (e.g., resource management, monitoring, infrastructure), the latter involve handling information that is "inaccurate or incomplete, with uncertainty and ambiguity, being ambiguous, volatile, poorly understood, and dynamic (e.g., educational policy, administration, decision-making, educational reforms)" (p. 66). This distinction highlights the limitations of purely technological approaches to complex educational challenges.

To address these limitations, Fawns (2022) proposes an "entangled pedagogy" that encapsulates "the mutual shaping of technology, teaching methods, purposes, values and context." In this view, pedagogy is not a driver of technology or vice versa, but rather a complex interrelationship in which all elements are interdependent and co-constitutive.

As educational systems evolve in response to changing societal demands, so too do economic frameworks that shape and are shaped by educational outcomes. The concept of ***smart economics*** has evolved significantly over the past several decades, reflecting broader shifts in economic theory, technological advancement, and changing societal priorities. While the term itself gained prominence in the early 21st century, its conceptual foundations can be traced to earlier economic paradigms that emphasized knowledge, innovation, and human capital as driving forces of economic development (Lundvall & Johnson, 1994; Powell & Snellman, 2004).

The initial formulations of smart economics emerged primarily from two distinct but converging trajectories: the technological revolution associated with digitalization and the increasing recognition of knowledge as a critical economic asset. As Kostyk et al. (2023) argue, modern economic development is primarily shifting toward a digital economy, with digitalization increasingly serving as the key factor driving global growth. This technological dimension of smart economics emphasized how digital technologies could transform production processes, enhance efficiency, and create new economic opportunities. This technological focus has been further embedded within broader urban development contexts, as the evolution of smart economics has been intertwined with the emergence of "smart cities" as an urban development paradigm. As Nastjuk et al. (2022) observe, "In the last two decades, the concept of smart cities has attracted significant research and policy attention. Despite its extensive discussion in literature, the term smart city is a fuzzy concept" (p. 1917). Within this broader smart city framework, the smart economy has been conceptualized as one of several interconnected dimensions of "smartness," alongside smart governance, smart environment, smart mobility, smart people, and smart living (Adiyarta et al., 2020).

The conceptual evolution of smart economics has continued with more recent iterations influenced by the transition from Industry 4.0 to Industry 5.0, as highlighted by Kostyk et al. (2023). This shift marks a move away from a predominant focus on connectivity, automation, and data exchange, toward a more human-centered approach that prioritizes sustainability, resilience, and the integration of human and machine capabilities. This evolution reflects an increasing awareness that technological advancement alone is insufficient to create truly "smart" economic systems, which must also be efficient, equitable, and aligned with broader societal goals. The multidisciplinary nature of smart economics is captured by Purnomo et al. (2021) in their bibliometric review, noting that "the smart economy concept is a synergy between computer science, social science, engineering, and business" (p. 3), highlighting the technological emphasis in much of the smart economy literature. Their classification of smart economy research themes – abbreviated as EASSEES (economy, application of smart cities, smart technology, smart cities character, economy of urban, education, and smart environment) – further illustrates how technology permeates diverse aspects of smart economics research.

While technological innovation remains a crucial component of smart economics, recent conceptualizations recognize that technologies are primarily enablers of knowledge creation, sharing, and application. The fundamental driver of economic value is the capacity for transformation – of resources, of capabilities, and of relationships themselves. This perspective aligns with empirical research on smart economies. For instance, Indrawati et al.'s (2018) emphasis on "a variety of employment opportunities with labor market flexibility, diversification that promotes entrepreneurship and innovation as well as more productivity through local, regional and global linkages" (p. 175) can be understood as creating conditions for the effective development and application of transformative knowledge within a network of economic relationships. Similarly, the direct impact of smart economy on all smart domains identified by Popova and Popovs (2022) reflects how knowledge-mediated relationships in the economic sphere can enhance capabilities and conditions in other domains, from smart people and smart society to smart environment and smart living. This perspective recognizes the interconnected nature of economic, social, and environmental systems, viewing smart economics not as an isolated domain but as an integrated component of a broader smart ecosystem.

Beyond technological and systemic perspectives, smart economics also encompasses important human capital and gender dimensions. The human capital perspective on smart economics underscores the importance of education and lifelong learning. Dulhunty's (2022) research on microcredit programs in India offers a nuanced view of the gender-focused aspect of smart economics. The study identifies a continuum of approaches, ranging from smart economics strategies to more holistic gender-focused programming, with the latter including broader elements such as gender training and community support. As one of Dulhunty's (2022) interviewees articulates, "Money is not everything, awareness is everything. If awareness levels are not high, they do not spend money in the right way. Education is the main thing – both school and informal education" (p. 1220). This highlights the critical role of knowledge and skill development in enabling individuals to engage meaningfully in economic systems.

Critical perspectives on smart economics provide important considerations for educational frameworks. As Prügl (2021) critically notes, smart economics often involves making "women fit for capitalist markets" (p. 471), focusing on aligning gender equality with economic growth rather than challenging the structural inequalities within the capitalist system. From this critique, we can conclude that a challenge for smart education is not merely to adapt to the skills demanded by smart economies but to foster the creative and transformative capabilities that can reshape economic relationships in ways that enhance human development.

The evolution of smart economics has increasingly emphasized relational and transformative dimensions that go beyond technology-centric and human capital approaches. Recent conceptualizations view economic interactions not merely as transactions of goods and services but as complex relationships mediated by knowledge and characterized by mutual transformation. Ramaswamy and Ozcan (2014) propose a co-creation perspective on economic value, suggesting that value emerges from the interactions between stakeholders rather than being embedded in products or services themselves. This relational view resonates with network-based economic models that emphasize the importance of connectivity, collaboration, and knowledge flows in determining economic outcomes (Castells, 2010; Benkler, 2006). In summary, the conceptualization of smart economics evolves from technology-centric approaches focused on digital infrastructure and automation to more comprehensive frameworks that emphasize knowledge creation, human capital development, and transformative relationships.

This convergent evolution of smart education and smart economics creates the context for our assessment framework, which draws upon several distinct theoretical traditions that provide its conceptual underpinnings. Central among these is transformative learning theory, particularly as articulated by Mezirow (1991, 2000), who conceptualizes learning as a process of perspective transformation. Mezirow's work emphasizes how educational experiences can fundamentally alter learners' frames of reference, enabling them to construct more inclusive, discriminating, and integrative perspectives. As further developed by Taylor (2007), Taylor & Cranton (2012), and Cranton (2016), transformative learning theory emphasizes how critical reflection on assumptions and experiences can lead to profound shifts in individuals' frames of reference, enhancing their capacity to navigate complex and ambiguous situations – capabilities that are increasingly valued in the smart economy. This aligns with our model's emphasis on measuring not just knowledge acquisition but transformative impacts on student qualities.

Complementing this transformative perspective, Schön's (1983, 1987) works on reflective practice provides another theoretical pillar, particularly his distinction between "technical rationality" and "reflection-in-action." Schön's critique of technical rationality as insufficient for navigating complex, uncertain, and value-laden professional contexts resonates with our framework's emphasis on meta-skills and adaptability. Schön advocated for educating "reflective practitioners" through experiential learning environments, preparing them to meet the multifaceted demands of practice that technical rationality alone cannot address. This corresponds to the demand-oriented approach our model seeks to undertake, recognizing that educational quality must be assessed in terms of how well it prepares individuals to navigate the complexities of professional practice.

The economic dimensions of our framework integrate human capital theory (Becker, 1964) while extending beyond its traditional formulations. Becker emphasized education as an investment that enhances productive capacity, a perspective that remains foundational to understanding the economic value of education. Expanding on this foundation, Lundvall and Johnson (1994) introduced the taxonomy of "know-what," "know-why," "know-how," and "know-who" knowledge types, emphasizing that different forms of knowledge contribute differently to economic value. The knowledge economy literature, particularly contributions by Powell and Snellman (2004), has emphasized how economic value increasingly derives from intellectual capabilities rather than physical inputs. Our framework's conceptualization of the smart economy as a *network of relationships between people mediated by knowledge* builds upon this tradition while emphasizing the relational and transformative aspects of knowledge production and application. This perspective recognizes that economic value emerges not just from possessing knowledge but from the ability to apply that knowledge to transform resources and relationships in ways that meet evolving needs.

A distinctive feature of our framework is its incorporation of market-based mechanisms for information aggregation, particularly the use of futures contracts on program rewards. This approach connects to research on prediction markets as tools for aggregating distributed information. Arrow et al. (2008) demonstrate how prediction markets can effectively harness collective intelligence to forecast complex outcomes, while Wolfers and Zitzewitz (2004) have shown how such markets provide an algorithm for aggregating diverse opinions and create incentives for information seeking and truthful revelation. These market-based approaches offer a potential solution to the challenge of assessment in educational contexts, where outcomes are complex, multidimensional, and often manifest over extended timeframes. By incorporating market mechanisms into our assessment framework, we harness the collective intelligence of diverse stakeholders, creating a dynamic system that can adapt to changing conditions and integrate new information as it becomes available.

To fully appreciate the contribution of our framework, it is important to consider the historical evolution and current limitations of educational assessment approaches. The evolution of educational assessment and evaluation has been marked by significant paradigm shifts that reflect changing understanding of learning processes and emerging technological capabilities. Traditional assessment approaches, rooted in psychometric traditions of the early 20th century, predominantly relied on standardized testing and summative evaluation methods (Gipps, 1994; Koretz, 2008). However, the latter half of the 20th century saw increasing recognition of assessment's formative role in learning, sparked by seminal works like Scriven's (1967) distinction between formative and summative evaluation.

More recent approaches have attempted to address the complex challenge of quantifying educational impact. Value-added modeling (VAM), as described by Raudenbush (2004), attempts to isolate the contribution of educational institutions to student outcomes by controlling for prior attributes. Competency-based assessment frameworks measure educational outcomes in terms of capabilities rather than content knowledge; Voorhees (2001) provides an overview of competency-based approaches, discussing "learning opportunities that are intensely focused and are populated by learners and employers who are chiefly interested in the shortest route to results" (p. 5), which align with our framework's representation of education as providing the most optimal path of student qualities' transformation. However, traditional competency frameworks often lack the dynamic, system-level perspective that quantitatively relates education with economy.

Despite these developments, current literature on educational assessment reveals several persistent limitations that our framework attempts to address. First, most assessment frameworks struggle to capture long-term impacts and adaptability, focusing instead on immediate outcomes (Caley et al. 2021; Boud & Falchikov, 2006). Second, there is the challenge of measuring meta-skills like adaptability and self-regulation, which are increasingly central to success in knowledge economies (Siburian & Mardiyanti, 2023). A significant gap remains in approaches to educational assessment that comprehensively integrate economic, transformative, and measurement perspectives. While scholars have theorized the connection between educational quality and economic systems (Lauder et al., 2012), there is need to further formalize this relationship in mathematical terms that enable quantitative assessment and prediction.

Perhaps the most challenging limitation of current assessment approaches is their vulnerability to what has been termed Campbell's Law – the tendency for measures to become corrupted when used as targets. While scholars like Koretz (2008) have documented this phenomenon extensively, proposed solutions typically involve expanding the set of measures rather than fundamentally rethinking measurement approaches. Our framework's incorporation of dynamic adjustment mechanisms and Bayesian updating processes represents a novel approach to this longstanding challenge, creating an assessment system that can adapt and evolve in response to changing conditions and emerging information.

In conclusion, this literature review reveals a significant evolution in both smart education and smart economics concepts, from technology-centric perspectives to more holistic frameworks that emphasize knowledge-mediated relationships and transformative capabilities. The definition of smart economics as *a network of relationships between subjects (human subjects, but perhaps also non-human subjects in the future) mediated by knowledge of transformation of objects and subjects themselves* provides a foundation for bridging technological, educational, and social dimensions of economic development – a perspective that aligns closely with our proposed framework for educational assessment.

METHODS

Our methodological approach is grounded in the construction of a rigorous mathematical framework that quantitatively assesses educational programs in alignment with the evolving demands of the smart economy. At the core of this framework is the ambition to integrate diverse stakeholder perspectives – including industries, educational entities, students, and academic experts – into a unified, convertible measure of educational effectiveness. This section delineates the foundational principles and criteria that guide our construction of the framework, ensuring both conceptual robustness and practical applicability.

To facilitate a comprehensive and comparable assessment across all types of stakeholders, the proposed framework introduces two interconnected "fields":

1. The field of student qualities (both personal and professional). This field represents a multi-dimensional space encompassing the diverse attributes of students, alumni, and employees. It can be conceptualized as a vector space where each dimension corresponds to a specific trait – encompassing skills, knowledge, competencies, and even soft skills and personal characteristics. Crucially, within this field, we posit the existence of a potential or value function. This function reflects the perceived worth or importance of specific quality combinations as valued by companies, society, and governmental bodies. The value function dynamically represents the evolving demands of the smart economy and the broader societal context. A graduate's "position" within this field, defined by their unique combination of qualities, can then be evaluated against this value function, providing a measure of their preparedness and potential impact.
2. The field of educational programs (curricula and pedagogical approaches). This field is also envisioned as a multi-dimensional space, where each dimension can represent a specific educational method, module, course, or pedagogical approach. Educational programs and courses are positioned within this field based on their constituent elements and pedagogical design. Furthermore, each position in this field is associated with costs – reflecting the resources required by educational institutions to develop and deliver these programs. Drawing upon the work of educational theorists who emphasized transformative learning and critical pedagogy, and building upon the principles of competency-based

education, this field acknowledges the diverse pedagogical landscape and the resource implications of different educational approaches. Moreover, aligning with contemporary research in learning analytics and educational data mining, this field can incorporate data-driven insights into the effectiveness and resource utilization of various educational programs.

A primary criterion in our approach is adherence to the insights of Goodhart's and Campbell's laws[1]. These principles warn that once a measure becomes a target, it risks losing its effectiveness, as it may be subject to manipulation or over-optimization. To counteract this, our framework incorporates dynamic adjustment mechanisms and forecasting elements that continuously recalibrate the value function underlying the assessment metric. By embedding feedback loops – such as Bayesian updating processes and market-based futures contracts – the system is designed to mitigate the risks of metric gaming and ensure that the measure remains a reliable indicator of genuine educational improvement, rather than a simplistic target for performance inflation.

The second key principle is the cross-stakeholder nature of our measurement unit. In our framework, educational programs and student attributes are represented within interconnected multi-dimensional fields – the field of educational programs and the field of student qualities. This dual representation allows us to establish a convertible unit of measurement that is applicable across all stakeholder groups. Specifically, a student's progress is captured by a quality vector, while educational programs are modeled as vectors of attributes, each with an associated cost function. A transformation matrix then formalizes the influence of program attributes on student qualities. This structure not only ensures that improvements in student capabilities are directly quantifiable in economic terms but also facilitates comparisons and aggregations across diverse stakeholder perspectives, making the measure universally applicable within the ecosystem of smart education and smart economy.

The third principle centers on the importance of a long-term, dynamic perspective. Recognizing that education is a process that unfolds over time, our framework is explicitly designed to evaluate both immediate outcomes and long-range trajectories. Forecasting is treated as an essential form of knowledge, with the accuracy of predictive models being qualitatively assessed. This temporal dimension is further enriched by incorporating market-based elements, such as futures contracts on program rewards, which provide real-time, forward-looking indicators of program impact. By assessing not only the current state but also the projected evolution of student capabilities, our methodology aligns short-term educational interventions with the long-term demands of a dynamic, knowledge-driven economy.

The implementation of our framework involves defining the dimensions of the capability space, representing educational programs as vectors, specifying transformation matrices, calibrating cost functions, estimating value functions, monitoring outcomes for Bayesian updating, and implementing futures contracts on program rewards. The mathematical formalism underlying these steps – including vector representations, Lagrangian approaches, and functional integrals – provides a foundation for educational assessment that integrates diverse perspectives, accommodates temporal dynamics, and remains resilient against metric corruption. Our framework offers an illustrative example of how the relationship between educational interventions and their impacts on student qualities and capabilities can be approached, with the goal of aligning educational systems with the evolving demands of the smart economy.

Constructing the Mathematical Model: An Upward Perspective

This section develops the mathematical framework, starting with the basic representation of students and programs and progressively introducing more complex elements. Let us represent a student's qualities by a vector $\mathbf{S}(t) \in \mathbb{R}^n$, where n is the number of qualities considered. Each component, $S_i(t)$, of the vector represents the student's proficiency or level in the i-th quality at time t. Similarly, an educational program is represented by a vector $\mathbf{P}(t) \in \mathbb{R}^m$, where m is the number of program attributes. These attributes might include aspects of the curriculum, pedagogical methods, or resource allocation. Both student qualities and program attributes can evolve over time.

The mechanism by which an educational program influences student qualities is captured by a transformation matrix, $M(\mathbf{P})$. This matrix acts as a linear operator, mapping program attributes to changes in student qualities. The updated student qualities, $\mathbf{S}'$, are given by:

$$\mathbf{S}' = M(\mathbf{P})\mathbf{S}. \tag{1}$$

A linear approximation can be adopted for the dependence of the transformation matrix on program attributes:

$$M(\mathbf{P}) = I_n + \sum_{j=1}^{m} P_j(t)\, T_j, \tag{2}$$

where I_n is the $n \times n$ identity matrix (representing no change), and $T_j \in \mathbb{R}^{n \times n}$ are fixed generator matrices. Each T_j represents how the j-th program attribute influences student qualities.

To evaluate the overall "value" of a student's qualities, we use a scalar function, $Q(\mathbf{S})$. A simple example is a linear function:

$$Q(\mathbf{S}) = \mathbf{w}^T\mathbf{S}, \tag{3}$$

where $\mathbf{w} \in \mathbb{R}^n$ is a weight vector. A more general quadratic form is:

$$Q(\mathbf{S}) = \mathbf{S}^T W\mathbf{S}, \tag{4}$$

where $W \in \mathbb{R}^{n\times n}$ is a symmetric, positive semi-definite matrix (to make Q positive number).

The cost of an educational program is represented by a cost function, $C(\mathbf{P})$. A quadratic form can be used:

$$C(\mathbf{P}) = \mathbf{P}^T K\mathbf{P}, \tag{5}$$

where $K \in \mathbb{R}^{m\times m}$ is a positive definite matrix. This form captures the idea that different program attributes have different associated costs.

Initially, we can express the relationship between program cost and the resulting change in student quality through a simple, non-differentiable expression:

$$L(\mathbf{S},\mathbf{P}) = C(\mathbf{P}) - \Delta Q(\mathbf{S}), \tag{6}$$

where $\Delta Q(\mathbf{S}) = Q(\mathbf{S}') - Q(\mathbf{S})$ and $\mathbf{S}'$ is the student's quality vector after applying the program's transformation. This equation represents the net benefit of the program, balancing the cost of implementation against the improvement in student qualities.

To incorporate time dependence and enable a more rigorous mathematical treatment, we introduce a differentiable Lagrangian function, $L(t)$, that depends on time-varying student qualities, $\mathbf{S}(t)$, and program attributes, $\mathbf{P}(t)$:

$$L(t) = C(\mathbf{P}(t)) - Q(\mathbf{S}(t)). \tag{7}$$

This differentiable form allows for the application of calculus of variations and facilitates the analysis of system dynamics. A rigorous mathematical treatment of these time-dependent functions is beyond the scope of this chapter; however, this formulation serves as a foundation for understanding the dynamic interplay between program costs and student quality improvements. The action, A, is then the time integral of the Lagrangian:

$$A = \int_{t_1}^{t_2} L(t)\,dt. \tag{8}$$

To account for the inherent variability and uncertainty in educational outcomes, we introduce stochasticity[2] into the transformation process. We model the transformation using a stochastic differential equation:

$$d\mathbf{S}(t) = M(\mathbf{P}(t), d\mathbf{W}(t))\,\mathbf{S}(t)\,dt, \tag{9}$$

where $d\mathbf{W}(t)$ is the increment of a Wiener process (Brownian motion), and the transformation matrix M now depends on both the program attributes $\mathbf{P}(t)$ and the stochastic process $d\mathbf{W}(t)$. This formulation captures the idea that the program's effect is not deterministic but is subject to random fluctuations.

To penalize excessive stochastic fluctuations, we introduce a penalty term into the Lagrangian. Since the variance of the Wiener process scales linearly with time, a natural choice for the penalty term is one that is proportional to the square of the stochastic component of the transformation. We modify the Lagrangian (Eq. 7) to include this penalty:

$$L(t) = C(\mathbf{P}(t)) - Q(\mathbf{S}(t)) + \lambda \mathrm{Tr}[\mathrm{Cov}(M(\mathbf{P}(t), d\mathbf{W}(t)))], \tag{10}$$

where:

- λ is a positive constant representing the "penalty coefficient" for stochastic fluctuations. A higher λ indicates a stronger aversion to variability.
- $\mathrm{Tr}[\mathrm{Cov}(M(\mathbf{P}(t), d\mathbf{W}(t)))]$ represents the trace of the covariance matrix of the stochastic transformation matrix. This term quantifies the "spread" or variability introduced by the Wiener process. The covariance is taken over the stochastic elements, and the trace sums the variances of the diagonal elements of the resulting transformation. This choice ensures that the penalty term has the same units as the other terms in the Lagrangian (cost and quality).

This modified Lagrangian now includes a cost for program implementation, a benefit from improved student qualities, and a penalty for excessive stochasticity. This reflects the principle that, all else being equal, educational programs with more predictable outcomes are preferred.

Going further, we consider an ensemble of programs, indexed by i. Each program has an associated action, $A^{(i)}$, defined as:

$$A^{(i)} = \int_{t_1}^{t_2} L^{(i)}(t)\,dt, \tag{11}$$

where $L^{(i)}(t)$ is the Lagrangian for the i-th program, as defined in Eq. 10. To determine the probability of selecting a particular program, we introduce a partition function:

$$Z = \sum_i \exp(-A^{(i)}/T), \tag{12}$$

where T is a "diversity temperature" parameter. The probability of selecting program i is given by the Boltzmann weight:

$$p_i = \frac{\exp(-A^{(i)}/T)}{Z}. \tag{13}$$

A low temperature favors programs with low action (high effectiveness), while a high temperature allows for a greater variety of programs to be selected, even if they are less efficient. This *exploration* of a wider range of programs is crucial, as programs that appear less effective in the short term may reveal hidden benefits or prove more adaptable in the long run.

As a next step, the reward for those who elaborated a new educational program should introduced. For that, let us start with defining the ensemble entropy, a measure of diversity, as:

$$S = -\sum_i p_i \ln p_i. \tag{14}$$

The "work" done by the ensemble on the students is the average improvement in student quality:

$$W = \langle \Delta Q \rangle = \sum_i p_i \Delta Q^{(i)}, \tag{15}$$

where $\Delta Q^{(i)} = Q(\mathbf{S}'^{(i)}) - Q(\mathbf{S})$ and $\mathbf{S}'^{(i)}$ is the student's quality vector after program i.

The change in the ensemble's average action is:

$$\Delta\langle A\rangle = \langle A\rangle^{(2)} - \langle A\rangle^{(1)}, \tag{16}$$

where the superscripts (1) and (2) refer to two different ensembles (e.g., before and after adding a new program).

Finally, we define a reward function for program creators:

$$R = -\Delta\langle A\rangle + \gamma T \max(0, \Delta S). \tag{17}$$

This reward function has two components:

1. Action Reduction. The term $-\Delta\langle A\rangle$ rewards programs that reduce the average action of the ensemble. This is the primary driver, incentivizing cost-effective programs that improve student qualities.
2. Diversity Bonus. The term $\gamma T\max(0,\Delta S)$ [3] provides a bonus for programs that increase the ensemble's entropy (diversity). The parameter γ controls the strength of this bonus, and the temperature T scales the entropy change. The $\max(0,\Delta S)$ ensures we only reward increases in diversity.

The reason for including the diversity bonus is to encourage exploration and prevent premature convergence to a single, potentially suboptimal, program. While minimizing the average action is the primary goal, maintaining a diverse set of programs allows for adaptation to changing circumstances and the discovery of potentially superior approaches that might not be immediately apparent. However, the diversity bonus should not be so large that it overshadows the action reduction; its purpose is to incentivize the exploration of a variety of promising approaches, not to reward diversity for its own sake. The parameter γ allows for fine-tuning the balance between these two objectives.

The introduction of a new educational program into the ensemble should be envisioned as a process analogous to dissertation defense or the peer review and acceptance of scientific findings. A proposed program, characterized by its attribute vector, $\mathbf{P}^{(new)}$, is presented along with an *expected* transformation matrix, $M_{\text{expected}}(\mathbf{P}^{(new)})$. This matrix embodies the program creators' *a priori* claims about how the program will influence student qualities. Critically, this proposal also includes an estimate of the uncertainty surrounding these expected transformations, formalized as the covariance matrix of the stochastic component of the transformation, as previously discussed. This comprehensive proposal is then subjected to scrutiny by a "scientific council" – a body that could be composed of experts, educational entities, educator, regulators, etc. This council evaluates (or approves the evaluation of its creators) the program's potential merit, considering both its anticipated impact and the associated uncertainties. Approval by the council signifies the program's acceptance into the ensemble, with the proposed M_{expected} and its covariance serving as the initial *prior* for subsequent refinement. To incentivize informed and diligent evaluation, council members who endorse the program might receive a portion of future rewards linked to its actual performance, creating a direct stake in the program's success.

Once a program is active and applied to students, its actual impact on student qualities will, in general, deviate from initial expectations. To incorporate this real-world feedback, we employ Bayesian statistics to iteratively update our understanding of the transformation matrix, $M(\mathbf{P})$. The core idea is to treat M as a random variable, characterized by a probability distribution that evolves as new data become available. The initial distribution, the prior, is based on the M_{expected} and covariance provided during the approval process. As students undergo the program, their changes in quality, $\Delta\mathbf{S}(t)$, are observed. These observations, combined with the prior distribution, allow us to compute a *posterior* distribution for M, reflecting a refined understanding of the program's true effect. This Bayesian updating process can be efficiently implemented using well-established techniques, which are suitable for linear-Gaussian systems, or more general methods, applicable to non-linear and non-Gaussian scenarios. The repeated application of Bayesian updating, incorporating new data as they become available, leads to a progressively more accurate representation of the program's transformation matrix, reflecting its actual, rather than projected, impact on student qualities.

However, relying solely on Bayesian updates is not sufficiently effective since it lacks a forecasting element. Given that we are operating in the domain of the smart economy, cognitive and mark-based mechanisms should be blended together. Tradable "futures" contracts on the share of program rewards can be introduced. These futures represent claims on a portion of the future reward, R, generated by a specific program's application. Upon approval, futures contracts are issued, with their initial price anchored to the expected action based on the current (that is, prior in Bayesian statistics) values of transformation matrix, M_{expected}. Critically, these futures are specific, pertaining to a particular educational program (**P**), a defined time period ($[t_1,t_2]$), and a specific student cohort or institution. Since the contract settles at the conclusion of this educational program. The actual reward, R, is computed using the *posterior* values of transformation matrix (reflecting the learned understanding of the program's impact). Holders of the futures contracts receive a payment proportional to their share of this actual reward. So, the profit or loss of the holder depends on the ability to predict the changes between prior and posterior values of the transformation matrix for specific educational program. This market mechanism serves a crucial function: it provides a real-time, aggregated assessment of the programs expected future value, incorporating all available information. This is a form of *market-based truth discovery*, incentivizing participants to gather information, conduct analyses, and make informed predictions about program effectiveness. The market price of the futures, therefore, acts as an independent, forward-looking indicator of the program's potential. Furthermore, the trading activity and price fluctuations themselves constitute valuable data. Discrepancies between the market price and the posterior estimates derived from observed data can signal either undiscovered information about the program (which the market is pricing in) or market inefficiencies and biases. These discrepancies, therefore, provide valuable signals for further investigation and may indicate potential manipulation of assessment of the results of educational process. The interplay between the market mechanism and the statistical learning process creates a system for aligning incentives, promoting transparency, and continuously improving the assessment and reward of educational programs within the smart economy.

Thus, in this section, we have progressed from the basic representation of students and programs as vectors through educational dynamics, and concluded with market-based mechanisms for program selection and reward. In the next section, we will approach the problem from the opposite direction, starting with a more abstract formalism for the smart economy and working our way down to educational aspects.

FROM SMART ECONOMIC DYNAMICS TO EDUCATIONAL VALUATION: A DOWNWARD PERSPECTIVE

This section adopts a top-down perspective, drawing inspiration from field-theoretic concepts to construct a general framework for understanding the dynamics of smart economy and its implications for educational assessment. We begin with an abstract formulation, progressively connecting it to the concrete elements introduced in previous sections.

Let us envision a system comprising a large number of *agents*. These agents represent individuals (students, employees, etc.), institutions, or organizations – any entity participating in the knowledge-mediated relationships that define the smart economy. Each agent's state is characterized by a vector of *qualities*, denoted by $\mathbf{Q}(t)$. This vector encompasses a broad spectrum of attributes, including skills, knowledge, resources, and even material possessions. Some components of $\mathbf{Q}$ can take on both positive and negative values, representing supply (a surplus or capacity) and demand (a deficit or need), respectively.

The interactions between agents are mediated by a *knowledge field,* denoted by $A(\mu)$. This field encompasses all forms of knowledge, information, and technology that influence how agents transform their qualities and interact with each other. It includes educational programs, technological tools, communication networks, and cultural norms. The knowledge field acts as a kind of generalized "force" or "potential," shaping the evolution of agent qualities. We draw an analogy here to a gauge field in physics, which mediates interactions between particles. In our context, the knowledge field mediates interactions between agents, influencing their transformations in quality space.

The dynamics of this system are described by a Lagrangian, $\mathscr{L}$, which is a functional of the agent quality vectors, $\mathbf{Q}(t)$, the knowledge field, $A(\mu)$, and their derivatives. We postulate a general form for the action, S:

$$S = \int dt \mathscr{L}[\mathbf{Q}, A, \ldots]. \tag{18}$$

The integral is taken over the relevant time period. The ellipsis (...) indicates that the Lagrangian may depend on first- and higher-order derivatives of $\mathbf{Q}$ and A, as well as potentially other auxiliary fields or constraint terms. These terms are crucial for ensuring the system's consistency and reflecting the full complexity of interactions within the smart economy. The Lagrangian in Eq. 18 is intended to encompass the maximum number of degrees of freedom that affect agents in any aspect relevant to our model. This includes all agents, their interactions, and the full range of factors influencing their evolution in long-term perspective.

A crucial component of the Lagrangian is the interaction term, which describes how the knowledge field mediates transformations of agent qualities. We can represent a general form of this interaction as:

$$\mathscr{L}_{\text{int}} = g \int dt \mathbf{J}(\mathbf{Q}) \cdot A, \tag{19}$$

where g is a coupling constant, representing the strength of the interaction between the agents' qualities and the knowledge field. $\mathbf{J}(\mathbf{Q})$ is a "current" that depends on the agent qualities, $\mathbf{Q}$, and possibly their derivatives. This current represents the way in which agents' qualities "source" or respond to the knowledge field. The specific form of $\mathbf{J}$ will depend on the details of the interaction. The dot product (•) indicates a contraction over appropriate indices, ensuring that the interaction term is a scalar.

The principle of least action dictates that the system evolves to extremize the action, S. This principle leads to equations of motion for the agent qualities and the knowledge field, governing their co-evolution.

To find the optimal trajectories of agents over long time perspectives, we employ the concept of a *functional integral*. This approach allows us to consider all possible paths of evolution, weighted by the exponential of the action. The functional integral is written as:

$$Z = \int \mathscr{D}\mathbf{Q}\mathscr{D}A \exp(-S[\mathbf{Q}, A]), \tag{20}$$

where $\mathscr{D}\mathbf{Q}$ and $\mathscr{D}A$ represent integration over all possible configurations of the agent quality vectors and the knowledge field, respectively; $S[\mathbf{Q}, A]$ is the action, as defined in Eq. 18.

Our goal is to identify the paths (trajectories of agent qualities and the knowledge field) that minimize the action over a sufficiently long time period to ensure the *sustainable development* of the system. This sustainability is defined as the full reproduction of the system, taking into account all relevant constraints. The boundary conditions of the functional integral are chosen to reflect the reproduction of the variety of agents' possibilities, ensuring a diverse and adaptable population. By solving the functional integral (or, more practically, finding its saddle points), we identify the most probable paths of evolution, representing the optimal transformations of agent qualities under the influence of the knowledge field.

Since interactions occur between specific qualities rather than complete quality vectors (for example, there occurs transformation of one qualities into another ones, interaction between demand and supply of the same quality) and the dynamics of these interactions can be different depending the type of the interacting quality, it could be appropriate to quantize the knowledge and qualities fields. For example, positive and negative values in the quality vector, **Q** can be interpreted as "quasiparticles" of supply and demand, respectively. The displacement of quality, whether towards supply or demand, can therefore be described in terms of the creation and annihilation of these quasiparticles. The knowledge field, A, mediates interactions that can be interpreted as the creation, annihilation, or transformation of these quasiparticles. A positive component of **Q** (supply) can be "annihilated" by interacting with a negative component (demand), representing a successful exchange or transaction. Conversely, the knowledge field can facilitate the "creation" of new supply or demand quasiparticles, reflecting the development of new skills or the emergence of new needs.

The exact form of annihilation and creation operators is not crucial here, but we retain the concept that ω_k represents an indicator of value, or "energy," for these quasiparticles. Different types of quasiparticles (corresponding to different components of the quality vector, **Q**) will have different values of ω_k. The relative values of ω_k for different quasiparticle types can then be used to establish "exchange rates" or relative valuations between different qualities. To ensure a degree of systemic constraint and to further ground the demand-supply dynamics, we posit that each agent initially possesses a reservoir of non-directed qualities, distinct from explicit supply or demand. These can be conceptualized as fundamental capabilities or resources, represented mathematically as particles without a direct antiparticle counterpart, or alternatively as non-signed components within the quality vector. Following a principle of conservation, agents can manifest demand by converting a portion of these non-directed qualities into demand entities (antiparticles). This conversion process introduces a necessary limitation on the total demand an agent can express, rooted in their finite initial endowment of non-directed qualities. The economic value or cost associated with supply entities (particles) is then intrinsically linked to the characteristics of their corresponding demand entities. Specifically, the system seeks to establish a balance, where the aggregate value attributed to demand quasiparticles is reflected in the collective valuation of supply quasiparticles of the same type, effectively forming an exchange rate between the value of demand and the quantity of supply.

The "annihilation" of a demand quasiparticle and a supply quasiparticle corresponds to a successful matching of needs and resources in the smart economy. The "creation" of quasiparticles corresponds to the development of new skills or the identification of new demands. The knowledge field, through its interactions, drives these processes, guiding the system towards a dynamic equilibrium.

To develop a practical framework for analyzing agent interactions within the smart economy, and specifically for assessing educational programs, we focus on an effective "Hamiltonian," H_{eff}, which describes the interaction of a specific agent with a subset of the knowledge field. This effective Hamiltonian is conceptually derived from the full Lagrangian by "integrating out" or averaging over the vast majority of degrees of freedom, retaining only those variables that are directly relevant to the specific agent and the interactions under specific case.

Let us consider the following form for the effective Hamiltonian:

$$H_{\text{eff}} = B_a(t) + \Phi(\mathbf{Q}) + \gamma\Gamma(t). \tag{21}$$

Let us examine each term in detail. Balance term ($B_a(t)$) represents the agent's instantaneous balance, $B_a(t)$, which is the running difference between the agent's supplied and demanded quasiparticles – a measure of the agent's net contribution to the system at time t.

Potential term ($\Phi(\mathbf{Q})$) is crucial for connecting the short-term dynamics of the individual agent to the long-term optimal trajectory of the entire system. It is a potential that depends on the agent's current qualities, $\mathbf{Q}$. This potential is derived from the long-term analysis – specifically, from the solution of the functional integral (Eq. 20) subject to the boundary conditions that ensure sustainability. $\Phi(\mathbf{Q})$ acts as a "guiding force," encouraging the agent's qualities to evolve in a direction that aligns with the long-term optimal path. It effectively encodes a "credit limit" or a set of constraints that the agent must adhere to in order to remain consistent with the overall system's goals. The balance term and the potential term have the same unit measures.

The potential $\Phi(\mathbf{Q})$ is obtained as follows:

(1) We consider the agent's quality vector, $\mathbf{Q}$, which represents a specific point in the multi-dimensional space of all possible qualities.
(2) We construct a quadratic form:

$$\Phi(\mathbf{Q}) = \mathbf{Q}^T M \mathbf{Q}, \tag{22}$$

where M is a symmetric matrix. This matrix M is derived from the solution of the long-term functional integral. It is chosen such that $\Phi(\mathbf{Q})$ reflects the maximum allowable negative balance (the "credit limit") for an agent with quality vector $\mathbf{Q}$. In other words, M and $\Phi(\mathbf{Q})$ are constructed to ensure that the agent's trajectory, driven by the short-term Hamiltonian, remains close to the optimal path determined by the full, long-term analysis. The specific entries of M will depend on the details of the long-term solution and the constraints imposed on the system. The vector $\mathbf{Q}$ in this context can represent a selection of an agent A qualities, denoted as $\mathbf{Q}(A)$.

The reward/penalty term ($\gamma\Gamma(t)$) represents an accumulated reward (if positive) or penalty (if negative) based on the agent's past contributions to the system's overall balance. The quantity $\Gamma(t)$ evolves according to:

$$\frac{d\Gamma}{dt} = \kappa B_a(t), \tag{23}$$

where κ is a scaling parameter. $\Gamma(t)$ integrates the agent's balance over time. A consistently positive balance leads to an increasing $\Gamma(t)$ (accumulated reward), while a consistently negative balance leads to a decreasing $\Gamma(t)$ (accumulated penalty). The parameter γ (or potentially κ) determines the weight of this accumulated reward/penalty for the positive/negative balance in the overall Hamiltonian.

To make the effective Hamiltonian constant (implicitly time-independent), we can redefine the potential term to incorporate the running balance and the accumulated reward/penalty.

$$\Phi(\mathbf{Q}) = \mathbf{Q}^T M\mathbf{Q} - B_a(t) - \gamma\Gamma(t). \tag{24}$$

This represents a kind of "available credit," that is, the difference between the agent's potential "credit limit" ($\mathbf{Q}^T M\mathbf{Q}$) and their current balance and accumulated reward/penalty. We require that the available credit be non-negative:

$$\mathbf{Q}^T M\mathbf{Q} - B_a(t) - \gamma\Gamma(t) \geq 0. \tag{25}$$

This constraint ensures that the agent does not exceed their allowed imbalance.

In other words, all changes in $B_a(t)$ and $\Gamma(t)$ are implicitly accounted for within the dynamics governed by $\Phi(\mathbf{Q})$ and the constraint, keeping the total effective Hamiltonian constant.

It is essential to note that while the functional integral (Eq. 20) is solved once for the entire system, the effective Hamiltonian (Eq. 21) applies to individual agents for specific cases, allowing for a tractable analysis of short-term dynamics. We will have a set of effective Hamiltonian equations, one for each agent under specific conditions.

Finally, we emphasize that the global constraint on the sum of all agents' current balances,

$$\sum_a B_a(t) = 0, \tag{26}$$

reflects the principle that within the closed system of the smart economy, the net exchange of contributions (the value of all provided and received "quasiparticles") of all agents must balance.

The abstract framework developed above can now be connected back to the concrete problem of assessing educational programs. The central idea is to link the *value* of an educational program to its impact on the agent's *potential*, which, in turn, is tied to the long-term dynamics and constraints of the smart economy.

Let **Q** represent the agent's quality vector before undergoing an educational program, **P**, and **Q**′ represent the quality vector after the program. The change in the agent's qualities is given by $\Delta\mathbf{Q} = \mathbf{Q}' - \mathbf{Q}$, which, in the deterministic case, is $\Delta\mathbf{Q} = M(\mathbf{P})\mathbf{Q}$, or is determined by the stochastic evolution (Eq. 9) in the more general, stochastic case.

We now define the value of the educational program, **P**, for a specific agent. We denote this program value as $V_{\text{program}}(\mathbf{P},\mathbf{Q})$. A natural definition for the program's value is the change in the agent's "credit limit" potential minus the cost of the program:

$$V_{\text{program}}(\mathbf{P},\mathbf{Q}) = \Phi_{\text{limit}}(\mathbf{Q}') - \Phi_{\text{limit}}(\mathbf{Q}) - C(\mathbf{P}). \tag{27}$$

Or, using explicit formula:

$$V_{\text{program}}(\mathbf{P},\mathbf{Q}) = \mathbf{Q}'^{T} M \mathbf{Q}' - \mathbf{Q}^{T} M \mathbf{Q} - C(\mathbf{P}). \tag{28}$$

where $C(\mathbf{P})$ is the cost of the program, as defined earlier (Eq. 5).

A positive program value, $V_{\text{program}}>0$, indicates that the program has moved the agent's qualities to a more favorable state with respect to the long-term constraints, and that this improvement outweighs the program's cost. A negative value indicates that the program either worsened the agent's standing with respect to the long-term constraints or that the improvement in qualities was not sufficient to justify the program's cost.

For a group of agents undergoing the same program, the average program value is:

$$\bar{V}_{\text{program}}(\mathbf{P}) = \frac{1}{N}\sum_{a=1}^{N}\left[\Phi_{\text{limit}}(\mathbf{Q}'_{a}) - \Phi_{\text{limit}}(\mathbf{Q}_{a}) - C(\mathbf{P})\right], \tag{29}$$

where N is the number of agents in the group. This provides a quantifiable measure of the program's overall effectiveness, taking into account both its impact on individual agents and its cost. Programs that consistently lead to positive average program values are considered valuable within the framework.

This definition of program value directly links the short-term impact of an educational program to the long-term dynamics of the smart economy, as captured by the potential $\Phi_{limit}(\mathbf{Q})$. It provides a concrete, quantifiable measure that can be used to assess and compare different programs, incentivizing the design and implementation of programs that are both cost-effective and aligned with the long-term needs of the system. The matrix M, derived from the long-term functional integral, plays a crucial role in determining this value, ensuring that the assessment is not based on arbitrary or easily manipulated metrics, but on a deep connection to the underlying dynamics of the system.

This field-theoretic approach to educational assessment aims to transcend traditional evaluation paradigms. Beyond simply offering a quantitative tool for assessment, the proposed framework is grounded in an ongoing reconceptualization of education's role in society: not as an isolated process of knowledge acquisition, but as an integral component in the complex, evolving network of relationships that constitutes the smart economy. This perspective invites educational stakeholders to view program design and assessment through a wider lens, one that acknowledges education's transformative power in shaping not just individual capabilities but the collective trajectory of social and economic systems.

DISCUSSION

The mathematical framework presented in this chapter represents a preliminary step towards a more holistic and quantitative approach to assessing educational programs within the context of a smart economy. The equations presented here are, in many cases, illustrative rather than definitive, serving to convey the core ideas rather than providing a fully operational model. It is crucial to acknowledge the limitations and many avenues for future refinement and development.

One significant area for future work lies in the development of more rigorous mathematical formulations. The equations presented in this chapter, while capturing the essential relationships between student qualities, program attributes, and system dynamics, were often formulated in a relatively loose manner. For instance, the precise form of the stochastic term of the transformation matrix, the details of the Bayesian updating process for the transformation matrix, and the functional integral over long-term trajectories all require more careful mathematical specification. This would involve, among other things, a more precise definition of the function spaces involved, a rigorous treatment of the stochastic elements, and a deeper exploration of the conditions under which the various approximations (e.g., the linear approximation for the transformation matrix) are valid. The connection to existing mathematical results in optimal control, stochastic processes, and statistical inference should be made more explicit.

Another limitation of the current framework is the lack of explicit distinction between individual agents and composite entities such as organizations or institutions. While the formalism treats all agents as abstract entities characterized by quality vectors, a more realistic model might need to incorporate hierarchical structures, internal dynamics within organizations, and the different roles and responsibilities of individuals and collectives. The challenge would be to do this in a way that maintains the tractability of the model while capturing the essential features of these more complex relationships.

A further, and perhaps harder to overcome, limitation lies in the treatment of demand as an external, fixed target. In the current framework, the demand for specific qualities is represented by a "potential" or "value function" that guides the evolution of student qualities. Educational programs are then seen as means to achieve this pre-defined target. However, in reality, the relationship between means and ends is often more reciprocal. Education not only empowers individuals with the skills and knowledge to achieve their goals; it also shapes those goals themselves. The very process of learning and acquiring new qualities can transform an individual's desires, aspirations, and understanding of what is possible. The current framework, with its emphasis on a fixed target, does not fully capture this dynamic interplay between means and ends. A more sophisticated model might need to incorporate a mechanism by which the "demand landscape" itself evolves in response to the changing qualities of the agents.

In the present framework, we introduced a stochastic term into the transformation matrices to capture the variability in outcomes produced by identical educational programs applied to students with the same initial quality vector. However, this approach may oversimplify the complexity inherent in educational processes. For instance, a particularly challenging program might, in some cases, motivate students while, in others, it might have a demotivating effect. Consequently, the range of outcomes may not be adequately represented by simple fluctuations around a mean value; rather, the program's impact could manifest as a set of qualitatively distinct possibilities – a notion we could refer to as "superpositional probabilism." It is an intriguing prospect to explore whether the framework might be extended or shifted from this model of stochastic probabilism toward one of "probabilistic determinism," in which multiple, discrete outcomes are deterministically bounded within a probabilistic structure. We acknowledge the speculative nature of these preliminary thoughts, and further investigation is necessary to assess their viability and implications for the overall model.

A further, subtle challenge for our proposed framework lies in the nature of what is *conserved* within the system. Traditional economic models, even those incorporating multiple goods and services, tend to operate on a principle of *exchangeable conservation*. Diverse qualities are reduced to a common, scalar metric (often monetary value), allowing for trade-offs and comparisons based on a single dimension. This approach, analogous to an L1 norm, virtually collapses the inherent multidimensionality of the system. Consequently, these models struggle to adequately represent phenomena like education, which are fundamentally about *transformation*, not just exchange. Education alters the *relationships* between an individual's qualities, not merely their aggregate quantity. Our framework attempted to address this by employing quadratic forms and transformation matrices, aiming to capture the interactions *between* different qualities. The underlying, though as yet unfulfilled, goal was to move towards a principle of *relational-structural conservation*. This concept, inspired by the distinction between energy (monopole moment) and momentum (dipole moment) conservation in physics, emphasizes the preservation not just of a total quantity of value, but also of the *structure* of relationships between different dimensions of value. While our use of quadratic forms represents a move in this direction, the framework, as it stands, lacks rigorous conservation principles that would demonstrably maintain this relational-structural conservation, a feature we hypothesize as crucial for a robust mathematical framework for the economy of smart education[4].

Another critical consideration is the potential for contradictions between the interests and demands of agents from a long-term perspective. Two distinct types of contradictions can arise: (a) intra-agent contradictions, where a single agent's long-term goals conflict with their short-term needs or desires, and (b) inter-agent contradictions, where the long-term goals of different agents are incompatible. The current framework, while emphasizing long-term sustainability, does not fully address these potential conflicts. Since the basis upon which the qualities are defined is not necessarily orthonormal, future work will need to explore methods for resolving these contradictions. This could involve developing a metric matrix that accounts for the non-orthogonal nature of the quality space, enabling a more nuanced assessment of trade-offs and compromises between different objectives.

Furthermore, the framework, in its quantitative emphasis, primarily focuses on qualities that are readily measurable and economically relevant. This raises a critical question regarding the inclusion and assessment of moral, ethical, and humanitarian qualities. Some of these qualities may be inherently difficult, if not impossible, to quantify objectively. Moreover, attempts to assess certain moral attributes could be ethically problematic, potentially intruding on personal autonomy and fostering a culture of surveillance. Future development should explicitly address the challenge of considering these less tangible yet crucial qualities, acknowledging both the limitations of the proposed methodology and the possibility that some aspects of human life should remain unfathomable and beyond instrumental valuation.

Finally, the parallels between the proposed framework and certain concepts in machine learning suggest a potentially fruitful avenue for future research. Specifically, the idea of using path-dependent functional integrals to optimize long-term trajectories could have connections to the training of artificial intelligence (AI) agents. For example, in "pure" transformer-based chatbots, semi-supervised learning is widely used. For AI reasoners, reinforcement learning is becoming increasingly important. It is conceivable that for AI agents that are capable of not just answering questions or performing tasks, but of producing and transforming things in the real world, a formalism based on path integrals, with appropriate boundary conditions and constraints, could provide a powerful optimization principle. This suggests a task of elaborating machine learning algorithms based on the core ideas presented here. Such algorithms might leverage the concepts of stochastic transformations, Bayesian updating, and ensemble averaging to create AI agents that are not only efficient in achieving specific goals but also adaptable and capable of contributing to a dynamic, knowledge-driven economy.

It is also vital to emphasize that education is fundamentally a *process of self-cultivation*, not solely a means to achieve high scores on an assessment system. While the proposed framework aims to provide a more comprehensive and externally aligned measure of educational quality, it is crucial to recognize the *limited significance of any assessment method.* Students should always prioritize their intrinsic motivation for learning and personal growth, understanding that all assessment systems, including this one, are inherently imperfect and subject to potential misuse. The intention behind this framework is not to further "economize" human beings, but rather to offer a system of assessment that is *intrinsically resistant to becoming an end in itself.* If successfully implemented, it could potentially *replace* some existing, more reductive forms of quantification in education, rather than adding to the existing burden of standardized testing and performance metrics. The aim is to shift the focus from narrow, easily gamed measures to a more holistic and dynamically adaptable assessment of long-term value creation.

CONCLUSION

The underlying motivation of this work is to contribute to a more holistic understanding of education, one that recognizes the epistemological, transformative, and economic dimensions of learning as fundamentally intertwined. Knowledge is not simply a static body of facts to be transmitted; it is a dynamic force that shapes individuals, mediates relationships, and drives economic activity. Education, therefore, is not merely a process of acquiring information but a process of becoming – of transforming one's capabilities, adapting to changing demands, and contributing to the collective knowledge base. The Smart Economy, in this view, is not just an economy driven by knowledge; it is an economy constituted by the ongoing, knowledge-mediated transformations of its participants.

A core premise of this framework is that the perceived dichotomy between economic and epistemological domains, on the one hand, and between market exchange and transformative processes, on the other hand, is not fundamental, but rather a consequence of the absence of a unifying mathematical apparatus. We hypothesize, albeit tentatively, that a formal calculus can be developed to achieve a convergence of these seemingly disparate spheres. The "market-based truth discovery" we alluded to earlier is not merely a metaphor; it reflects a belief that market dynamics, properly structured, can be integrated with the process of knowledge acquisition and validation. The development of such a calculus represents a step towards a more holistic and integrated understanding of the smart economy, one where education, cognition, valuation, exchange, and transformation are viewed as interwoven aspects of a single, dynamic system.

It is crucial to acknowledge the significant gap between theoretical abstraction and practical implementation. The ideas presented here are intended to paint a broad picture of core principles, rather than to offer an implementable blueprint. Translating this framework into a real-world assessment system would require addressing numerous practical challenges and would likely involve a blending of different approaches, incorporating elements of our framework alongside existing assessment tools and methodologies. Indeed, given the complexity of educational systems and the inherent uncertainties involved, we ourselves cannot claim full confidence in the intrinsic correctness and viability of the proposed framework. In sharing these ideas, our hope is not to present a definitive solution but to invite dialogue, critique, and collaborative refinement.

REFERENCES

Adiyarta, K., Napitupulu, D., Syafrullah, M., Mahdiana, D., & Rusdah, R. (2020). Analysis of smart city indicators based on prisma: Systematic review. [IOP Publishing.]. *IOP Conference Series. Materials Science and Engineering*, *725*(1), 012113. DOI: 10.1088/1757-899X/725/1/012113

Arrow, K. J., Forsythe, R., Gorham, M., Hahn, R., Hanson, R., Ledyard, J. O., Levmore, S., Litan, R., Milgrom, P., Nelson, F. D., Neumann, G. R., Ottaviani, M., Schelling, T. C., Shiller, R. J., Smith, V. L., Snowberg, E., Sunstein, C. R., Tetlock, P. C., Tetlock, P. E., & Zitzewitz, E. (2008). The promise of prediction markets. *Science*, *320*(5878), 877–878. DOI: 10.1126/science.1157679 PMID: 18487176

Bajaj, R., & Sharma, V. (2018). Smart education with artificial intelligence based determination of learning styles. *Procedia Computer Science*, *132*, 834–842. DOI: 10.1016/j.procs.2018.05.095

Becker, G. S. (1964). *Human capital: A theoretical and empirical analysis, with special reference to education*. University of Chicago Press.

Benkler, Y. (2006). *The wealth of networks: How social production transforms markets and freedom*. Yale University Press.

Boud, D., & Falchikov, N. (2006). Aligning assessment with long-term learning. *Assessment & Evaluation in Higher Education*, *31*(4), 399–413. DOI: 10.1080/02602930600679050

Caley, L., Williams, S. J., Spernaes, I., Thomas, D., Behrens, D., & Willson, A. (2021). Frameworks for evaluating education programmes and work related learning: A scoping review. *Journal of Workplace Learning*, *33*(6), 486–501. DOI: 10.1108/JWL-09-2020-0157

Campbell, D. T. (1979). Assessing the impact of planned social change. *Evaluation and Program Planning*, *2*(1), 67–90. DOI: 10.1016/0149-7189(79)90048-X

Castells, M. (2010). *The rise of the network society* (2nd ed.). Wiley-Blackwell.

Cranton, P. (2016). *Understanding and promoting transformative learning: A guide to theory and practice* (3rd ed.). Routledge.

Demir, K. A. (2021). Smart education framework. *Smart Learning Environments*, *8*(1), 29. DOI: 10.1186/s40561-021-00170-x

Díaz-Parra, O., Fuentes-Penna, A., Barrera-Cámara, R. A., Trejo-Macotela, F. R., Ramos-Fernández, J. C., Ruiz-Vanoye, J. A., Zezzatti, A. O., & Rodríguez-Flores, J. (2022). Smart Education and future trends. *International Journal of Combinatorial Optimization Problems and Informatics, 13*(1), 65.

Dulhunty, A. (2022). The microcredit continuum: From 'smart economics' to holistic programming. *Journal of International Development, 34*(6), 1213–1226. DOI: 10.1002/jid.3625

Fawns, T. (2022). An entangled pedagogy: Looking beyond the pedagogy – technology dichotomy. *Postdigital Science and Education, 4*(3), 711–728. DOI: 10.1007/s42438-022-00302-7 PMID: 40477872

Gipps, C. (1994). *Beyond Testing: Towards a Theory of Educational Assessment* (1st ed.). Routledge.

Hwang, G. J. (2014). Definition, framework and research issues of smart learning environments – a context-aware ubiquitous learning perspective. *Smart Learning Environments, 1*(1), 4. DOI: 10.1186/s40561-014-0004-5

Indrawati, A., N., & Amani, H. (2018). Indicators to measure smart economy: An Indonesian perspective. In *ICBIM'18: Proceedings of the 2nd International Conference on Business and Information Management* (pp. 173–179). Association for Computing Machinery. DOI: 10.1145/3278252.3278278

Koretz, D. M. (2008). *Measuring up: What educational testing really tells us*. Harvard University Press. DOI: 10.4159/9780674039728

Kostyk, Y., Tiuleniev, S., Goi, V., Kovalenko, O., & Pochernina, N. (2023). The National Model of the Smart Economy for Achieving the Goals of Innovative Development. *Revue d'Economie Financiere, 21*, 622–632.

Lauder, H., Young, M., Daniels, H., Balarin, M., & Lowe, J. (Eds.). (2012). *Educating for the knowledge economy? Critical perspectives*. Routledge. DOI: 10.4324/9780203817698

Lundvall, B. Å., & Johnson, B. (1994). The learning economy. *Journal of Industry Studies, 1*(2), 23–42. DOI: 10.1080/13662719400000002

Martín, A. C., Alario-Hoyos, C., & Kloos, C. D. (2019). Smart Education: A Review and Future Research Directions. *Proceedings, 31*(1), 57.

Mezirow, J. (1991). *Transformative dimensions of adult learning*. Jossey-Bass.

Mezirow, J. (Ed.). (2000). *Learning as transformation: Critical perspectives on a theory in progress*. Jossey-Bass.

Popova, Y., & Popovs, S. (2022). Impact of smart economy on smart areas and mediation effect of national economy. *Sustainability (Basel), 14*(5), 2789. DOI: 10.3390/su14052789

Powell, W. W., & Snellman, K. (2004). The knowledge economy. *Annual Review of Sociology, 30*(1), 199–220. DOI: 10.1146/annurev.soc.29.010202.100037

Ramaswamy, V., & Ozcan, K. (2014). *The co-creation paradigm*. Stanford University Press. DOI: 10.1515/9780804790758

Raudenbush, S. W. (2004). What are value-added models estimating and what does this imply for statistical practice? *Journal of Educational and Behavioral Statistics, 29*(1), 121–129. DOI: 10.3102/10769986029001121

Schön, D. A. (1983). *The reflective practitioner: How professionals think in action*. Basic Books.

Schön, D. A. (1987). *Educating the reflective practitioner: Toward a new design for teaching and learning in the professions*. Jossey-Bass.

Scriven, M. (1967). The Methodology of Evaluation. In Tyler, R. W., Gagne, R. M., & Scriven, M. (Eds.), *Perspectives of Curriculum Evaluation Chicago* (pp. 39–83). Rand McNally.

Siburian, J., & Mardiyanti, L. (2023). Prediction of Meta-Skills Based on Metcognition. *Jurnal Penelitian Pendidikan IPA, 9*(12), 11053–11059. DOI: 10.29303/jppipa.v9i12.5683

Strathern, M. (1997). 'Improving ratings': Audit in the British University system. *European Review (Chichester, England), 5*(3), 305–321. DOI: 10.1002/(SICI)1234-981X(199707)5:3<305::AID-EURO184>3.0.CO;2-4

Taylor, E. W. (2007). An update of transformative learning theory: A critical review of the empirical research (1999–2005). *International Journal of Lifelong Education, 26*(2), 173–191. DOI: 10.1080/02601370701219475

Taylor, E. W., & Cranton, P. (2012). *The handbook of transformative learning: Theory, research, and practice*. John Wiley & Sons.

Voorhees, R. A. (2001). Competency-based learning models: A necessary future. *New Directions for Institutional Research, 2001*(110), 5–13. DOI: 10.1002/ir.7

Wolfers, J., & Zitzewitz, E. (2004). Prediction markets. *The Journal of Economic Perspectives, 18*(2), 107–126. DOI: 10.1257/0895330041371321

Zhu, Z. T., & He, B. (2012). Smart education: New frontier of educational informatization. *E-Education Research*, *12*, 1–13.

Zhu, Z. T., Yu, M. H., & Riezebos, P. (2016). A research framework of smart education. *Smart learning environments*, 3, 1–17.

ENDNOTES

1 Goodhart's law states, "When a measure becomes a target, it ceases to be a good measure" (Strathern, 1997, p. 308). Campbell's law: "The more any quantitative social indicator is used for social decision-making, the more subject it will be to corruption pressures and the more apt it will be to distort and corrupt the social processes it is intended to monitor" (Campbell, 1979, p. 85).

2 In future iterations of this framework, we may consider replacing the stochastic process representation with explicit probability distributions to more comprehensively capture the range of possible educational outcomes.

3 In analogy to statistical mechanics, the introduction of a new educational program, akin to a new energy level, maximizes its contribution to the ensemble's entropy when its action value lies within the range of existing educational programs' actions. As the action approaches infinity, the entropic contribution of the new program asymptotically approaches zero. Conversely, when the action of the newly introduced program is significantly lower than that of existing programs, it results in a negative entropy change.

4 It is important to note that this notion of relational-structural conservation remains, at this stage of development, a speculative assumption, and further rigorous investigation is required to fully validate and reject its applicability within our proposed framework. However, even if the concept of relational-structural conservation, as presented here, proves ultimately untenable or inconsistent, the underlying challenge persists: to identify and implement formal approaches that effectively safeguard the inherent multi-dimensionality of qualities from reductive scalar valuation.

Chapter 9
Exploring Factors Influencing Digital Inclusion Within Modern Pedagogical Practices Prevalent Amongst Schoolteachers:
Smart Students, Smarter Teachers?

Rabia Abhay
https://orcid.org/0000-0002-2888-6523
Christ University, India

Sruthi Suresh
https://orcid.org/0000-0002-2505-9872
Christ University, India

ABSTRACT

Teachers serve as torchbearers to the vast expanse of knowledge, leaving a lasting impact on the minds of their students through their teaching styles and pedagogical practices. With the advancement of technology, the integration of digital technologies has increasingly been observed in pedagogical approaches. The current study seeks to explore various subjective factors related to teachers, such as their teaching styles, age, classes taught, educational qualifications, and years of experience, in relation to their individual digital competencies and the extent of digital inclusivity

DOI: 10.4018/979-8-3373-7922-7.ch009

in their pedagogies. Three psychometric scales were employed for data collection (TSI-Q, DiPeCoS, and TDiCoS). Data were gathered from 215 school teachers in the United Arab Emirates (UAE) and subsequently analysed. The results revealed a significant influence of teaching styles ($p = 0.045$), classes taught ($p = 0.001$), age ($p = 0.006$), and years of teaching experience ($p = 0.016$) on the digital inclusivity of the teachers' pedagogical practices.

INTRODUCTION

It is said that there is no profession more noble than teaching. The dissemination of the knowledge acquired by the elders through experience and research amongst the upcoming generations is a revered and honored profession around the world. The roots of knowledge sown by these 'teachers' into the minds of the younger generations form the foundation stone for all the developments and progress taking place in a region. It is the teacher's duty to ensure that the concepts taught by them have been effectively understood and learned by the students. To this end, teachers should adopt the necessary educational methods and strategies to increase the impact of their teachings on their students' inquisitive minds.

Pedagogy is the field that looks into the techniques and strategies used by teachers in their educational practices. It studies the various techniques and innovations brought in by the teachers to make their classes more impactful in disseminating the intended knowledge. The art of passing on accumulated knowledge by the older generation to the younger generation has taken many forms throughout different cultures throughout history.

The Past

Humans, in the early stages of evolution, lived a nomadic life of hunting and gathering. Devoid of any long-term settlement and in a state of constant movement, their reduced tendencies to study the nature around them hindered the passing down of such knowledge to the younger generation and what information was passed on was mainly confined to oral traditions.

upon the advent of agriculture and the change from a nomadic lifestyle into a more sedentary settled lifestyle at the early stages of the Holocene epoch (around 11000-12000 years ago) (Waters et al., 2016), humans started to give more attention to studying their surroundings and unravelling the mysteries around them. This gave rise to the initial quantification of the properties present within the abstract and material worlds (Eskelson, 2020). This gave humans the ability to codify this information and preserve it to be shared with the next generation. With the advent of civilization, fueled by population growth due to agricultural practices and the giving up of the nomadic lifestyle, more formalized institutions were set up to impart the accumulated knowledge to the uninitiated younger generation. This was the dawn of formalized education.

The oldest of the ancient civilizations, the Mesopotamian civilization, also had a rich history of teaching and education. It has been found that this civilization had schools. The early compilation and imparting of knowledge was done with the help of symbols that did not resemble the modern-day alpha-numeric method of communication. It has been found that Proto-cuneiform texts began to appear in Mesopotamian civilization at around 3400 B.C (Trigger, 1998; Englund, 2004). The advent of Scribal schools within the Mesopotamian civilization is another instance of formalized education in ancient civilizations. Within these schools, the children learned literacy, mathematics and bureaucratic skills necessary to contribute to society effectively. These scribal schools were called 'Edubba' (George, 2005). The subjects of literacy and mathematics were not just new and refined concepts developed by humans, but they also exerted great influence in transforming the nature of education as a whole.

As time passed, education became controlled by the elite section of the society. Its roots can be traced even back to Egyptian civilization. It has been found that the education system in Egyptian civilization was predominantly controlled by priests, who were powerful elites in Egyptian society (Szyliowicz et al., 2025). Egyptians had developed two types of formal schools for privileged youth under the supervision of governmental officials and priests, one for educating the scribes and the other for training priests. A student entered these schools at the age of five and underwent rigorous training in reading and writing until the age of 16-17 (Szyliowicz et al., 2025).

Ancient India, with its universities such as Takshashila, Nalanda, and Vikramashila, dating as back as the sixth century BC, was one of the beacons of education in the ancient world. These universities, for the first time, represented a global education system focused on imparting practical knowledge to students coming from all across the world. Subjects such as grammar, mathematics, astronomy and medicine were taught here (Ghonge et al., 2020). Vedic scriptures also talked about the Gurukul system. It was an in-situ educational system where children from wealthy and upper castes stayed with the family of the 'guru' for a prolonged period of time and learned the knowledge (usually about Vedas and ancillary topics) that was passed onto them by the guru usually in an oral manner (Dutt, 2017). It was a holistic education where the students partook in the everyday life of the Gurukul along with their educational routine.

The first instance of a pedagogic practice was seen in the time of the Greek thinker Socrates. He had come up with a teaching style, which later came to be known as the "Socratic method". This method did not work in increasing the knowledge of the individual, but it helped strengthen the logic behind the knowledge that already existed (Delic & Becirovic, 2016). This method involved the usage of open-ended questions that enabled an individual to explore a topic and develop critical thinking skills. This method still has practical relevance in the modern world.

The start of the new millennium (around 500 AD) saw the Dark Ages, where the religious views and outlooks of the Catholic church, along with myriad superstitions, ran rampant. Education in these times was limited and was strictly monitored by religious organizations, thus curtailing any improvements that could have been brought into the then-existing educational system (Spaulding, 2022). The Christian scholar Tertullianus is known for stating that all pagan education systems should be abhorred since he thought they were false roads that led to the misleading information of ancient philosophers (Compayre & Payne, 2015).

The situation changed with the advent of the 'Renaissance', which was a period starting from the early 15th century and lasting till the mid-19th century when passion for knowledge and educational reforms saw a re-emergence. It is said to have begun in Florence, Italy (Burke, 1998). It then spread widely to other European countries. This period saw the origin of the scientific method of exploring nature. The pedagogical structure existing within educational practices at that period inclined towards not just teaching but also focusing on expanding the knowledge of an individual. The ideology of humanism was one of the pillars around which the educational pedagogical practices of the Renaissance were moulded (Team, n.d.). The schools created by the humanists of this period strived to instil moral values and plant seeds of curiosity within the minds of the individuals through their teaching practices. This helped them to expand their knowledge as well as fuel intellectual development, making the teaching practices prevalent in the Renaissance the predecessors to modern-day pedagogical practices.

The Present

With the advent of the 21st century, monumental changes have been observed in various fields of knowledge, including pedagogy (Lavryk & Yaresko, 2015). If the past was dominated by the behaviourist school of thought, where teaching methods were made such that education just meant moulding students into socially confirmed roles without any regard to their individuality, the present is more or less dominated by the student-centered learning where the individuality of learners is catered to and the pedagogies are structured in such a manner that it aids in fostering lifelong learning and self-actualization of one's highest potential. In the current era where digital and technological progress is advancing at breakneck speeds, its inclusion in the educational setup is inevitable. In a highly interconnected world such as today, the knowledge that one seeks is available now at their fingertips. In such an environment, it is of importance to develop pedagogical structures that seamlessly attempts to integrate these technological advancements into the teaching practices such that it enables a more efficient and effective way of imparting knowledge in the classrooms.

The Digital Explosion

The explosion of the EdTech industry has invaded numerous learning spaces worldwide. Across the world, the teaching-learning experiences offered are incorporating digital tools to enhance and personalize learning. Some of the services available in this regard include online learning platforms and materials as well as learning support and enhancement tools. These tools promise enhanced accessibility, interactivity, availability, collaboration, and quality of teaching-learning experiences (Trang, 2024). The term ed-tech is said to have become popular around 2015 and boomed during the COVID-19 pandemic (West, 2023). Among the OECD countries, almost three-quarters of the households have access to at least one smartphone (Schleicher, 2024), whereas in the United States, 78% of the households own a computer or tablet, and 97% of individuals own a cellphone. However, there are still gaps in the availability and access to technology, especially in rural areas and among those who are less educated (Golinkoff et al., 2025).

For educators, the digital explosion has acted as a boon and a bane simultaneously. These digital tools help promote student engagement through the use of game-based learning strategies but also reduce student responsiveness when interacting through virtual meeting platforms. It encourages students to be attentive through simulations but makes them zone out when viewing presentations or videos. Additionally, familiarity with the digital world and digital competencies influenced teacher's ability to harness these digital tools in the classroom. Several factors, such as the size of the class, the teacher's digital competencies, availability of digital tools, student familiarity with the platforms, and much more, influence the adoption of these digital tools in the classroom. Keeping these points in mind, it is important to note that it is not the mere presence of digital tools but how they are used that contributes effectively to the teaching-learning environment.

The Middle East

The rich and vibrant history of teaching and learning within the Middle East from ancient times makes it an area of interest for this study. The seamless integration of technology into educational and teaching practices in the current era also makes it an ideal place for this study to be conducted. This region has a long-lasting legacy of contributing to education and scientific exploration. It was the Semitic-speaking people of ancient Egypt who, for the first time, transmuted the pictorial style of writing reflected in hieroglyphs into a different form of alphabets (Albright, 1998), which later on paved the way to the development of the first consonantal alphabets. This later on developed into the alphabet that we know today. Perhaps the most important contribution this region has given to the field of educational exploration is the invention of the Arabic numeric system (Dana, 2021).

In the current era, this region has made great strides in technological progress, especially in the United Arab Emirates (UAE). UAE has been termed as the "Most Technologically ready" Arab nation. On the global scale, UAE ranks 37th in the UN Technology Index. Even in the field of education, the UAE has made great strides. In 1971, at the time of its formation, adult literacy rates stood at 54%, which later became 98.29% in 2022 (MACROTRENDS, n.d.). This trend is particularly high compared to other financially advanced countries. For instance, the United States of America (USA), with a GDP of 30.33 trillion, which is approximately 70 times higher than the GDP of the UAE, has an average adult literacy rate of 79% (Ingridhaynesphd, 2024). This drastic jump in literacy rates could be attributed to the increased government expenditure in this field. As of 2022, the UAE spends around 16% of its total federal funding on education (*Federal Budget | the Official Portal of the UAE Government*, n.d.). Thus, the UAE, with its advances in technology and enhancements in the field of education, sits in a unique place to satisfy the demands of the current study to check the digital inclusion into teaching practices prevalent in the education system.

INSIGHTS FROM PAST LITERATURE

The concept that education is a process through which an individual is shaped based on unchanging variables independent of the time and place surrounding them is one of the core concepts that has been passed on to the modern zeitgeist revolving around teaching practices by the process of pedagogical thoughts spanning many centuries (Suchodolski, 1961).

Pedagogical Approaches

Various philosophers and thinkers have contributed to the evaluation of educational and pedagogical practices. As stated in the earlier section, the first documented instance of a particular pedagogical structure dates back to the time of the Greek philosopher Socrates, who developed his method of teaching, where individuals asked questions to examine their existing knowledge and enhance their critical thinking. This kind of teaching style usually comes under the 'dialogic teaching style' in which the learning takes place through dialogues where individuals put forward claims and arguments based on validity and logic to support their deductions (Kincheloe & Horn, 2007). His successor, Plato, also held education in high esteem. He had felt education as the road through which one's soul attained its fulfilment (Compayre & Payne, 2015). The Roman philosopher Marcus Fabius Guntillanius published his approach towards pedagogy in *Institutio Oratoria,* where he stated education was a gradual affair and placed certain responsibilities on the teacher. Saint Jerome, an early Christian priest, emphasized the need for experiential learning and the importance of positive reinforcement.

Herbartianism is usually referred to as the educational movement based on the ideologies proposed by John Friedrich Herbart, a German philosopher and psychologist. He, in his ideologies, points fingers towards the existing correlation between individual development and the resulting benefits to society. He was of the view that humans felt a sense of fulfilment once they felt that they had established themselves as productive citizens, aiding in the development of their communities. To enable an individual to attain this goal, he asserted that all teaching pedagogies must contain the five attributes of preparation, presentation, association, generalization and application (Kenkiles, 2012). John Dewey, an American psychologist and philosopher in his 1906 book, "The Child and the Curriculum", advocated for a pedagogical structure that focused on striking a balance between delivering knowledge while also catering and fostering the interests and experiences of the students. He felt that the purpose of teaching should not just pivot around gaining pre-determined skill sets but should also help in realizing one's full potential and ability and enable them to use the same for the greater good (Dewey et al., 1897). These thoughts and concepts have had a far reaching impact in shaping the modern day pedagogical structures and practices.

There are numerous approaches towards pedagogy and its underlying practices. The behaviourist pedagogical approach sees education as a tool to enforce a child on the track of training and knowledge. The individuality of a learner is not given much consideration under this approach. Individual learners were seen fit to be moulded by the conventional ideas or the interests of teachers and parents (Joshi, n.d.). This trend in educational practice has seen to change, thanks to pioneering works of education thinkers and philosophers such as Rousseau and Montessori. They claimed that educational practices should be moulded in such a manner where it must focus on bringing out the learners' own intellectual and moral capacities to the highest extent possible (Joshi, n.d.).

The evidence-based approach is based on the educational principle which states that educational practices should be based on the best and most recent available scientific evidence rather than hinging on the traditional, personal, and prejudiced approaches of the past. This approach revolves around evidence-based education, evidence-based learning and school effectiveness. This approach correctly encapsulates the modern-day tendency to revise the educational tools, practices and content being taught with the availability of the most recent scientific evidence. This helps in keeping the educational system up to date with the current knowledge frontier and enables students to contribute to expanding this frontier even further.

Perhaps one of the most well-advocated pedagogical structures in recent times would be the 'student-centered learning' approach which is also often called 'learner-centered education'. This approach entails taking up teaching practices by the educators that shift the focus from teachers to students. Such teaching practices aim to develop learner autonomy and independence by putting the responsibility of the learning path and the pace of learning into the hands of the students (Jones, 2007; Pederson & Liu, 2003; Hannafin & Hannafin, 2010). A strong focus on the inculcation of lifelong learning tendencies and independent problem-solving skills among students is another characteristic of student-centred learning (Young & Paterson, 2007). The majority of the pedagogical structures around the world also follow the above-said trend. The New Educational Policy (NEP) enacted in India, fosters the concept of lifelong learning. The policy highlights the necessity of kindling activities that foster lifelong learning at all educational levels. The policy's inculcation of adult education as one of its pillars shows the importance it gives to lifelong learning opportunities (Joshi, n.d.).

Modern-day pedagogy is founded on a plethora of paradigms such as humanistic, knowledge-oriented, culturally-inclined, societal and anthropocentric (Bordovskaya, 2000). All these paradigms are inevitably associated with appropriate approaches. The approach and method together provide the instrumentality of the paradigm, for example, the knowledge-based paradigm rests on the systemic nature of educational facilities which allows presenting the educational system as a set of sequentially arranged components (teacher, objectives, principles etc.) connected by backward and forward linkages. With the advancements in the field of technology, a trickle-down effect has been seen in the field of education. Teaching practices in recent times have witnessed an increased integration of digital technologies within it, giving rise to a new era of pedagogies.

Smart Pedagogy

Smart pedagogy refers to the use of technology in the teaching-learning environment. It uses tested innovative strategies including flipped classrooms, crowdsourcing, collaborative learning, and gamification (Uskov, 2022). It harnesses the services offered by a range of digital platforms beginning from the smartphone to the use of smartboards and simulators. These technologies are tailored to promote higher-order skills such as adaptivity, sensing, inferring, anticipation, self-learning and self-organization skills among the students (Uskov et al., 2018). The smart pedagogy framework has three parts – the SMART key elements model, the curriculum design method, and the detailed teaching strategy. The SMART key elements model includes situated learning (S), mastery learning (M), adaptive learning (A), reflective learning (R) and thinking tools (T) (Meng et al., 2020). The use of such technology-enhanced teaching and learning strategies can be articulated as both the art and science of teaching pedagogy. It can play a key role in enhancing access to education and better educational facilities. However, digital literacy is imperative to make the best use of these pedagogical tools (Daniela, 2021).

The thinking style of an educator is also of importance, as it dictates which methodology and techniques will be utilized by them in imparting knowledge to their students. Sternberg (1995) in his theory of mental self-government, states that the preference for certain mental operations and abilities is a personal variable which is termed as 'thinking style'. The central tenet of the theory is the need for individuals to control how everyday activities are carried out. These thinking styles are not permanent and are socialized, i.e., they can be modified over time depending on the needs that may arise (Zhang, 2001; Zhang & Sternberg, 2005, 2006). The educators also feel the need to adopt pedagogical strategies that they deem fit for imparting knowledge in an educational setup. Correlation between teaching approaches and thinking styles within teaching activity has been seen throughout various studies (Zhang, 2001).

Teachers who had global and liberal thinking styles were seen to structure their pedagogy to foster conceptual changes allowing students to attain intellectual autonomy whereas teachers who had local and conservative thinking styles were seen to adopt pedagogical practices that just focused on conveying information, without instilling a fundamental change in the student (Zhang, 2001) Thus the teaching styles adopted by teachers that reflects in their pedagogical structure is found to mirror their thinking styles. Four major teaching styles have been postulated based on thinking styles held by teachers (Palos & Maricutoiu, 2013). They are teaching styles that stimulate - (a) reproductive abilities (recognizing and reproducing information that is taught), (b) analytical abilities (analysing the information provided and assessing what is being learned), (c) creative abilities (inventing and exploring new ways to solve various problems), and (d) practical abilities (application of what is being taught in class into everyday activities).

Digital competency is a term that is used to denote an individual's ability to utilise digital resources in the day-to-day activities of their lives. Thus, the digital competencies of teacher play a pivotal role in helping them to use such digital resources to expedite their teaching protocols, evidently leading them to have a more digitally inclusive pedagogical structure. It has been found that teachers with enhanced digital competencies do a better job in assisting and encouraging their students to learn more efficiently and enable them to enjoy their school work more by using digital methods (Caena & Redecker, 2019). Thus the current study attempts to investigate the role of teacher sociodemographic and teaching style in influencing a teacher's digital competency and their tendencies for digital inclusion within their pedagogical structures.

METHOD

Research Problem

The present study aims to investigate the factors influencing the inclusion of digital technologies within the pedagogical practices used by teachers in the UAE. It examines the primary data collected on teachers such as teaching styles, age, experience, classes taught, subjects taught, and how it would seemingly influence their digital competencies and their overall tendencies regarding digital integration within their teaching practices. It is important to obtain insights into this booming area of EdTech to assess the existing patterns of how teachers use digital tools in their classrooms. Further, identifying the teaching styles that incorporate these digital tools in the classroom will help all stakeholders in the field of education tailor EdTech tools and train teachers accordingly. The results of the study will improve present understanding of teacher's digital and digital pedagogy competencies, thereby encouraging the development of appropriate training and professional development programs. Finally, schools can benefit from the findings of this study by recognizing the grades and teachers' who are most likely to benefit and effectively use digital technologies. Hence, the present study will strive to build an initial empirical understanding of existing patterns of digital technology usage by teachers in classrooms to guide the future development and customization of these tools.

Objectives

- To examine how subjective factors pertaining to teachers such as their teaching styles, experience, age, classes taught, and educational qualification influence their individual digital competencies.
- To examine how subjective factors pertaining to teachers such as their teaching styles, experience, age, classes taught, and educational qualification influence digital inclusion within their pedagogies.

Research Design

A survey is usually the best method used to obtain a quantitative description of opinions, attitudes, and trends or to test for associations among variables in a population. Such a design helps answer descriptive, relationship, and predictive relationship-related questions about the chosen variables (Creswell & Creswell, 2017). A quantitative survey method was used to collect data on teaching styles, digital competency and digital pedagogical competency through a Google form.

Sample

A convenience sampling technique was used to collect data from 215 school teachers working in two schools in the Middle East. A majority of the participants were female (94.9%) and between the ages of 30 and 39 years (48.8%). Most of the teachers taught primary grade levels (grade five and below) (57.7%) and had one to five years of teaching experience (40.9%).

Table 1. Sociodemographic Profile of Participants

Variable	Number	Percentage
Gender	204	94.8%
Female	10	4.6%
Male	1	0.6%
Prefer not to say		
Age	21	9.7%
20 to 29	105	48.8%
30 to 39	69	32.1%
40 to 49	19	8.8%
50 to 59	1	0.6%
60 and above		
Teaching Level(s)	132	61.4%
Primary School (grade 5 and below)	42	19.5%
Middle School (grade 6 to 8)	28	13%
Secondary School (grade 9 and 10)	13	6%
High Secondary School (grade 11 and 12)		
Teaching Experience (years)	6	2.7%
Less than 1 year	88	40.9%
1-5 years	66	30.6%
6-10 years	26	12.1%
11-15 years	15	6.9%
16-20 years	14	6.8%
More than 20 years		

continued on following page

Table 1. Continued

Variable	Number	Percentage
Educational Qualification		
Undergraduate	57	27%
Postgraduate	143	67.7%
MPhil/PhD	11	5.3%
Subjects Taught		
Language	48	22.3%
Non-Science	75	34.8%
Science	92	42.9%

Research Tools

Participant Sociodemographic Profile. The sociodemographic profile is used to collect general participant data such as gender, age, and familiarity with digital devices. Additionally, the details of the participant's educational qualification and teaching experience are collected through the profile.

The Teaching for Successful Intelligence Questionnaire (TSI-Q) (Paloú & Maricutoiu, 2013). It is a 20-item scale that enables the identification of the teachers' preferences for teaching styles relevant to stimulating the students' creative, analytical, practical and reproductive abilities. Items are framed to promote self-reflection among the participants such as, "In my teaching activity, I favour and value the use of memorizing in the learning process". Participants are required to respond on a six-point scale ranging from very strong disapproval (1) to very strong approval (6). It has good internal consistency with a Cronbach alpha of 0.78.

Teachers' Digital Competence Scale (TDiCoS) (Ergül & Taşarb, 2023). It consists of nineteen items asking participants how often they do or do not do specified actions in the learning-teaching environment or have no knowledge at all regarding the action. Participants are required to respond on a scale of never (1) to always (5) or choose not appropriate (NA). A sample item of the scale is "I create learning environments suitable for students' individual differences by using digital technologies". It was validated on a sample of 288 in-service teachers from different regions of Turkey, and Cronbach's alpha value was determined to be .949.

The Digital Pedagogy Competency Scale (DiPeCoS) (Rawat et al., 2022). The DiPeCoS is an eight-item scale that assesses school teachers on their digital inclusion competencies in their pedagogical structure. The scale consists of eight scenario-based items such as, "According to your, which of the following is the most critical element for a successful online class?". Participants are required to choose one out of the three to four responses that are given. It has good psychometric properties, reporting a reliability value of 0.62, and has been validated on a sample of 1315 teachers in India.

Data Analysis

The data collected was analysed using descriptive and inferential statistics. Descriptive statistics such as mean, median, and standard deviation were used to understand the spread of the data. A Welch's T-test was used to examine the factors influencing the digital competency, and digital pedagogical competencies prevalent among primary, middle, and high school teachers.

Ethical Considerations

A detailed informed consent form was provided to the participants at the beginning of the survey used for data collection. The form outlined all the information about the study, including the voluntary nature of participation and participants' right to withdraw. Further, participants were informed that all data collected would be stored securely, accessed only by the researchers, and would only be used anonymously for research and publication purposes. Finally, participants were provided with the contact details of the researchers so that they could reach out with any further queries or concerns. Participants were provided with the option to provide their consent or refrain from participating in the study at the end of the informed consent form.

RESULTS

There was a significant difference ($t(5)=2.443, p=0.045$) in the teaching styles used by teachers in their classes and how digitally inclusive were their pedagogical practices. It was seen that the teachers who fostered the practical abilities of the students were the ones who used the most digitally inclusive pedagogies (M=5, SD= 1.65) followed by teachers whose teaching styles fostered creative abilities (M=4.6, SD= 1.69), and analytical abilities (M=4.6, SD= 2.01). The teachers whose teaching styles catered to the reproductive abilities of the students had the lowest digitally inclusive pedagogies (M=3.5, SD=1.82). An interesting find was that the teachers who used a combination of two or more types of teaching styles had the second highest DiPeCoS score, reflecting their integration of digital aids within their pedagogies (M=4.86, SD=1.53) (Figure 1).

Figure 1. Mean Scores across Digital Pedagogy Competency Scale (DiPeCoS) vs Teaching Styles

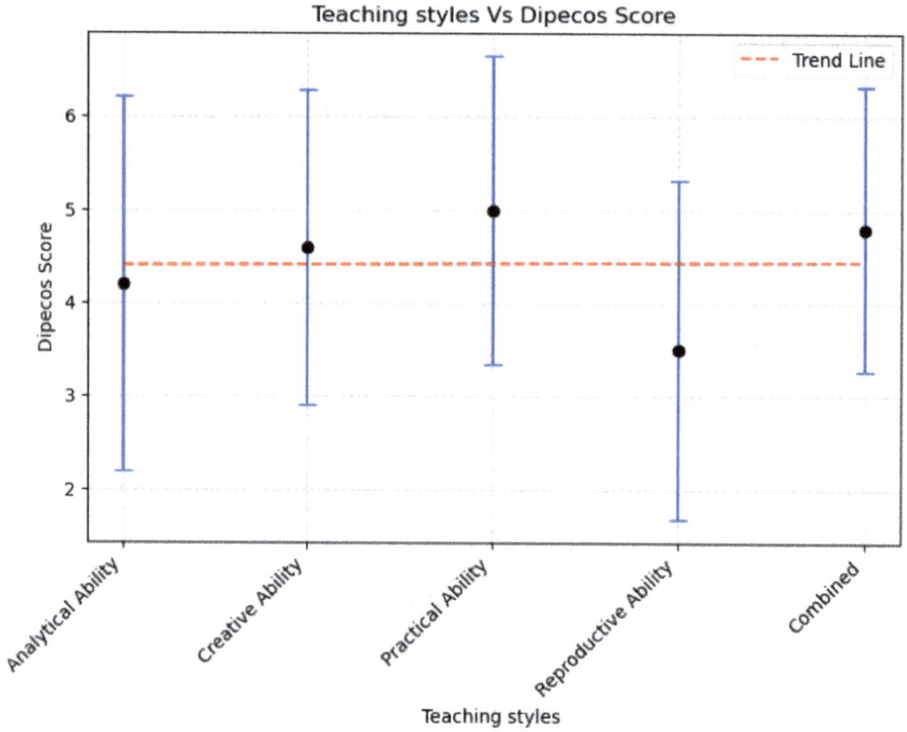

The individual digital competencies of teachers (TiDeCOs results) did not show any significant difference based on their teaching styles (t(5)=0.245, p=0.9) (Figure 2). Though these results were not significant, it was seen that teachers that taught science subjects were seen to have more digitally inclusive pedagogies (M=4.9, SD=1.5) than the teachers who taught non science subjects (M =4.41, SD= 1.74) and languages (M=4.39, SD=1.84). Even the digital competencies of teachers did not show any significant difference based on the subjects that they taught (t(2)=0.517, p=0.597).

Figure 2. Mean Scores of Teachers Digital Competency Scale (TDiCoS)

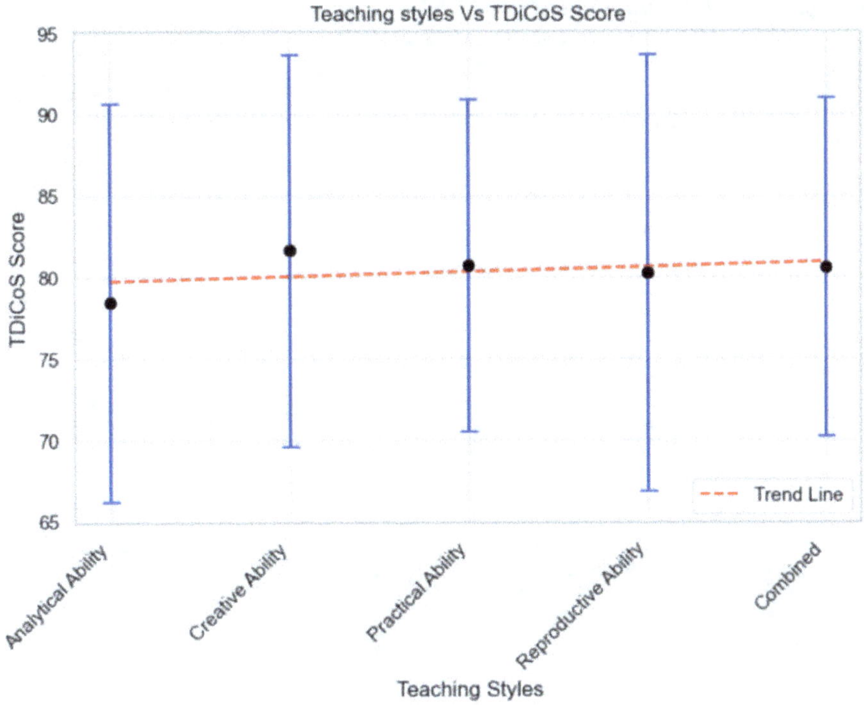

Looking into the classes taught by the teachers, it was seen that there indeed existed a significant difference in trends relating to digital inclusion within pedagogies depending on which strata of the classes were taught (t(5)=5.142, p=0.001) (Figure 3). A post hoc test (Games-Howell) revealed that the teachers who taught secondary school students had a significantly higher digitally inclusive pedagogy (M=5.71, SD=1.18) than teachers who taught middle school (M=4.17, SD= 1.566, P<0.001) and the teachers who taught primary school (M=4.54, SD=1.76, p=0.001). The digital competencies of the teachers were also significantly different (t(5)=3.087, p=0.021) based on the classes that they taught (Figure 4).

Figure 3. Mean Scores of Digital Pedagogy Competency Scale (DiPeCoS) vs Sociodemographic Factors

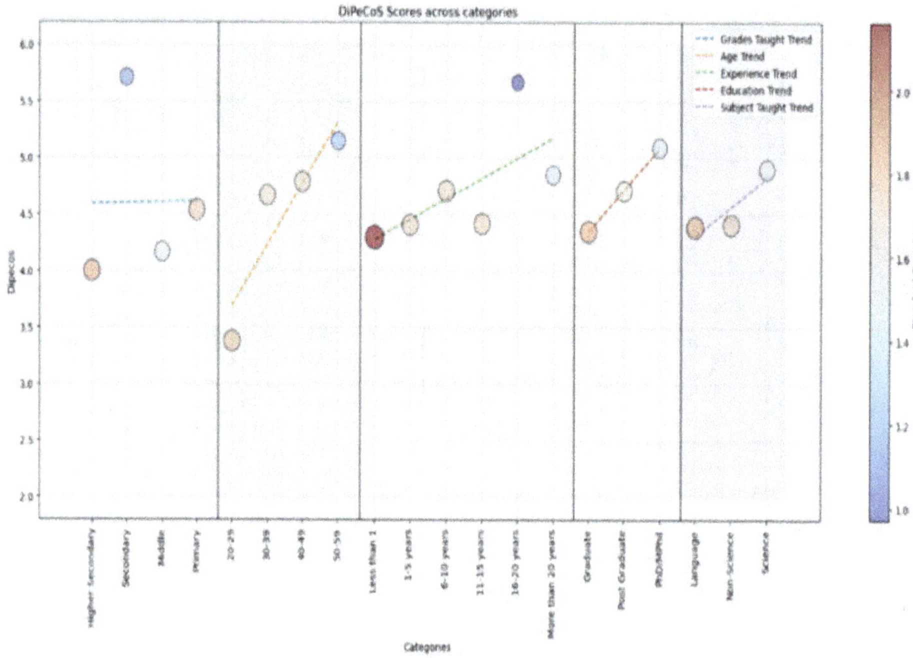

Figure 4. Mean Score of Teachers Digital Competency Scale (TDiCoS) vs Factors

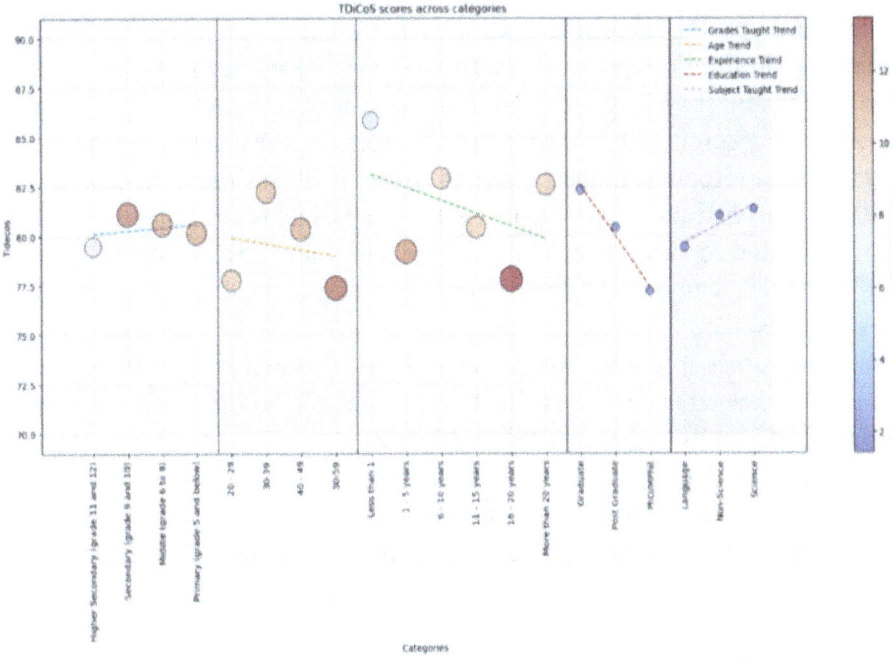

It was seen that the age of the teacher also had an effect on the digital inclusiveness of their pedagogies. With only one participant in the above 60 category, that age category was not included in the analysis. Teachers within the age group of 50-59 years had the highest digitally inclusive pedagogies (M=5.15, SD= 1.25) followed by teachers in age group of 40-49 (M=4.79, SD=1.65), 30-39 (M=4.65, SD=1.66) and the teachers in age group of 20-29, i.e., the youngest cohort had the least digitally inclusive pedagogies (M=3.38, SD=1.8). These differences were significantly different as well (t(5)=4.553, p=0.006). A post-hoc test (Games-Howell) revealed that the digital inclusiveness of teachers of the age group 20-29 was significantly different from that of teachers in the age group 30-39 (p=0.028), 40-49 (p=0.016) and 50-59 (p=0.005). There was no statistically significant difference seen for the digital competencies of teachers based on their age.

When looking into the teaching experience of teachers (in years), there was a significant influence of their teaching experience on digital inclusiveness within their pedagogies (t(5)=3.309, p=0.016). A post-hoc test (Games-Howell) revealed that teachers with teaching experience of 1-5 years (M=4.4, SD=1.73) significantly varied in their digital inclusion within their pedagogies when compared to teachers with teaching experience of 16-20 years (M=5.66, SD=0.975) (p=0.004). Teaching experience did not significantly influence the individual digital competencies of teachers.

Four participants did not have an educational qualification (diplomas, TTC, etc.) that fell into the three categories that were studied (undergraduate, postgraduate and MPhil/PhD), thus they were excluded from the analysis. Though not significant, the educational background of teachers was also seen to influence their tendencies to include digital resources in their teaching methods. It was seen that teachers with PhD/MPhil qualifications had the highest digitally inclusive pedagogies (M=5.1, SD=1.44) followed by teachers with postgraduate qualifications (M=4.7, SD= 1.6) and the teachers with undergraduate qualifications had the least digitally inclusive pedagogy (M=4.35, SD=1.85).

When it comes to digital competency, the trend seems to be reversed. TiDeCOs results showed that, teachers with undergraduate qualification seemed to be more technologically competent (M=82.3, SD=10.74) followed by teachers with postgraduate qualification (M=80.4, SD=10.95) and the teachers with PhD/MPhil qualifications were the least technologically competent (M=77.18, SD=11.95). These insights, though pointing to a possible trend, were not significant.

DISCUSSION

The study looked into how a teacher's individual digital competency and digital inclusiveness within their pedagogies is influenced by factors such as teaching styles, experience, and educational qualification. Teaching styles are broadly divided into four categories. They are teaching styles that cater to a) practical abilities (PA) b) creative abilities (CA) c) reproductive abilities (RA) and d) analytical abilities (AA) of the students. Significant differences were seen in the digital inclusiveness within the pedagogical structures of the teachers depending on what teaching style they used in their classes. It was seen that teachers with teaching styles that catered to enhancing the practical abilities (PA) of their students through their teaching methods used the most digitally inclusive pedagogies. It is always a usual question that is asked by the student to their teachers on where and how they can apply what they have learned in a real-life scenario. A teaching style that stimulates the practical abilities of the students is usually the best remedy to such queries. Such a teaching style essentially helps them in applying the theoretical knowledge that they have gained in the class in various everyday practical scenarios. In such scenarios, digital tools could be a great help in stimulating real-life scenarios where the students can be engaged to apply the theoretical knowledge that they have gained in the class. For instance, digital tools that mimic gravity in an artificial graphical interface could be used to show how different values of the gravitational constant would impact differently on different objects. Digital tools thus provide an alternate way to simulate real-life scenarios where the students could practically apply their theoretical knowledge and gain a more in-depth and fundamental understanding of what they have learned. Thus, it is of sound logic that teachers that cater to stimulate the practical abilities of the students to have a more digitally inclusive pedagogy.

It was also seen that teachers with teaching styles that catered to enhancing the reproductive abilities of the students had the lowest digitally inclusive pedagogies. Reproductive capacities of students refer to the extent to which they can recognize what they have been taught and re-enact it out of their memory. Memorization usually increases in its efficiency when no external help is provided and the memorizer comes up with their own tactics to memorize the information. Studies have shown that when learners create their own retrieval paths or study strategies, learning is more durable and impactful (Butler et al., 2014). Thus reducing the aid for the students to remember, forcing them to come up with their own mental representations to remember what is being taught more efficiently might be the possible explanation as to why the teachers with teaching styles catering to the student's reproductive abilities have the least digitally inclusive pedagogies.

Individual digital competencies of the teachers were not seen to be significantly influenced by their teaching styles. This might be owing to the fact that individual digital competencies of the teachers might be influenced by many personal and subjective factors such as their overall exposure to digital tools in the surrounding environment, past experience of using such tools and much more, which may not have any direct influence on their belief on how to teach in a classroom. It is possible that unless they feel the need to include digital tools within their teaching routine they may not have the incentive to gain more knowledge on such digital tools, invariably not influencing their individual digital competencies.

The teaching stratas were divided into four, depending on the classes that were being taught by the teachers. They were divided into (a) primary school (grades 5 and below), (b) middle school (grades 6 to 8), (c) secondary school (grades 9 and 10) and (d) higher secondary (grades 11 and 12). The results showed the existence of a significant difference in how digitally inclusive the pedagogies of the teachers were depending on the classes that they taught. It was seen that teachers teaching secondary school students had the most digitally inclusive pedagogies when compared to teachers teaching middle and primary school students. One reason for this trend could be the fact that when information being taught in the classes becomes increasingly abstract, digital resources help the teachers to show such abstract concepts through simulations and digital visual aids, helping the students to grasp the concepts in a more foundational manner. It is a well-documented fact that abstraction of what is being taught in the school linearly increases along with the grades that are being taught. Piaget in his 1972 paper "Intellectual Evolution from Adolescence to

Adulthood" mentions that students between the age of seven to 11 have a more concrete operational thought process, which enables them to learn what is concrete and objective in the world around them. Formal operational thought begins around the age of 12 and up when they can delve on more abstract concepts (Piaget, 2016). Modern-day curriculum is also structured in such a manner that there is a gradual progression in conceptual and abstract complexity over time (Billett, 2006). Thus this increasing conceptual and abstract thinking across the grades in the schools could be a possible reason for increased digital inclusiveness within the pedagogies of teachers teaching secondary school when compared to middle and primary school.

A counterargument can be placed here, stating the need for increased digitally inclusive pedagogies for primary and middle school students. The statement arises from acknowledging the fact that students in these strata are young and have a highly imaginative and creative mind and digital resources are best utilized here, to augment their creativity helping them to think out of the box. Another argument that gives credence to using more digitally inclusive pedagogies in primary and middle school strata would be the fact that this is the age where theoretical foundations for many topics that will be studied in depth at the secondary and higher secondary classes are placed into the minds of the students. Digital tools could go a long way in strengthening these foundations by engaging the students in a more interactive manner making the concepts and facts get deeply rooted within their minds. The findings from the study thus advocate the necessity for having more digitally inclusive pedagogies for primary and middle school teaching stratas.

Individual digital competencies of teachers were also seen to be significantly different in spending on what teaching strata they taught. This may be due to the fact that when dealing with certain strata of students, the teachers would have come to realize the need to include various digital aids in their pedagogies to increase the efficiency of the teaching process. This has been found true from the above results showing that teachers who teach the high school stratum had the most digitally inclusive pedagogies. So in order to facilitate this need to include digital aids in the teaching environment, they (the teachers) need to be well-versed with the digital tools first, thus this could lead them to have a better understanding of the digital tools and aids available, invariably increasing their individual digital competency as a teacher. It could be thus said that the recognition of the need for more digital inclusiveness within the pedagogy based on the stratum of the class which is being taught could play a factor in positively influencing the individual digital competencies of the teachers.

When looking into the age groups of the teachers, a significant trend was seen in the level of digital inclusiveness within their teaching structure. Contrary to the perceived notion of the younger generation being technologically more in tune than the older generation, it was seen that older teachers had more digitally inclusive pedagogies than younger teachers. The results showed that teachers in the age bracket of 50-59 years had the most digitally inclusive pedagogies and on the contrary teachers in the youngest age bracket of 20-29 had the least digitally inclusive pedagogies. The occurrence of such a discrepancy in digital inclusiveness within the pedagogies of the so-assumed technologically savvy younger generation teachers when compared to the older generation teachers might result from the fact that the older generation teacher, through their experience, have a better understanding of the requirements the students might have for a better understanding of the concept being taught. Older teachers would have had years of experience to fine-tune their pedagogies to ensure the most effective manner of teaching, whereas teachers who are in the early stages of their teaching career would be still figuring out methods that can be used to deliver a more impactful teaching experience. Thus with experience, the teachers would have found out that a more digitally inclusive pedagogy would be best suited for an effective delivery of teaching concepts and thus adopt such teaching methods. Furthermore, the analysis of the teaching experience (years) of teachers and its influence on their digital usage within their pedagogies also came to the same conclusion that, with more experience, the pedagogies of the teachers became more digitally inclusive. It was seen that teachers with a teaching experience of 16-20 years had the most digitally inclusive pedagogies and on the contrary teachers with less than one year of experience had the least digitally inclusive pedagogies. Thus, it can be seen that older teachers with more experience had the most digitally inclusive pedagogies as compared to younger cohorts with comparatively less experience. Schools thus can arrange mentoring sessions where older teachers could give insights and innovative methods to their younger cohorts through which they could include more digital aids within their pedagogies enabling them to better convey their teaching lessons to their students.

Though not significant, the digital inclusiveness within pedagogies seemed to follow a trend based on the field of subjects taught by the teachers. For the study, the subjects taught by the teachers were divided into three categories, (a) science subjects (physics, environmental studies, biology, etc.), (b) non-science subjects (social studies, moral science, etc.), and (c) language subjects (Arabic, Hindi, Malayalam, English, etc.). Teachers who taught science subjects were seen to have a more digitally inclusive pedagogical structure and the teachers who taught language subjects had the least digitally inclusive pedagogies. The reason behind this trend is an evident one. Science being a subject that evokes a much higher cognitive load with a lot of abstract concepts and logic, external interventions such as digital aids could be of great help in solidifying the foundation for these concepts within the minds of the students. Digital aids could help in simulating abstract science topics (gravity, cloud formation, light diffraction, etc.) being taught in the class and help the teachers explain the actual reasoning behind these concepts in a more concrete manner enabling the students to remember the facts and understand them in a more efficient manner.

When educational backgrounds and the qualifications of the teachers were looked into, despite lacking a significant difference, a trend was visible. It was seen that the higher the educational qualification of the teacher, more digitally inclusive were their pedagogical practices. Teachers with PhD/MPhil qualifications were seen to have the highest digitally inclusive pedagogy followed by teachers with post-graduation and the least digitally inclusive pedagogies were seen for the teachers with under graduation. However, when the individual digital competencies of teachers were measured using TDiCoS, an inversion of the above-seen trend. Teachers with undergraduate degrees were seen to have the highest individual digital competency followed by teachers with postgraduate qualification. Teachers having PhD/MPhil qualifications were seen to have the lowest individual digital competency. A possible explanation as to why the educational qualification of the teacher could be positively correlated with the inclusion of digital technology within their pedagogy, may arise from the fact that the education qualification of the teacher also often translates to an increased time spent by the teachers as students themselves. A higher qualification means more time spent as a student, and this would have helped them to gain insights on how to make learning more enhancing and interesting through their learning practices and the methodologies that their teachers must have used to teach them. This concept is termed as 'the apprenticeship of observation' which states teachers' prior experiences as students shape their beliefs and practices in teaching (Borg, 2004). This could have given them insights on coming up with more innovative pedagogical structures that they might feel are the best method to convey the topic that they are teaching, thus leading to more digitally inclusive pedagogies. Another explanation may be the fact that with increasing educational qualifications, teachers may have been compelled to use more digital tools to aid them in their studying courses. For instance, on average a PhD/MPhil student needs to learn various software for data analysis, visualization, and graphical representations of the concepts to name a few. This knowledge that is acquired by the teachers, when they were students could enable them to set up more efficient pedagogies suitable for conveying the concepts that are being taught in a more fundamental and easy-to-understand manner.

Reduced age and lack of experience (in years) could be a possible reason behind why the results seen while looking into the individual digital competence of the teachers, where the teachers had the lowest educational qualification (undergraduate) had the highest individual digital competence did not translate into an increase in digital inclusiveness within their pedagogies. Teachers with undergraduate qualifications were also the cohort with the least number of teachers having more than 10 years of experience (15%), which in comparison was 26% for teachers with postgraduate qualifications. Furthermore, the undergraduate cohort also had the youngest stratum of teachers, with around 65% of them falling between 20-39 years of age (it was 58% for the postgraduate cohort). A 2013 study done by the Pew research Centre found that younger teachers were more technologically confident than older teachers (Purcell et al., 2024).

Thus, taking these factors together it can be said that teachers belonging to the undergraduate cohort, being younger compared to the other two cohorts would be in a much better position to be aligned with the current technological zeitgeist, allowing them to be more digitally competent than the other two cohorts, but their lack of experience in teaching prevents them from translating this competency into a more digitally inclusive pedagogical structure.

Implications

The study highlights a pivotal point that the experience of the teacher may have a positive influence on their attitude towards digital inclusion within their pedagogical structure. Despite younger teachers having better individual digital competencies, older teachers with more experience were seen to have a more digitally inclusive pedagogy. This gap between increased individual digital competency of younger teachers not translating into a more digitally inclusive pedagogical structure can possibly bridged by organizing training sessions where the older more experienced teachers can give insights on how they integrated digital aids into their pedagogical structure and give tips from personal experience to the younger teachers on how they can successfully integrate digital aids within their teaching practices for assuring a more interactive and effective learning experience for their students. An interactive session between younger teachers who have low teaching experience and older teachers with more teaching experience can be conducive for both parties. Younger teachers who are more digitally competent than older teachers on an individual level can introduce new and advanced digital aids to them and in return older teachers who have a more digitally inclusive pedagogies can provide their younger cohorts with insights on how they can translate their enhanced individual digital competencies into a more digitally integrated teaching practices (Figure 5).

Figure 5. Proposed Knowledge Transfer

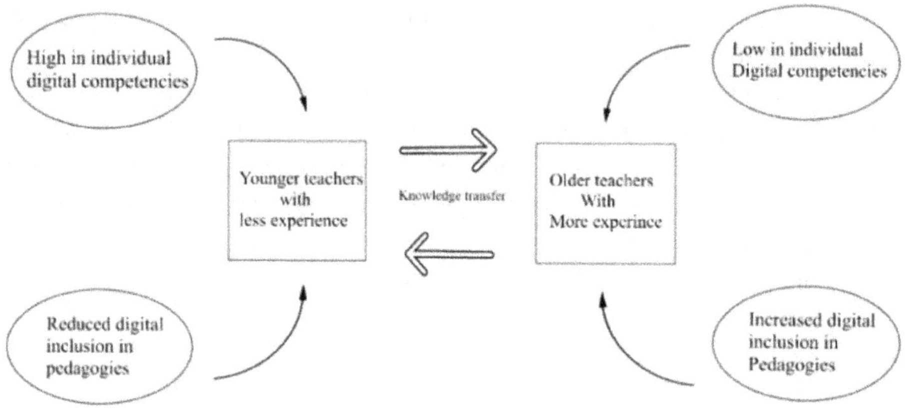

Another interesting finding from this study is the fact that the higher the educational qualification of the teachers, the more digitally inclusive their pedagogies were and the less competent they were in terms of individual digital competencies. This seemingly contradicting fact where higher digital competencies of teachers with lower educational qualifications did not translate into more digitally inclusive pedagogies, but teachers with higher educational qualifications but having lower individual digital competencies did have more digitally integrated pedagogical practices gives credence to the fact that experience plays a pivotal role in enhancing digital integration with pedagogical practices.

An interesting implication that can be deduced from this is a possible symbiosis between younger teachers with increased individual digital competencies and elder teachers who have more digitally inclusive pedagogical practices. School authorities could arrange for knowledge-sharing sessions where an interaction between younger less experienced teachers and older more experienced teachers could enable the elder teachers to learn more about new digital tools that are available and the younger teachers could get a better insight into how to integrate their increased digital competence into their pedagogical practices. This exchange of knowledge could enhance the overall inclusion of digital aids within the pedagogical structures of the teachers within an educational institution, thus improving the teaching standards of that institution.

Insights from the current study can also be used as a sounding board for discussing policy implications at the management level of an educational institution and coming up with customized ways, catering to different strata of teaching staff to expedite enhanced digital integration within the pedagogical practices practiced by them.

Limitations

The study attempts to explore the factors that influence the inclusion of digital aids within the pedagogical practices of teachers. The Middle East (Abu Dhabi) was the geographical area that was chosen for the study. This could be seen as a limitation as Abu Dhabi is a technologically advanced country and thereby digital integration into education would be in a more enhanced manner than what might be seen in a less technologically advanced country. Thus factors influencing digital inclusion within the pedagogies used in an educational institution might vary across geographical regions. Future researchers could conduct a cross-sectional study across various geographical regions and explore the underlying factors that might influence digital inclusiveness within pedagogical structures.

The self-reported survey method also contributes to the limitations of the study. Although it is a quick and easy method of data collection, it lends itself to numerous biases. Participants would have been influenced by the social desirability bias while responding to the question, especially as they were contacted through their schools. Additionally, participant beliefs and attitudes towards the EdTeach boom could have also influenced their responses to the survey. Future research can use different forms of triangulation to combat the limitations of using just one research method.

CONCLUSION

The current study attempts to explore the factors that could influence a teachers tendency to enhance their pedagogical practices by integrating digital aids within it. It was found that there indeed existed a significant influence of the teaching style used by the teachers in their classes on their pedagogical practices. Similarly, the age and classes taught by the teachers and their experience (in years) were also seen to exert significant influence on determining how digitally inclusive their pedagogical practices were. Although lacking a significant impact, a trend was observed based on the educational background of the teachers on their digital inclusiveness within their pedagogies. Despite studies being conducted on digital inclusion within teaching practices prevalent in higher education institutions, a comparative study looking into multiple strata (primary, middle, secondary, and higher secondary) of learning and individual factors pertaining to teachers such as their age, experience, and teaching styles have not been done. The study aims to be an initial exploratory step in filling this gap. With the breakneck speeds in which technological advancements are being made on a daily basis in the current era, it is of importance to enable these advancements in aiding everyday learning process of the students to strengthen their foundational knowledge about the topics that are being taught to them. A student with strong foundational knowledge would in turn have a better chance of contributing back to society in substantial ways.

ACKNOWLEDGEMENTS

The authors sincerely thank the school authorities for giving the permission to conduct this study and the school teachers who took part in this study.

REFERENCES

Albright, W. F. (1998, August 23). *Ancient Middle East | History, cities, civilizations, & Religion*. Encyclopedia Britannica. https://www.britannica.com/place/ancient-Middle-East/Science-and-law

Billett, S. (2006). Constituting the workplace curriculum. *Journal of Curriculum Studies, 38*(1), 31–48. DOI: 10.1080/00220270500153781

Bordovskaya, N. V. Pedagogy. Manual for High Schools / N.V. Bordovskaya, A.A. Rean. SPb.: Publishing house "Peter", 2000. 304 p.

Borg, M. (2004). The apprenticeship of observation. *ELT Journal, 58*(3), 274–276. DOI: 10.1093/elt/58.3.274

Burke, P. (1998). *The European Renaissance: centres and peripheries*. Blackwell Publishers.

Butler, A. C., Marsh, E. J., Slavinsky, J. P., & Baraniuk, R. G. (2014). Integrating cognitive science and technology improves learning in a STEM classroom. *Educational Psychology Review, 26*(2), 331–340. DOI: 10.1007/s10648-014-9256-4

Caena, F., & Redecker, C. (2019). Aligning teacher competence frameworks to 21st century challenges: The case for the European Digital Competence Framework for Educators (Digcompedu). *European Journal of Education, 54*(3), 356–369. DOI: 10.1111/ejed.12345

Compayré, G., & Payne, W. H. (2015). *The history ofpedagogy*. Routledge. Daniela, L. Smart Pedagogy as a Driving Wheel for Technology-Enhanced Learning. Tech

Daniela, L. (2021). Smart Pedagogy as a Driving Wheel for Technology-Enhanced Learning. *Know Learn, 26*(4), 711–718. DOI: 10.1007/s10758-021-09536-z

Delić, H., & Bećirović, S. (2016). Socratic method as an approach to teaching. *European Researcher. Series A*, (10), 511–517.

Dewey, J., Small, A. W., & Dutton, S. T. (1897). *My pedagogic creed. and The demands of sociology upon pedagogy*. No Title.

Dutt, T. (2017). Education in Vedic and Post Vedic Era. *Journal of National Development, 30*(2), 159–166.

Englund, R. K. (2004). Proto-cuneiform account-books and journals. Creating economic order: Record-keeping, standardization and the development of accounting in the ancient Near East, 23-46.

Eskelson, T. C. (2020). How and Why Formal Education Originated in the Emergence of Civilization. *Journal of Education and Learning, 9*(2), 29–47. DOI: 10.5539/jel.v9n2p29

Federal budget | The Official Portal of the UAE Government. (n.d.). https://u.ae/en/information-and-services/finance-and-investment/budget/federal-finance#:~:text=Federal%20Budget%202023&text=is%20as%20follows:-,AED%2023.9%20billion

George, A. (2005). In search of the é. dub. ba. a: The ancient Mesopotamian school in literature and reality.

Ghonge, M. M., Bag, R., & Singh, A. (2020). Indian education: Ancient, medieval and modern. In *Education at the Intersection of Globalization and Technology.* IntechOpen.

Golinkoff, R. M., Hassinger-Das, B., Kucirkova, N., Myers, L., & Dore, R. A. (2025). The Explosion of EdTech: Can Its Promise Be Fulfilled? In Christakis, D. A., & Hale, L. (Eds.), *Handbook of Children and Screens.* Springer., DOI: 10.1007/978-3-031-69362-5_75

Hannafin, M. J., & Hannafin, K. M. (2010). Cognition and student-centered, web-based learning: Issues and implications for research and theory. *Learning and instruction in the digital age*, 11-23.

Ingridhaynesphd. (2024, March 7). Literacy Statistics 2024- 2025 (Where we are now). *National Literacy.* https://www.thenationalliteracyinstitute.com/post/literacy-statistics-2024-2025-where-we-are-now#:~:text=On%20average%2C%2079%25%20of%20U.S.,below%205th%2Dgrade%20level)

Jones, L. (2007). *The student-centered classroom* (Vol. 32). Cambridge university press.

Kincheloe, J. L., & Horn, R. A., Jr. (2007). *The Praeger Handbook of Education and Psychology. Volume 1.* Praeger. Available from: Greenwood Publishing Group, Inc. PO Box 5926, Portsmouth, NH 03802-6926.

Kireet Joshi. (n.d.) https://kireetjoshi.in/books/philosophy/philosophy-of-indian-pedagogy/philosophy-of-indian-pedagogy-2

Lavryk, T. V., & Yaresko, E. V. (2015). The Development of the Foundations of Modern Pedagogy: Paradigmal and Methodological Aspects of Research. *European Journal of Contemporary Education, 12*(2), 150–157. DOI: 10.13187/ejced.2015.12.150

MACROTRENDS. (n.d.). *UAE literacy rate 1975-2025.* https://www.macrotrends.net/global-metrics/countries/ARE/uae/literacy-rate

Meng, Q., Jia, J., & Zhang, Z. (2020). A framework of smart pedagogy based on the facilitating of high order thinking skills. *Interactive Technology and Smart Education, 17*(3), 251–266. DOI: 10.1108/ITSE-11-2019-0076

Palos, R., & Maricutoiu, L. P. (2013). Teaching for Successful Intelligence Questionnaire (TSI-Q)-a new instrument developed for assessing teaching style. *Journal of Educational Sciences and Psychology, 3*(1).

Pedagogy: Innovative Teaching and Learning Strategies in Engineering Education, 2018 IEEE World Engineering Education Conference (EDUNINE), Buenos Aires, Argentina, pp. 1-6, DOI: 10.1109/EDUNINE.2018.8450962

Piaget, J. (2016). Intellectual evolution from adolescence to adulthood1. In *Cognitive and moral development, academic achievement in adolescence* (pp. 1–12). Routledge.

Purcell, K., Buchanan, J., & Friedrich, L. (2024, April 14). *How teachers are using technology at home and in their classrooms.* Pew Research Center. https://www.pewresearch.org/internet/2013/02/28/how-teachers-are-using-technology-at-home-and-in-their-classrooms/

Schleicher, A. (2024, March 18). Digital dreaming: the role of technology in expanding education access. https://www.unesco.org/en/articles/digital-dreaming-role-technology-expanding-education-access

Spaulding, W. (2022, December 28). *Education in the Middle Ages | Schools, History & Facts.* Study.com. https://study.com/academy/lesson/education-middle-ages-schools-history-facts.html

Sternberg, R. J., & Grigorenko, E. L. (1995). Styles of thinking in the school. *European Journal for High Ability, 6*(2), 201–219. DOI: 10.1080/0937445940060211

Suchodolski, B. (1961). Education for the future and traditional pedagogy. *International Review of Education,* •••, 420–431.

Szyliowicz, J. S., Lauwerys, J. A., Bowen, J., Browning, R., Thomas, R. M., Vázquez, J. Z., Mukerji, S., Nakosteen, M. K., Chen, T. H., Naka, A., Swink, R. L., Lawson, R. F., Shimahara, N., Gelpi, E., Anweiler, O., Graham, H. F., Ipfling, H., Moumouni, A., Scanlon, D. G., . . . Arnove, R. F. (2025, January 19). *Education | Definition, Development, History, Types, & Facts*. Encyclopedia Britannica. https://www.britannica.com/topic/education/Education-in-the-earliest-civilizations

Team, F. (n.d.). Education during the Renaissance period. *Fairgaze*. https://fairgaze.com/educationnews/education-during-the-renaissance-period.html

Trang, H. (2024, October 22). Digital Transformation in Education: Benefits, Challenges and Solutions. Tokyo Tech Lab. https://tokyotechlab.com/blogs/digital-transformation-in-education

Trigger, B. G. (1998). Writing systems: A case study in cultural evolution. *Norwegian Archaeological Review*, *31*(1), 39–62. DOI: 10.1080/00293652.1998.9965618

Uskov, V. L. (2022). Smart Pedagogy: Concepts, Components, Outcomes. In Uskov, V. L., Howlett, R. J., & Jain, L. C. (Eds.), *Smart Education and e-Learning - Smart Pedagogy. SEEL-22 2022. Smart Innovation, Systems and Technologies* (Vol. 305). Springer., DOI: 10.1007/978-981-19-3112-3_1

Uskov, V., Bakken, J. P., Aluri, L., Rachakonda, R., Rayala, N., & Uskova, M. (2018, March). Smart pedagogy: Innovative teaching and learning strategies in engineering education. In *2018 IEEE World Engineering Education Conference (EDUNINE)* (pp. 1-6). IEEE.

Waters, C. N., Zalasiewicz, J., Summerhayes, C., Barnosky, A. D., Poirier, C., Gałuszka, A., Cearreta, A., Edgeworth, M., Ellis, E. C., Ellis, M., Jeandel, C., Leinfelder, R., McNeill, J. R., Richter, D. B., Steffen, W., Syvitski, J., Vidas, D., Wagreich, M., Williams, M., & Wolfe, A. P. (2016). The Anthropocene is functionally and stratigraphically distinct from the Holocene. *Science*, *351*(6269), aad2622. DOI: 10.1126/science.aad2622 PMID: 26744408

West, M. (2023). *An ed-tech tragedy? Educational technologies and school closures in the time of COVID-19*. UNESCO., DOI: 10.54675/LYGF2153

Young, L. E., & Paterson, B. L. (Eds.). (2007). *Teaching nursing: Developing a student-centered learning environment*. Lippincott Williams & Wilkins.

Zhang, L. F., & Sternberg, R. J. (2006). The nature of intellectual styles. Mahwah, NJ: Lawrence Erlbaum.

Zhang, L., & Sternberg, R. J. (2005). A Threefold Model of Intellectual Styles. *Educational Psychology Review*, *17*(1), 1–53. DOI: 10.1007/s10648-005-1635-4

Zhang, L. F. (2001). Do thinking styles contribute to academic achievement beyond self-rated abilities? *The Journal of Psychology, 135*(6), 621–637. DOI: 10.1080/00223980109603724 PMID: 11931003

Zhang, L. F. (2001). Approaches and thinking styles in teaching. *The Journal of Psychology, 135*(5), 547–561. DOI: 10.1080/00223980109603718 PMID: 11804007

Creswell, J. W., & Creswell, J. D. (2017). *Research design: Qualitative, quantitative, and mixed methods approaches*. Sage publications.

KEY TERMS AND DEFINITIONS

Digital competencies: The skills required to harness the power of digital tools in the classroom.

Digital tools: Any form of technology that offers strategies or techniques to improve the teaching-learning process.

EdTech: Any technological innovation that aims to facilitate or improve the teaching and learning processes.

Smart pedagogy: The use of digital tools and techniques to create and conduct effective teaching-learning experiences.

Teaching styles: The strategy of teaching and learning offered by a teacher.

Chapter 10
Predictive Analytics as a Catalyst for Disruption in the Fashion Ecosystem

Naadia Ghalib Sheriff
https://orcid.org/0009-0005-3881-1364
Christ University, Bangalore, India

Pragya Bhattacharjee
https://orcid.org/0009-0009-0354-2247
Christ University, Bangalore, India

B. R. Gagan Kumar
https://orcid.org/0009-0004-9095-4971
Christ University, Bangalore, India

ABSTRACT

The fashion industry is one of the most dynamic industries, with trends that change constantly being moulded by the changing preferences of consumers. In this dynamic environment it is crucial for companies to adjust to the current market demand, failing which companies might plunge into losses and even face shutdowns. Predictive analytics plays an important role in trend prediction, demand forecasting, optimizing the supply chain and inventory management, personalization, dynamic pricing, waste minimization and prediction of sustainable materials. This, of course, is not without challenges. Data privacy concerns and over reliance on AI algorithms leading to inhibition of human creativity stand out as the most concerning challenges that need to be addressed to ensure that predictive analytics is utilised to its fullest potential in

DOI: 10.4018/979-8-3373-7922-7.ch010

the fashion industry. Predictive analytics and AI can ensure that the fashion industry is not just profitable but also consumer centric and sustainable in its operations thus ensuring long term growth and sustenance of businesses in this industry.

1. INTRODUCTION TO PREDICTIVE ANALYTICS IN FASHION ECOSYSTEM

Predictive analytics has changed the entire landscape of the fashion industry by allowing brands to make data-driven decisions, ensuring efficiency in all operations (Anitson, 2023). By analysing past trends and current market data, along with specific consumer data from social media analytics, companies can predict the demand of products, consumer preferences, optimise inventory management and ensure overall profitability and sustainability.

1.1 Evolution of Data Driven Decision Making in Fashion

Data driven decision making has been adopted by various businesses, across industries today, and has been an undeniable changemaker in enhancing the profitability and long-term sustainability of businesses (Jančiarová, 2024). So, what exactly is data driven decision making? It is, as the term itself suggests, the process of making crucial business decisions on the basis of reliable and relevant data. It is only after collecting the necessary data and analysing it that brands tend to take decisions in a data driven decision making process. It has been found by a McKinsey study that brands using this strategy are more likely to be profitable and retain their old customers besides acquiring new ones (Harreis et al., 2023). The same applies to the fashion industry that is rapidly shifting to data analytics to estimate and meet consumer demands for quick deliveries, and to ensure sustainability and profitability(Nwobodo, 2024). Traditionally in the fashion industry decisions were based mostly on intuition, previous experiences and the demand of the readily available population to a particular brand. However, with globalisation, catering to the global demands whilst being situated in just one place did pose a challenge. Besides this, earlier designers relied solely on their personal insights, creativity and industry expertise to predict upcoming trends, which was rather flawed, as trend agencies made informed guesses on the next big trends by observing runway shows, fashion week and celebrity influences, and this catered just to the elite few(Howland, 2021). The unpredictability of consumer preferences, especially with expansion, often led to misjudgements which further resulted in either overproduction or missed opportunities, both resulting in losses (Alina, 2023). And this is where data analytics changed the entire scenario. Zara is often considered a pioneer in using data analytics to drive its fast fashion model and meet the consumer demands efficiently. This approach to decision making comes with several benefits, namely, better forecasting to ensure right products are manufactured to meet consumer demands, improving efficiency through supply chain by analysing the inventories and demand fluctuations ensuring efficiency in internal operations, sustainability by ensuring that waste is minimized and trimming excess stock to prevent losses for the companies (Anitson, 2023).

1.2 Predictive Analysis, from Past to Present in Fashion Industry

Predictive Analytics in the fashion industry started off initially with trend forecasting agencies that would analyse historical trends and cultural movements (WGSN, n.d.). They did so by observing runway shows, fashion week and celebrity influences. They relied heavily on expert analysts who studied the fashion cycles, customer preferences and broader cultural shifts in order to forecast future trends (Howland, 2021). Besides observing runways, they also examined past trends, art movements, economic influences and global events and based their decisions of predicting future trends on these factors. Some examples of publications playing a major role in influencing and predicting trends were WGSN and Pantone's colour forecasts (WGSN, n.d.). These agencies provided detailed reports on emerging trends of patterns, colour palettes, fabric and even cultural sentiments that influences preferences thus helping brands align collections with anticipated demand, basically playing the role of what predictive analytics does today with much less of a hassle. Today predictive analytics helps fashion business predict trends with the help of historical and current data in a matter of minutes by identifying essential correlations which may go unnoticed by humans (Leitão, 2024). Artificial intelligence plays an important role in predicting trends today utilising machine learning models to analyse hug data, including social media trends, search keywords and sales figures (Shi & Lewis, 2020). The insights from these analyses are used to fine tune the companies' respective marketing campaigns, focus on meeting the demand and optimise supply chain management. As mentioned earlier, with the advent of globalisation, understanding every consumer's attitude is hard but crucial, and this is possible today with the help of predictive analytics, thus making it indispensable (ZYOD, 2023). By delving into customers' digital footprints, like browsing habits, purchase history and social media interactions, brands tailor marketing strategies ensuring that they retain their customers whilst also acquiring new ones. AI predictive analytics does not only help in trend prediction but also streamlines supply chain operations by analysing sales trends, inventory stocks and production schedules and thus helping companies to make eco-conscious and profit-conscious choices by reducing unnecessary waste and ensuring cost savings (Siddique, 2024). This results in a more efficient and eco-friendly production and operation cycle, that not only fits consumer demands and enhances the company's financial performance but also helps protect the planet.

1.3 Role of Artificial Intelligence and Machine Learning in Predicting Fashion Trends

The global fashion industry, valued at roughly $2 trillion is experiencing a huge technological makeover, and it is Artificial Intelligence and Machine Learning that serve as the primary cause for this (UN-HJDM, 2024). With the advent of AI and ML, trend forecasting has become highly accurate thus ensuring that the fashion industry is not just profitable by meeting the necessary demand but also sustainable by ensuring wastage of resources is reduced (Admin, 2024). This is extremely important as the fashion industry is responsible for 4% of the world's emissions. AI analyses not just historical sales and trends but also current consumer preferences, live market data and even inventory stock levels thus aligning the production with the demand efficiently. Earlier trend agencies would base decisions on future trends by observing fashion shows and celebrity influences but AI today does not only this but also analyses millions of social media posts each day, spotting new trends in pictures, videos and company sales figures (Shi & Lewis, 2020). Trends can be predicted up to a year ahead thus contributing to more efficiency (Chakraborty et al., 2020). A Paris based company, Heuritech has developed AI models that analyse millions of images posted on social media every day, classifying users as either "edgy", "trendy" or "mainstream", post which these are coded from a list of 2000 attributes and tracked in real time to allow the AI model to determine the popularity of a specific attribute in clothing (Heuritech, 2020). This of course comes with the concern that all designs of different brands might become too similar due to the similar algorithms and the only way to address this is by ensuring the right mix of technology and human creativity (Howland, 2021). AI will undoubtedly boost the fashion industry's profits and cut down on costs by aligning with what the consumers are looking for, but this does not mean human intervention should be completely cut off (Reporter, 2023). Data is accuracy whilst the human touch adds in that pinch of emotion that makes brands stand apart. Market leaders like McKinsey foresee huge profitability for brands that use AI creatively (Harreis et al., 2023). Besides helping in predicting trends with predictive analytics, AI also improves the shopping experience for consumers by providing virtual try-on tech that lets the consumers see how the clothes would look on them, leading to more satisfaction and lesser purchase returns (StartUs Insights, 2022). AI also plays a crucial role in supply chain management as will be elaborated upon later leading to better inventory control.

Even designers today use AI tools like Generative Adversarial Networks and Style Transfer Models (UN-HJDM, 2024).

Generative AI also plays a crucial role, slightly different from that of a traditional AI model, by generating new content, simulating complex scenarios and providing deeper insights with the help of advanced data synthesis (K et al., 2024). Often companies face the challenges of incomplete or unstructured data and this is where Gen-AI plays the most important role by filling in the gaps with high quality synthetic data and by improving data completeness it enhances the overall accuracy of the model in predicting trends (Harreis et al., 2023). Besides this, Gen- AI models help simulate future scenarios like inflation, stock market changes etc, allowing analysts to make informed decisions to ensure that companies work seamlessly even in cases of external contingencies, Gen-AI is also primarily useful as it can interpret data not just from texts, but also videos, images and various other formats, thus allowing for a more holistic approach ensuring efficiency in predictions. These models also help reduce human biases as predictions are based only on the data available and the patterns recognised in it without any intentional subjective influences (K et al., 2024). Thus, it helps in improving objectivity and strategic-decision making.

Machine Learning complements AI in all of its functions (Labs, 2024). By using ML for trend prediction, the fashion industry aims to cut down on the fabric wastage, which roughly amounts to 92 million tonnes annually. It also reduces overproduction as about 40% of the goods produced are not sold at the full price and 25% are never sold (Alina, 2023). ML models are used to analyse consumer preferences by using both first-party and third-party data and hence meet consumer needs effectively (Nadeem et al., 2022). By lessening overproduction, increasing efficiency and reducing waste, ML plays an important part in keeping the industry profitable and sustainable.

2. FASHION TREND FORECASTING USING PREDICTIVE ANALYTICS

All the fashion brands in today's ever evolving world, use predictive analytics to predict upcoming trends, sometimes even up to a year, which in turn helps them refine their products, in order to meet the current demand, providing a competitive edge in the industry. With the help of data analysis, derived from consumer purchase data, social media interactions and live sales data, brands ensure profitability, efficiency and sustainability (Gong, 2024).

2.1 Leveraging Data Science for Trend Prediction and Demand Forecasting

In today's constantly evolving marketplace, predicting trends is of utmost importance to ensure profitability for businesses. Trend prediction in turn leads to demand forecasting which is nothing but the art of assessing the current demand of the market and tailoring the production accordingly (Waterston & Choudhary, 2024). And we've transitioned from reliance on gut instincts to spreadsheets to finally Data Science that helps businesses leverage the power of accurate data for useful insights that help in strategic decision making.

There are several methods and techniques of demand forecasting which include the moving average that calculates the average demand over a specific predefined period (Raj, 2024). The only drawback of this is seasonality. Exponential smoothing is another technique that assigns weightage to past data points, providing a higher weightage to more recent data. This adapts better to trend fluctuations than the moving average method. The ARIMA model, that stands for Autoregressive Integrated Moving Average analyses trends, seasonality and random error in the time series data, which is helpful in capturing complex patterns but also requires statistical expertise to be implemented efficiently (Waterston & Choudhary, 2024). Customer Surveys and Sales force composite are other methods of data collection. Sales force composite basically refers to leveraging the opinions of salespeople but comes with the problems of individual biases and optimism. Barometric Analogy assumes a relationship between broader economic factors like GDP and future trends, but the strength of each relationship may vary (Raj, 2024). The steps involved in data preparation include data collection from internal and external sources, data cleaning which ensures there is no missing data that would impact the results, data transformation that is basically creating new variables that might help in the study and data exploration and visualisation to identify pattern, trends and potential relationships between variables to predict accurately (Waterston & Choudhary, 2024).

The various Machine Learning techniques used for demand forecasting include regression techniques, decision trees, random forests, support vector machines and neural networks (Chakraborty et al., 2020). The evaluation metrics for trend prediction and demand forecasting models include Mean Squared Error(MSE) which measures the average squared difference between predicted and actual values and a lower MSE indicates better accuracy; Mean Absolute Error(MAE) which measures the average absolute difference between predicted and actual values and is easier to interpret than MSE; Mean Absolute Percentage Error (MAPE) which expresses the error as a percentage of the actual demand values and is useful for comparing forecasting performance across different categories; and Root Mean Square Error(RMSE) which is nothing but the square root of MSE and provides a measure of the error in the same units as the predicted and actual values (Waterston & Choudhary, 2024). Beyond these basic techniques, advanced models like casual modelling, ensemble learning and deep learning exist today that are shaping the industry as never imagined before. Casual Modelling models casual relationships between all the factors affecting the demand, leading to a more robust forecast. It often involves techniques like SEM (structural equation modelling) to understand the cause-effect relationship. Ensemble learning on the other hand combines multiple forecasting models to leverage the strengths of each potentially leading to higher accuracy (Zhu & Zhan, 2022). Deep Learning used deep neural networks that can capture complex non-linear relationships in data, in large unstructured datasets, leading to high accuracy. The future of Data science approaches in trend and demand forecasting is very bright and some of the achievable possibles are real-time forecasting, incorporation of external data like weather data etc., and explainable AI that would help understand how models arrive at their predictions fostering trust and interpretability, thus moving beyond just guesswork and gaining a clear picture of future demand (Waterston & Choudhary, 2024).

2.2 Social Media and Sentiment Analysis for Fashion Forecasting

Social media has become a widely used tool to predict fashion trends, influencing consumer preferences and thus moulding the strategies that companies use to promote their products (Koren, 2024). All social media platforms like Twitter, Instagram, TikTok and Facebook are real time sources of trend data with users discussing, sharing, commenting and engaging with fashion related content. Apart from this, celebrity influencing on social media also plays a very important role in determining the next popular trend. According to research conducted, interactions on social media like likes, comments, and shares indicate strongly the emerging trends (Koren, 2024). A study highlighted the integration of data from social media platforms in forecasting models, showcasing that the engagement metrics from different platforms provide invaluable insights into consumer behaviour thus helping companies base their marketing strategies for effective promotion of products whilst also ensuring that they catered to the demands of the market effectively (Shi & Lewis, 2020). Unlike traditional sales data, which is most often historical, social media content ensures that early-stage trends are captured before they reach the retail market, thus helping to anticipate the demand more effectively. Another study explored the role of Twitter in fashion forecasting, which showed that keywords, hashtags and user discussions related to fashion, could be mined using natural language processing (NLP) which would help in detecting trending styles, colour palettes, fabrics etc

(Koren, 2024). Brands have thus been able to track real time consumer sentiments and cater accordingly, leading to more profitability.

Sentiment analysis helps in enhancing the emotions and opinions expressed on social media related to fashion and thus helps companies identify the pain points of consumers and change their products accordingly to better fit needs (GhorbanTanhaei et al., 2024). Using ML and NLP techniques, consumer discussions are categorized into positive, negative or neutral sentiments which further provides better insights into preferences. By integrating sentiment analysis with trend forecasting models, the accuracy of prediction could be highly improved and thus identify if a particular trend is gaining traction or backlash (Chakraborty et al., 2020). Positive sentiments generally indicate the strong likelihood of the particular aspect being measured of turning into a trend. This helps companies refine their products and marketing strategies ensuring they meet the current demand and enhance profitability of the overall business.

2.3 Case study of Brands Leveraging Predictive Analytics for Trend Prediction: Adidas

Adidas has used predictive analytics to ensure their products meet the current trend and are customised according to consumer requirements. They introduced interactive digital platforms on their website and mobile app, powered by AI that enable consumers to select colour schemes, pattern, type of material and even add personalised text (Idara, 2023). This AI component collects user preferences and provides real time visual-mock-ups ensuring the end product is exactly according to preferences. Adidas also coupled design software with machine learning in order to detect design clashes and ensure the feasibility of each customization. In terms of personalization, for the uncertain consumers, the AI platforms use data collected from similar consumer profiles to suggest styles that might resonate. For marketing strategies, the brand's marketing team uses certain AI algorithms and categorizes consumers based on purchase history, browsing behaviour, geographic location and social media interactions (Idara, 2023). Adidas also deploys AI driven predictive analytics to forecast the outcomes of marketing campaigns before launch. Demand and trend forecasting is done by analysing sales data, social media trends, economic indicators and even weather patterns thus predicting future demand with impressive accuracy (Idara, 2023). Adidas also uses predictive analytics to implement dynamic pricing with the help of Artificial intelligence in some markets and hence adjusting prices based on real time factors like stock levels, pricing of products offered by competitors and current demand. By using predictive analytics and AI to predict trends and forecast demand, besides also assisting in personalization, Adidas positions itself as a leader ensuring brand loyalty, operational excellence, profitability and eco-conscious strategies.

3. DEMAND FORECASTING AND INVENTORY MANAGEMENT

Inventory management and demand forecasting are crucial in the fashion industry to prevent stockout or overstocked situations which would otherwise result in dissatisfied consumers, reduced brand loyalty, missed opportunities to gain profits, in the case of a stock out or excessive storage costs and wastage of products in the scenario of an overstock (Team, 2023). Demand forecasting ensures optimized supply chain operations leading to smooth productions and distribution processes (Mahler, 2024). This will allow brand to maintain a competitive advantage, have a loyal and satisfied customer base, increased brand loyalty and eco-conscious operations.

3.1 Optimizing Demand Forecasting with Historical Sales Data

Demand forecasting refers to predicting or foreseeing the future demand for a particular product or service, ensuring that business utilize all resources effectively and waste is minimized (Raj, 2024). Demand forecasting is done by analysing the historical sales data and taking into consideration the various factors that may influence demand of a product such as economic activities, competitor pricing etc. ("Optimizing demand forecasting", n.d.)

Accuracy in demand forecasting ensures companies do not overproduce or underproduce and meet the demands of the consumers effectively, thus ensuring profitability for the business and sustainability for the environment. Businesses can plan their production operation more effectively and even be prepared for potential contingencies such as disruptions in the supply chain, natural disasters etc. Overall accurate demand forecasting would help in improving the decision-making process, reducing the cost of production and increasing profitability for long term sustenance of the business ("Optimizing demand forecasting", n.d.). There are several methods to demand forecasting which include historical data methos, demand sensing method, external macro forecasting methos, market research and Delphi methos, and the predictive sales analytics method (Raj, 2024). Delving deeper into what the historical sales data methos is, it includes analysing, identifying and segmenting the data. By analysing historical trends, companies can figure out trends or patterns that can influence possible future demand for a product or service .("Optimizing demand forecasting", n.d.). Identification of seasonality is another important step, which involves recognizing seasonal patterns so that business can align their inventory and production cycles accordingly. This is followed by the segmentation of data by various categories such as consumer demographics, product type etc. This is done to identify demand variations based on different factors thus enabling accuracy in forecasting for specific products, regions or consumer segments. It is imperative to consider external factors such as economic conditions, competitors' sales and pricing and market trends to enhance the accuracy of prediction. AI and ML as already mentioned earlier, play a crucial role in enhancing the accuracy of demand forecasts and are of prime importance for the same ("Optimizing demand forecasting", n.d.). Demand forecasting however comes with a few challenges such as data quality and availability, rapid market changes and forecasting biases, each of which can be overcome by implementing robust data management practices and investing in good technologies, by constantly monitoring trends and adjusting forecasts as needed, and by relying on data-driven analysis and objective insights respectively (Raj, 2024).

Delving into the other methods of demand forecasting, the market research and Delphi method involves direct market research through surveys but is very time-consuming. In the demand sensing method, ML is used to monitor and capture real time variations in purchase behaviour, ML is also used in the Predictive Sales Analytics Method along with advanced IoT to determine relationships between various factors and make accurate predictions based on the relationships Raj, 2024). External Macro forecasting method is as the name suggests analysing trends in the broader economy to understand how they would affect the companies' goals. Some real-life companies using demand forecasting include Johnson & Johnson that uses demand forecasting to manage the complexities of a healthcare supply chain ensuring product availability at all times ("Optimizing demand forecasting", n.d.). With the unpredictable, volatile and fast changing trends of the current market, demand forecasting plays a crucial role in ensuring that profitability is maintained and wastage is reduced.

3.2 Preventing Stockout and Overstock with Predictive Analytics

With demand forecasting that is accurate, predictive analytics helps companies determine the required quantity of a product to ensure that it is neither understocked or overstocked hence preventing the cases of both unsold inventory and missed sales opportunities from occurring respectively (Safna, 2025; Team, 2023). Using data analytics, AI and ML, analysis of past sales history, current market trends, economic factors and other influencing factors is conducted to make informed decisions on purchasing, storing and distributing inventory (Mahler, 2024). The key benefits of predictive inventory management include precision, cost efficiency, customer satisfaction, improved decision making and streamlined supply chain operation which help in reducing disruptions whilst also keep companies prepared for any contingencies (Safna, 2025). Predictive Inventory Management requires huge amounts of quality, error-free data to make accurate predictions. The data includes historical sales data, seasonal trends, economic factors, market conditions etc. Apart from directly related data, external data like changing climate conditions should also be included as they have been proven to influence consumer preferences. It is very important to choose the right predictive analytics software that must align with the quantity of data the company wishes to feed in and also with the specific features that a business would require to succeed (Safna, 2025). A real-world application of utilisation of predictive analytics for inventory management is Zara, that employs predictive analytics to maintain optimal stock levels thus reducing waste and ensuring that their most popular products are readily available to consumers at all times. Zara with the help of predictive inventory management is able to respond swiftly to market changes thus ensuring that it maintains itself as a global leader in fast fashion (Mahler, 2024).

3.3 Role of Predictive Analytics in Supply Chain Optimization

Predictive analytics for supply chain management uses data, statistical algorithms and ML techniques to identify the probability of various future outcomes (Raj, 2023). The benefits of using predictive analytics for supply chain include the accurate determination of market demand, efficiency in inventory management, reduction of stock outs and elimination of overstocked situations and improved logistics management (Vavilala, 2023). It also helps in identifying potential disruptions in the supply chain thus helping in smooth operations with no delays and downtimes (Prajapati, 2024). Predictive analytics can be used in supply chain management for demand forecasting as elaborated upon earlier. It can also be used for capacity planning which is a critical aspect of supply chain management involving the determination of the right number of resources required to meet future demand (Vavilala, 2023). With accurate demand forecasts, supply chain management professionals can make informed decisions about the staffing, resource allocation and production levels. With demand forecasting, predictive analytics also helps ensure that the risks of stockouts and overstocking are reduced (Omojola, 2025). Predictive analytics is especially useful in retail fashion and for all the above functions and to also identify potential bottlenecks, ensure stores have the right store layouts, staff levels and promotional activities. Even in the manufacturing process, predictive analytics play an important role by predicting equipment failures, identifying bottlenecks and hence improving the overall efficiency (Prajapati, 2024). It can also help companies manage supplier performance and reduce risks by using metrics to identify delays or quality issues. Thus, supply chain experts can use the data and results to renegotiate contracts, diversify supplier portfolios or implement contingency plans (Raj, 2023). Predictive analytics is also vital for the logistics and transportation management as it helps in optimizing transportation routes and schedules by considering various internal and external factors that would affect the efficiency of transportation. It also helps in real-time tracking and monitoring of shipments and several companies today have integrated predictive analytics with IoT devices and sensors for real-time data collection on the location, condition and status of shipments (Vavilala, 2023). This helps to proactively address potential delays or disruption in the supply chain. All of this leads to an increase in the efficiency of supply chain management and overall customer satisfaction (Prajapati, 2024).

4. PERSONALIZATION AND CUSTOMER BEHAVIOUR PREDICTION

Predictive analytics plays a crucial role in ensuring that consumer experiences are enhanced with the seamless delivery of recommendations that are personalised and also helps in the prediction of consumer purchase behaviour which further helps in demand forecasting (GhorbanTanhaei et al., 2024). By analysing user preferences, past purchase history and their browsing patterns which are tracked by analytics to offer personalised recommendation which in turns helps in customer satisfaction and retention (Yalcin, 2024).

4.1 Personalised Recommendations with Predictive Models

Personalised recommendations refer to the practise of tailoring product recommendations and marketing messages and experiences to individual consumers based on their interest, preferences and behaviours (Yalcin, 2024). This is done by collecting data from various sources such as website analytics, social media profiles of consumers and surveys and feedbacks. The analysis of this data is then carried out using advanced algorithms and machine learning techniques to identify pattens and insights about each individual consumer (Shirkhani et al., 2023). This is done in order to create a seamless consumer experience that will in turn enhance brand loyalty and increase sales. AI and ML are extensively used to analyse consumer data and make personalised product recommendations based on each individual's preferences (Sridevi et al., 2020). They are also used to make predictions about future behaviour by identifying consumers most likely to make a purchase and targeting them with personalised marketing messages (TimesTech, 2024). Chatbots and virtual assistants, using AI and ML also provide personalised recommendations and assistance to customers. Personalised recommendations today are not limited to just online shopping; AI-powered in store experiences supported by smart mirrors suggest outfits based on the items tried on by a consumer creating an interactive shopping experience (TimesTech, 2024). AI is also improving the efficiency of personal stylists by analysing client preferences and current trends and assisting stylists in making personalised wardrobe selections for clients. A very good example of this is ASOS, one of the most recognizable e-commerce platforms that has created a reputation for itself by building highly personalised experiences for its more than 26.4 million active consumers. The recommendation engine of the company is powered by real-time behavioural data whose key tactics include hyper personalised homepages, dynamic style matching, for example if an individual bought a pair of sneakers, the ASOS website would not recommend just a pair of socks but a whole outfit curated personally for the individual consumer (Yalcin, 2024). They use diverse data sets, ranging from browsing behaviours, regional performances, weather patterns etc. While this is undoubtedly changing the landscape of the fashion industry, it is not bereft of concerns. The most important ethical concern is the collection and use of personal data of consumers (Fashion Law Journal, 2023). All companies must ensure that their systems handle the data securely, efficiently, transparently and operate within the ethical guidelines. By making personalised recommendations accessible to a larger audience, the fashion industry can ensure that consumers from all walks of life enjoy the experience of a tailored shopping experience whilst ensuring profitability for the company itself (TimesTech, 2024).

4.2 Predicting Customer Purchasing Behaviour and Preferences.

Data plays a very important role in predicting consumer behaviour and their responses to products (The Fullstory Team,2024). With the help of data collected from different sources like social media interactions, online purchase history, browsing history etc. Using AI and ML tools companies can derive results and relations that play an important role in influencing consumer behaviour which may otherwise be missed by human observation (GhorbanTanhaei et al., 2024). Some of the most common applications of using data to predict consumer behaviours include the estimation of customer lifetime value, identification of potential churn risks, discovery of cross-selling and upselling opportunities and forecasting revenue and sales (Krysiak-Adamczyk, 2024). Customer journey mapping on online fashion sites is of prime importance in understanding the consumer. It basically refers to recording in detail the path from the initial step to the final purchase step and sometimes even post purchase services to identify and improve the company's service to consumers (The Fullstory Team, 2024). Each interaction of a brand with a consumer is known as a touchpoint and implementing an analysis of these touchpoints allows brands to map the entire journey of the consumer from the consumer's point of view and helps the company to empathise with the consumers, identify pain points and address these in order to increase consumer loyalty and retention. The various stages that a consumer goes through before making a purchase constitute the conversion funnel, and mapping each stage of the funnel like awareness, interest, decision and action and using the results obtained to ensure a smooth progression for the consumer (Krysiak-Adamczyk, 2024). ML and NLP play an important role in predicting consumer behaviour by automating the recognition of the purchase behaviour thus helping in making personalised recommendations and interpreting consumer sentiments giving specific factors to predicting behaviour, respectively (GhorbanTanhaei et al.,2024). Some of the emerging behavioural metrics today are engagement scores that measure user interaction levels to predict purchase behaviour and social listening tools that keep tabs on brand popularity and influencer tractions. Predicting consumer behaviour is of prime importance for businesses in the fashion industry as it helps in ensuring that inventory is aligned to the current market demand thus reducing wastage of material promoting a cleaner and greener fashion industry (The Fullstory Team, 2024).

5. □DYNAMIC PRICING WITH ANALYTICAL STRATEGIES

Dynamic Pricing in simple terms is to change the particular price of a product according to the market conditions (Wang & Chen, 2024). When the demand is high, fashion companies often increase the price of particular products, for example, the price of branded sunglasses are likely to increase during the summer, or brands tend to often launch a new product like for their summer season or during the winter season that may be priced according to the demand. Data analytics aims to streamline this procedure by providing a vast array of customer data, which makes it easier and efficient for the brand to price their products strategically (Sánchez-Cartas et al., 2021). Throughout this section we will go in depth as to how.

5.1 Optimizing Pricing with Predictive Analytics

Predictive analytics plays a revolutionary role in the fashion ecosystem, from giving data driven insights to reshaping creativity, consumer behaviour and business strategy, it's causing disruption in today's day and age (ZYOD, 2023). Predictive analytics improves short-term sales, helps with forecasting consumer behaviour and providing flexibility for the company to not incorporate seasonal changes according to the preferences of the customers but also manage their supply chain to reduce wastage overall, making sure that a right estimation is made while stocking the stores (Alina, 2023).

Artificial Intelligence (AI), is nowadays used to identify patterns, trends from a vast pool of databases which include social media, online search histories, in order to predict future demand accurately. Pricing is also a major part of this procedure (Demera, 2024). Machine learning is used to analyse historical data, consumer patterns and market trends, enabling the retailers to offer optimal pricing points for their products according to the consumers that they are catering to, which maximizes revenue and increases customer satisfaction (Vorozhtsov, 2023). Broader market movements are used to make a proactive pricing decision, which is only possible due to real-time data utilization. Due to the amalgamation of technology in the field of fashion are businesses now capable of continuously learning, adapting, ensuring that pricing strategies remain optimized according to the customers and evolve according to the market conditions (Demera, 2024). Usage of AI, minimizes the scope of missing important details within trends and purchase patterns that a human analysis might. Leading to smarter and faster pricing decisions. The best thing about using predictive analytics? It's scalable across all product lines, markets, allowing retailers to maintain competitive pricing strategies globally (Vorozhtsov, 2023).

5.2 Real-time Pricing Adjustments based on Demand and Trends

With the help of algorithms, the fashion industry as a whole can analyse the market conditions, inventory levels and offer competitive pricing which adjust dynamically and optimize revenue (Sánchez-Cartas et al., 2021).

The Bayesian Model, utilizes historical pricing data as a prior belief and updates it dynamically with new data to adjust the prices in real time (Liu et al., 2020). Predictive analytics is a major boon, because it learns optimal pricing strategies by analysing the recent customer demands according to the seasonality, and competitor pricing in order to maximize revenue. Modern algorithms right now use web scraping tools to gather competitor prices and stock levels in real-time, we can get immediate pricing adjustments once we feed this data into pricing models (Demera, 2024). Another feature is personalized pricing, which means that the pricing is adjusted based on an individual customer's purchasing behaviour, which can be used to curate tailored offers.

Supply-Demand Balancing, is the kind of dynamic pricing which adjusts prices to incentivize suppliers during high demand periods, similarly adjusts the pricing when the demand is low (wang & Chen, 2024). The Phase Diagram Modelling is used to demonstrate conditions under which, dynamic pricing schemes succeed or fail to maintain balance.

We see today that E-commerce platforms like Amazon, Flipkart use automated web crawlers which monitor competitor prices, stock availability and keep an eye on the customer reviews and make adjustments accordingly. For example, Amazon changes its prices over 2.5 million times a daily in order to remain competitive and maximize its profits (Sánchez-Cartas et al., 2021). For companies leveraging these models effectively, Dynamic pricing is crucial, it has been shown to increase sales up to 5% (Demera, 2024).

Seasonality affects drastically, as trends tend to change almost instantaneously, with the help of predictive analytics, we can use historical data to combine it with seasonal trends to not only project demands but to adjust prices accordingly (Wang & Chen, 2024). Revenue Dashboards are used to visualize how demand forecasts impact financial metrics like gross profit and profit margins which also enables macro-level changes across product categories based on the aggregated demand projections.

Machine learning algorithms play a crucial role in analysing customer data, to determine the most effective price points as stated, however that is not where the advantages of using predictive analytics limits itself (Vorozhtsov, 2023). Data-Driven Decision Making, is something you will see in a lot of textbooks, but what it simply means is adjusting price after analysing market conditions by leveraging historical sales data, retailers identify seasonal demand fluctuations and adjust pricing accordingly. This strategy is particularly effective for increasing the sales volume during off-peak periods, which makes it possible to ensure consistent sales throughout the year.

Price Elasticity is a concept that retailers use to offer standard prices to customers who value immediate purchases and discounts to customers who do not mind waiting (Percherla, 2024). Often retailers offer discounts to clear the excess stock and stimulate the sales, which in turn also reduces wastage and gives value to the effort put in making the product. Let's explore an interesting case study now.

5.3 Case study of Dynamic Pricing in Fashion Retail: Zara

Zara is a major example about how a company can rightfully leverage data analytics to adjust to prices dynamically to respond to real-time demand, competition and market trends (productdatascrape, 2024). This approach allows Zara to maintain profitability while enabling it to stay competitive in the fast fashion industry. Zara also employs a targeted discount strategy, as quoted above which means offering discounts ranging from 15% to 60%, these discounts are strategically applied to clear the inventory while making sure that the brand value remains intact. Higher levels of discounts are reserved for specific items or promotional periods (productdatascrape, 2024).

Zara is a brand that is known for tailoring its pricing strategies according to the market conditions, which takes into account regional economic conditions, consumer purchasing power, pricing strategies and competitor pricing (productdatascrape, 2024). This type of flexibility allows Zara to remain competitive across global markets, oftentimes giving it a competitive edge over other similar companies which cater to the same audience. As mentioned before, with the help of Predictive Analytics, Zara adapts to the current market dynamics. This kind of agility in the supply chain which allows the inventory to increase and decrease as per customer demands, reduces the need for excessive markdowns.

Zara has positioned itself as an affordable luxury brand in the market, offering high fashion choices at accessible prices for the masses. This strategy has attracted style conscious consumers while also balancing profitability through efficient supply chain management and just in inventory models (productdatascrape, 2024). During the economic downturn like the financial crisis in 2008 and the COVID-19 pandemic, Zara had adjusted its pricing strategy by offering promotions and discounts to maintain its sales. This type of malleability has allowed the brand to not only retain its customers, by adapting to the ever-changing consumer behaviour but also offer competitive pricing in order to stay ahead of its competitors, in this dynamic world of fashion.

Another strategy is deliberately limiting stock availability for most items, which creates a perception of scarcity, which in turn drives the consumer's urgency to purchase at the full price rather than waiting for a discount (productdatascrape, 2024). This strategy minimizes markdowns while maximizing profits.

6. SUSTAINABILITY AND PREDICTIVE ANALYTICS

Sustainability is meeting the current needs without compromising on the needs of future generations. How does the fashion industry ensure that they stay sustainable? Starting from using sustainable materials to accurate production planning and stocking the inventory up with what is required, avoiding over-stocking, and using supply chain management procedures that support a circular economy within the fashion industry (Akram et al., 2022). These are just a few examples of how sustainability can be practiced in the world of fashion through small yet implementable steps with the help of Predictive Analysis.

6.1 Minimizing waste with Accurate Production Planning

Demand forecasting is used to align production with consumer needs, as mentioned before, historical data is used in order to reduce overproduction and ensure that only products which are likely to sell are manufactured, minimizing waste and optimizing human effort (Alina, 2023). Digital testing of colourways, and styles, in order to determine which options are most likely to succeed, provides the designers with valuable insights, which in turn helps them determine which options are most likely to succeed, and prevents the production of unpopular items which often cost a lot of manual labour as well as take up unnecessary stocking space, this aids in reducing inventory waste and chemical usage in processes like garment dyeing. Data-driven line planning has made it even more efficient for overstocking to be managed. AI systems analyse the production processes identifying inefficiencies and optimizing resource allocation in real time (e.g., water, energy and other such resources used during the printing process). This ensures a reduction in manufacturing waste and enhances sustainability efforts across the entire supply chain (Akram et al., 2022). Accurate dyeing processes are used with the help of predictive analytics, which improves the efficiency of dyeing by selecting optimal colour assortments based on consumer preferences, reducing water usages and minimizes chemical wastage in the process as well, this makes dyeing more sustainable in the long run while managing to save costs. Minimizing waste has always been an integral part of sustainable planning, because it ensures that the manufacturing process itself does not produce waste, as only the products which customer demands for, gets made, therefore preventing the process of waste management in its initial stages itself (Alina, 2023).

6.2 Predicting Sustainable Material Needs

AI algorithms analyse the product design and suggest sustainable materials based on the predefined parameters such as natural fibres, ethical sourcing and suppliers' compliance history (Akram et al., 2022). This ensures that brands not only use environmentally friendly materials but also ensure ethical sourcing of those materials while reducing reliance on non-sustainable options. Predictive analytics tools like Preware monitor suppliers for any violations in sustainability by scanning their online information related to environmental, labour and governance practices. This helps the brands to assess their supply chain's sustainability efforts and ensure adherence to ethical standards. Fashion businesses now use data analytics to evaluate the potential suppliers based on metrics like material quality, environmental impact and cost efficiency. This helps businesses make informed and data-driven decisions that prioritize sustainable and ethically produced materials (Gong, 2024).

Also helps reduce overproduction of unsustainable materials and ensures alignment with demands for eco-friendly products by optimally utilizing predictive analytics for material demand forecasting. These predictive models help identify opportunities to reuse or recycle materials within a circular economy framework, reducing waste and dependency on virgin resources while at the same time promoting sustainable practices in material sourcing (Akram et al., 2022).

6.3 Role of Artificial Intelligence in Advancing the Circular Fashion Economy

A circular economy is one in which aims to minimize waste and resource depletion by utilizing resources for as long as possible, by either recycling it, replenishment, refurbishment or repairing products for further usage. With the help of Artificial Intelligence (AI) designers in the garment-making industry use materials, patterns, which can later aid the process of working towards circularity (Akram et al., 2022).

AI also suggests designs that are easier to recycle or reuse, extending product lifecycles and supporting sustainable practices. Through the integration of AI algorithms, implementation of circular economy is possible which facilitates opportunities for repair, reuse or recycling based on data related to product returns, customer preferences, this creates a circular closed-loop system that reduces waste (Akram et al., 2022). Combined with blockchain technology which enhances transparency in the supply chain by tracking the lifecycle of garments from raw materials sourcing to recycling (StartUs Insights, 2022).

Circular design strategies include making of garments that can easily be disassembled and worn for multiple purposes, it can be reused and recycled according to the consumer needs, giving us dynamic options of how to wear a single piece of clothing in multiple different ways, increasing its product life cycle. "Refashion" projects use AI-aided designs processes to create multifunctional fabric blocks that can be transformed into various styles, this also does a great deal in promoting zero-waste production within the industry (Akram et al., 2022). Therefore, more of these projects should be funded and encouraged.

Reverse Logistics is the process of moving goods backwards in the supply chain back from the point-of-sale to the manufacturer, distributor for purposes like reusing and recycling. It encompasses a vast number of activities including managing unsold goods, customer returns and end-of-life production as well. This process heavily promotes the development of closed-looped systems within the fashion industry. Where the garments that are rejected somehow find purpose again, and are created into new products, conserving the resources with which it is made and giving justice to the labour out into making it in the first place (Percherla, 2024) .

Personalization of fashion material plays an important role even when we discuss sustainability, as the customers are more likely to retain and reuse a piece of garment if they feel personal attachment towards it (Yalcin, 2024). With the help of predictive analytics, retailers can now provide a personalized shopping experience like never before. By analysing the customer preferences, and suggesting durable and sustainable clothing options, made with material that are preferred by the customers according to the seasonality, they also manage to make a fashion statement in the process. This encourages the consumers to retain their garments for longer and value them, it also manages to reduce the frequency of replacements (TimesTech, 2024).

Ethical practices ensure and promote consumer trust in sustainable brands. Personalized AI-powered recommendations are given in order to extend product use, which is heavily used during personalized shopping experiences that reduce returns and unsatisfactory purchases, encouraging consumers to value and retain the items longer, this extends the lifecycle of garments and reduces waste (Shirkhani et al., 2023).

7. CHALLENGES AND ETHICAL CONSIDERATIONS

We know that "with great powers, comes great responsibility", similarly the use of AI in this industry is a boon, if used the right way. If not, it may lead to even more problems that it would end up solving, as data leaks, invasion of sensitive data of the customers are not only intrusive practices but are offensive. In this section we will explore how privacy remains a major concern in this domain as well, like any industry that uses AI to collect information about their consumers ("What are the data privacy concerns", 2024). This section will also address how the human mind constantly needs to be stimulated for the most creative outputs, unlike the ways in which today the use predictive analytics to constantly helps fashion companies to come up with creative solutions and often forget that the reason a personal brand exists is because of the uniqueness of the individual who created it. Afterall fashion is all about self-expression and creativity. Once that element is eliminated in the name of profitability and efficiency, are we really heading in the right direction?(Howland, 2021).

7.1 Data Privacy Concerns in Fashion Analytics

Fashion brands often share consumer data with third-party service providers for analytics and marketing purposes. This raises various concerns about consumer privacy protection, data control and security, once it is transferred as third parties may misuse data or fail to protect it adequately against breaches (Shewaramani, 2024).

AI applications in fashion are known to collect extensive and comprehensive data including their purchasing habits, style preferences and biometric data, which in turn makes it easier for the retailers to have a descriptive buyer persona, however, this raises multiple ethical concerns about how this data is analysed and used, and whether it can potentially lead to discriminatory practices and manipulation in marketing strategies ("What are the data privacy concerns", 2024).

Big data applications can lead to multiple privacy and security risks (Gong, 2024). The use of big data analytics in the fashion retail scene often exposes consumers to privacy invasive harms which can include exposure to sensitive information, price discrimination and manipulation of consumer choices (Fashion Law Journal, 2023). Additionally, this may also include multiple security vulnerabilities leading to data breaches and cyber-attacks. Under the compliance of the GDPR (General Data Protection Regulation in Europe, a regulation under the European Union and the CCPA (California consumer Privacy Act) which is a state law, in the United States of America, some fashion retailers have to face complex regulatory landscapes (Fashion Law Journal, 2023). These regulations impose strict requirements on data handling. And non-compliance can result in significant fines and reputational damage (Shewaramani, 2024). Many times, the consumers are unaware of the extent to which their data is being collected and distributed through various channels by the fashion brands. Therefore, ensuring that the consent of the consumers is taken to be transparent about data usage and making the consumers aware of this process is crucial for maintaining trust in the long run ("What are the data privacy concerns", 2024). This also aids the process of complying with legal standards. These concerns still persist today, and will remain an issue, however they can be minimized through proper regulatory measures and other consistent efforts to respect the privacy of consumers (Shewaramani, 2024).

7.2 Ethical Implications of AI Driven Fashion Predictions

AI systems often learn from historical data, as has been mentioned throughout this discussion, which itself may contain biases. Which in turn can often lead to limited representation and perpetuation of harmful stereotypes (Reporter, 2023). The biases may manifest themselves in favouring a certain body type, skin tone, or age groups, while excluding others, which gives rise to various concerns about inclusivity in the fashion industry as a whole, because whatever gets measured gets retained. Ethical and rightful usage of AI is exemplified when the fashion industry would use these resources in order to get a larger sample size of a more diverse population, promoting diversity and inclusivity in their designs, and size ranges

(UN-HJDM, 2024). AI can play an instrumental role in the attempt to mitigate biases for specific body types, cultures, sizes, etc, if used right.

The impact of AI in any industry is multifaceted, the adoption of AI in fashion design and supply chain management can also raise concerns about job displacement (Akram et al., 2022). Automation increases efficiency but can reduce the need for human workers, impacting livelihoods, creating ethical dilemmas regarding workforce sustainability. Through this we also see how the adoption of AI impacts the labour force and employment. Another concern it raises is one about intellectual property, AI-generated designs can blur the lines of ownership, as the algorithms can replicate or create designs that resemble existing works without proper attribution or consent (Fashion Law Journal, 2023). This can give rise to multiple legal and ethical questions about intellectual property rights and originality in the field of fashion design. The balance between creativity and automation is a tightrope that one needs to walk on carefully. While AI enhances and provides designers with real-time data and valuable insights for the most innovative designs, over-reliance on AI, restricts human creativity, the artistic expressions and their signature designs that the designers started out with (Howland, 2021). Leading to a mis-representation of their brand. Fashion companies must ensure AI serves as a tool to augment creativity rather than replace it entirely (Reporter, 2023).

7.3 Over Reliance on Algorithms vs Creativity Intuition

Risk is something that makes a journey interesting and intuition is built in an individual after multiple failed attempts, after which a calculated risk-averse decision leads to success. These are intuitive things that make humans experienced and train their minds to think on their feet and be creative problem solvers. There is a loss of this very crucial risk factor that often pushes fashion designers to put out their best work, with an over-reliance on AI algorithms and machine learning based models (Howland, 2021). In turn leading to a loss of creative, strategic risk taking in the fashion industry (Reporter, 2023). Designers today increasingly follow data-driven decisions rather than trusting their innate intuitions, resulting in repetitive and uninspired designs. This shift in society has weeded out individuals with strong fashion sense and real intuition from the industry. Generative AI tools like GANs (Generative Adversarial Networks) can produce innovative designs by analysing historical trends and customer preferences. These tools are most effective when they are guided by human creativity and customer preferences. They can be used as an example of how AI aids rather than replaces artistic vision.

Because so many of the fashion companies use similar technology to increase their efficiency and predictability, with the usage of predictive models, it also increases the risk of Homogenization in Fashion (Howland, 20210. We can see how many companies previously known to have their unique signature styles and iconic patterns are all starting to slowly follow similar templates to increase profitability just to meet the latest trends rather than give us an authentic taste of their brand identity, which given enough time in the market could have a brand value and audience of its own, even if it is not perceived initially as a hit. This diminishes diversity, and innovation in the fashion industry as a whole, highlighting the need for human input in order to maintain originality (Reporter, 2023). AI can never replicate the emotional and cultural nuances that human designers bring to their work, therefore a balanced approach is essential. Currently the AI-enabled design systems rely on human-in-the-loop (HITL) models, where the designs collaborate with AI tools to refine outputs and ensure originality. This approach preserves the artistic essence of fashion while leveraging AI for efficiency and trend forecasting (Howland, 2021).

8. CONCLUSION: FUTURE OF PREDICTIVE ANALYTICS IN FASHION

As we have seen throughout the paper, the future of fashion is going to be an amalgamation of creativity, innovation and technology (Magazzo, 2022). The field of fashion is ever evolving and dynamic, in ways which are rapidly changing, and the incorporation of technological tools like Artificial Intelligence, with the help of Predictive Analytics has only sped up the process (ZYOD, 2023). In this section we have quoted several examples of how integration of technology, namely predictive analytics has helped major brands to cater to the right audience, adopt smart inventory management procedures and make data driven decisions, the fruits of which the company will bear in the long run.

8.1 Emerging Technologies shaping fashion analytics

As exemplified by this discussion, AI is instrumental in revolutionizing fashion analytics by predicting consumer preferences and enabling a hyper-personalized shopping experience (Shirkhani et al., 2023). An example of this would be how Stitch Fix (an online personal styling service in the United States) uses AI to recommend styles based on the customer data like their preferences, size and purchase history, while a giant like Amazon integrates machine learning to tailor their recommendations in both their online and physical stores. Technologies like Augmented Reality (AR) and Virtual Reality (VR) are used to enhance the customer experience and customer engagement, through this, customers can virtually try-on several products, this in turn provides an immersive shopping experience (Magazzo, 2022). We see today with brands like Nike, and even Lenskart in an Indian context of how these tools have created a virtual showroom for their customers, where the customers try on accessories and specific clothing, to see how it looks on them (StartUs Insights, 2022). Another key tool used here is Blockchain Technology, which is widely adopted to improve supply chain transparency, allowing brands to track the lifecycle of garments, from procurement of raw materials, sourcing, to recycling the products to avoid any future wastage (Kochar, 2024). It is intended that ethical procedures are followed throughout the whole supply chain, which in turn supports sustainability goals in the process. Smart Inventory Management is possible today because of the Internet of Things (IoT). Which with the help of sensors track stock levels in real time across global supply chains (StartUs Insights, 2022). This in turn provides an efficient stock replenishment, reduces overproduction and minimizes wastage.

Trend forecasting is a term that has been used heavily throughout this discussion, let us go in detail about what it actually means with regards to practical implementation. Heuritech, is an advanced AI tool that helps major brands like Adidas, Dior and Prada, to analyse vast datasets from social media and e-commerce platforms and real-time-sales data (Heuritech, 2020). The insights gained from these platforms help the brand make educated decisions to adapt themselves to the changing market conditions faster and predict emerging trends, enabling faster adaptation to market changes, giving them a competitive edge (Admin, 2024).

Often designers need a dummy version of their product to test and experiment with, with the help of 3-D printing, the designers can now not only create intricate designs with minimal waste, but only make sure that only the required amount of material is utilized (Kochar, 2024). Designers are able to meet on demand production, circling back to the topic of reducing overstocking and promoting sustainable purchasing habits, through these types of technology. Major brands like Nike, use this type of 3-D printing to push their creative boundaries and set an example of how environmental damage can be minimized by utilizing these advanced technologies the right way (Magazzo, 2022).

Another example would be how Zara uses real time sales data to adjust their production quickly, ensuring that the available inventory aligns with the rapid customer demands (productdatascrape, 2024).

8.2 Staying Competitive in a Data Driven Fashion Industry

The industry today is all about making the right decisions as quickly and efficiently as possible to gain a competitive edge over the competitor companies and brands, as we have learnt, the use of predictive analytics in fashion achieves just that (Anitson, 2023). Starting from analysing the buyer behaviour to providing personalized recommendations and alteration to fit their preferences, to adapting itself to seasonality-based changes, everything is more efficient. Integration of online and offline channels to provide a seamless, data-driven shopping experience. We have also seen how the use of predictive analytics optimizes supply chain and inventory management, it also streamlines the development process, minimizing waste in production (Omojola, 2025). AI integration even in the talent acquisition process also ensures that a team with the right skills are employed in the industry, if used right. Therefore, investment in AI indirectly leads to an investment in the infrastructure of the organization and selecting the right talent which essentially builds a company.

Business Intelligence (BI) is a term for the process of using data and analytics in order to gain actionable insights and create strategies accordingly that helps increase profitability within an organization. Many businesses thrive due to the proper utilization of BI in their companies (Jančiarová, 2024). Data has shown that the fashion companies using advanced BI tools achieve 54% higher operating margins, as compared to their competitors who have limited adoption of this technology. BI can analyse KPIs like sales, consumer behaviour, and sales which can later be used to make educated decisions about the future of the company (Anitson, 2023). Apart from that, Next-gen ERP (Next-Generation Enterprise Resource Planning) is used to centralize vast sets of scattered data, from supply chains, different kinds of sales channels to style parameters streamlining them for easier interpretation by the business analysts. This provides a "single-source of truth" and aids the decision-making process by backing it with quantitative evidence.

Processing these vast sets of data was once a challenge, however, Cloud-based data lakes and real-time data-streaming platforms have enabled rapid processing of petabytes of data per hour (Gong, 2024). This has played an instrumental role in speeding up the process of adaptation to rapid market shifts, on the go.

Radio-Frequency Identification (RFID) and Point-Of-Sales (POS) data integration is further used to enhance visibility across all channels (Kochar, 2024). RFID is used in real-time as radio waves are used in order to identify and track garments, which makes inventory management very efficient. Whereas POS is similarly utilized to manage inventory, enabling the retailers to streamline their operations. All of these things combined is what provides a competitive edge in this dynamic and ever-changing world of fashion.

REFERENCES

Admin. (2024, June 7). Artificial intelligence in fashion forecasting and trend analysis - Venuez.dk. Venuez.Dk - Your Strategic Partner In The Sports Clothing Industry. https://www.venuez.dk/artificial-intelligence-in-fashion-forecasting-and-trend-analysis/

Akram, S. V., Malik, P. K., Singh, R., Gehlot, A., Juyal, A., Ghafoor, K. Z., & Shrestha, S. (2022). Implementation of digitalized technologies for fashion industry 4.0: Opportunities and challenges. *Scientific Programming*, *2022*, 1–17. DOI: 10.1155/2022/7523246

Alina. (2023, June 28). How Fashion Analytics is Saving the Industry (2025). Retalon. https://retalon.com/blog/fashion-analytics

Anitson (2023, June 5). How can fashion businesses use data analytics. WFX FASHION & APPAREL BLOG. https://www.worldfashionexchange.com/blog/how-data-analytics-is-revolutionizing-the-fashion-industry/

Chakraborty, S., Hoque, S. M. A., & Kabir, S. M. F. (2020). Predicting fashion trend using runway images: Application of logistic regression in trend forecasting. *International Journal of Fashion Design, Technology and Education*, *13*(3), 376–386. DOI: 10.1080/17543266.2020.1829096

Demera, C. (2024, April 16). Harnessing the potential of price optimization with machine learning. Netguru. https://www.netguru.com/blog/price-optimization-with-machine-learning

Fashion Law Journal. (2023, September 18). Fashion tech and data privacy: Legal challenges in wearable technology and smart fashion. Fashion Law Journal. https://fashionlawjournal.com/fashion-tech-and-data-privacy-legal-challenges-in-wearable-technology-and-smart-fashion/

GhorbanTanhaei, H., Boozary, P., Sheykhan, S., Rabiee, M., Rahmani, F., & Hosseini, I.GhorbanTanhaei. (2024). Predictive analytics in customer behavior: Anticipating trends and preferences. *Results in Control and Optimization*, *17*, 100462. DOI: 10.1016/j.rico.2024.100462

Gong, Y. (2024). The impact of big data on the fashion industry. Advances in Economics. *Management and Political Sciences*, *143*(1), 190–196. DOI: 10.54254/2754-1169/2024.GA18973

Harreis, H., Koullias, T., Roberts, R., & Te, K. (2023, March 8). Holger Harreis. McKinsey & Company. https://www.mckinsey.com/industries/retail/our-insights/generative-ai-unlocking-the-future-of-fashion

Heuritech. (2020, September 3). Trend forecasting with AI: Fashion's way forward. Heuritech. https://heuritech.com/trend-forecasting-fashion-ai/

Howland, D. (2021, April 29). Algorithms vs. humans: Who's better at predicting fashion trends? Retail Dive. https://www.retaildive.com/news/algorithms-vs-humans-whos-better-at-predicting-fashion-trends/599050/

Idara, W. N. (2023, November 6). Adidas sales analysis: Unveiling insights through data-driven exploration and visualization. Medium. https://medium.com/@williamsnseidara/adidas-sales-analysis-unveiling-insights-through-data-driven-exploration-and-visualization-aa77383c6411

Jančiarová, B. (2024). PlatformE. PlatformE. https://www.platforme.com/post/data-driven-decision-making-in-fashion-how-platforme-is-enabling-brands-to-make-better-informed-decisions

Kochar, S. (2024, January 2). Top 10 technology trends reshaping the fashion industry. Techpacker Blog. https://techpacker.com/blog/design/top-09-fashion-technology-trends/

Koren, M. (2024). The influence of social network on forecasting in fashion industry. *Journal of Textile Science & Fashion Technology*, *11*(1). Advance online publication. DOI: 10.33552/JTSFT.2024.11.000752

Krysiak-Adamczyk, A. (2024, August 16). Predicting customer behavior: Leveraging analytics for strategic insight. Survicate. https://survicate.com/blog/predicting-customer-behavior/

Labs, K. (2024, December 4). The role of AI & ML in trend forecasting! LinkedIn. https://www.linkedin.com/pulse/role-ai-ml-trend-forecasting-kreedalabs-pjrhf/

Leitão, M. (2024, July 29). Trend forecasting with big data and analytics. Izertis. https://www.izertis.com/en/-/blog/trend-big-data-analytics

Liu, J., Pang, Z., & Qi, L. (2020). Dynamic pricing and inventory management with demand learning: A bayesian approach. *Computers & Operations Research, 124*, 105078. DOI: 10.1016/j.cor.2020.105078 PMID: 32836690

Magazzo, S. (2022, June 20). 6 emerging technology trends in the fashion industry. Mondo Staffing Agency. https://mondo.com/insights/top-technology-trends-fashion-industry/

Mahler, C. (2024, October 30). Why Analytics is Essential for Fashion Inventory Forecasting. Prediko. https://www.prediko.io/blog/analytics-for-fashion-inventory-forecasting

Nadeem, K., Ahmad, M., Javed, Z., & Habib, M. A. (2022). Development of a machine learning model for prediction of colour trends in fashion industry. *2022 International Conference on Frontiers of Information Technology (FIT)*, 296–301. DOI: 10.1109/FIT57066.2022.00061

Nwobodo, L. K. (2024). The importance of data analytics in modern fashion industry. Global Journal of Economic and Finance Research, 01(05). https://doi.org/01e5DOI: 10.55677/gjefr/04-2024-vol

Omojola, S. (2025, February 18). Optimize supply chains using predictive fashion analytics. Woven Insights Blog. https://woveninsights.ai/site-blog/optimize-supply-chains-using-predictive-fashion-analytics/

Optimizing demand forecasting with historical data insights. (n.d.). EOXS- Where Steel Meets Technology. Retrieved March 17, 2025, from https://eoxs.com/new_blog/optimizing-demand-forecasting-with-historical-data-insights/

Percherla, A. K. (2024). Dropshipping strategies in luxury and fashion retail - Unveiling the potential. *Journal of Business and Strategic Management*, *9*(2), 52–60. DOI: 10.47941/jbsm.1896

Prajapati, N. (2024, September 6). Predictive analytics: Enhancing supply chain management. Softweb Solutions. https://www.softwebsolutions.com/resources/role-of-predictive-analytics-in-supply-chain.html

productdatascrape. (2024, October 14). Scrape Zara product pricing data to understanding Zaras pricing strategy. IndiBlogHub. https://indibloghub.com/post/scrape-zara-product-pricing-data-to-understanding-zaras-pricing-strategy

Raj, A. (2023, December 6). What is Supply Chain Predictive Analytics and how does it work? ThroughPut Inc. https://throughput.world/blog/predictive-analytics-in-supply-chain/

Raj, A. (2024, January 15). Top 5 demand forecasting methods in 2025. ThroughPut Inc. https://throughput.world/blog/demand-forecasting-methods

Reporter, G. staff. (2023, October 1). 'You've got to be data-driven': The fashion forecasters using AI to predict the next trend. The Guardian. https://www.theguardian.com/technology/2023/oct/01/ai-artificial-intelligence-fashion-trend-forecasting-style

Safna, H. (2025, February 25). How Predictive Inventory Analytics is Reducing Overstock & Stockouts for Retailers. Retailcloud. https://retailcloud.com/predictive-inventory-analytics/

Sánchez-Cartas, J. M., Tejero, A., & León, G. (2021). Algorithmic pricing and price gouging. Consequences of high-impact, low probability events. *Sustainability (Basel)*, *13*(5), 2542. DOI: 10.3390/su13052542

Shewaramani, M. (2024, August 30). Ensuring data security and compliance in fashion retail with PIM. Credencys Solutions Inc. https://www.credencys.com/blog/ensuring-data-security-and-compliance-in-fashion-retail/

Shi, M., & Lewis, V. D. (2020, May 3). Using artificial intelligence to analyze fashion trends. arXiv.Org. https://arxiv.org/abs/2005.00986

Shirkhani, S., Mokayed, H., Saini, R., & Chai, H. Y. (2023). Study of ai-driven fashion recommender systems. *SN Computer Science*, *4*(5), 514. Advance online publication. DOI: 10.1007/s42979-023-01932-9

Siddique, Z. (2024, August 14). Predictive analytics in fashion: Forecasting trends and optimizing inventory. Ceba Solutions. https://www.cebasolutions.com/articles/predictive-analytics-in-fashion-forecasting-trends-and-optimizing-inventory

Sridevi, M., ManikyaArun, N., Sheshikala, M., & Sudarshan, E. (2020). Personalized fashion recommender system with image based neural networks. *IOP Conference Series. Materials Science and Engineering*, *981*, 022073. DOI: 10.1088/1757-899X/981/2/022073

StartUs Insights. (2022, September 6). Top 8 fashion technology trends in 2023. StartUs Insights. https://www.startus-insights.com/innovators-guide/fashion-technology-trends/

Team, S. C. B. (2023, April 13). Can analytics prevent stockouts and overstocks? SPS Commerce. https://www.spscommerce.com/blog/in-a-ship-from-anywhere-era-can-analytics-prevent-stock-outs/

The Fullstory Team. (2024, April 17). Predicting customer behavior with data. Fullstory. https://www.fullstory.com/blog/predicting-customer-behavior/

TimesTech. (2024, July 8). How AI is Revolutionising Personalised Fashion. Times-Tech. https://timestech.in/how-ai-is-revolutionising-personalised-fashion/

Trend forecasting with data analysis and AI. (n.d.). WGSN. Retrieved March 17, 2025, from https://www.wgsn.com/en/wgsn/data

UN-HJDM. (2024, October 7). AI in fashion: How AI fashion trends shape the future of style. JD Institute of Fashion Technology. https://www.jdinstitute.edu.in/ai-in-fashion-how-ai-fashion-trends-shape-the-future-of-style/

Vavilala, N. (2023, May 3). Predictive analytics in supply chain. Admin. https://dilytics.com/predictive-analytics-in-supply-chain/

Vorozhtsov, A. (2023, May 3). Machine Learning for price optimization. Habr. https://habr.com/en/articles/692270/

Wang, X., & Chen, Q. (2024). Stochastic optimal control for dynamic pricing and production in fashion retailing: An economic sustainability approach. Discrete Dynamics in Nature and Society, 2024(1). DOI: 10.1155/ddns/4854557

Waterston, S., & Choudhary, N. (2024). Demand trends decoded: Data science approach. Pickl.AI. https://www.pickl.ai/blog/decoding-demand-the-data-science-approach-to-forecasting-trends/

What are the data privacy concerns with AI in fashion customer relations in 2024? (2024, June 2). Blog.Stateless.Nyc. https://blog.stateless.nyc/what-are-the-data-privacy-concerns-with-ai-in-fashion-customer-relations-in-2024/

Yalcin, M. K. (2024, December 24). Personalization in apparel e-commerce with predictive analytics. Vizio.AI. https://www.vizio.ai/blog/personalization-in-apparel-e-commerce-with-predictive-analytics

Zhu, M., & Zhan, S. (2022). Research and implementation of a trend prediction model based on trend similarity for the changing trends of fashion elements in clothing. *Proceedings of the 2022 6th International Conference on Electronic Information Technology and Computer Engineering*, 1460–1466. DOI: 10.1145/3573428.3573685

ZYOD. (2023, November 1). AI and predictive analytics: Revolutionising the fashion industry. https://www.linkedin.com/pulse/ai-predictive-analytics-revolutionising-fashion-industry-zyod-zjkyf/

Compilation of References

Abdurazakov, M. M., Korotenkov, Yu. G., & Mukhidinov, M. G. (2016). Educational space represent a tion in cyberspace. SHS Web of Conferences. T. 29. P. 01001. http://dx.doi.org/DOI: 10.1051/shsconf/20162901001 /20162901001

Abdurazakov, M., Gadjiev, D., Volkova, S., & Lyaginova, O. (2019). The Concept Of Electronic Learning With The Application Of Digital Technologies. // In: Jo˘ze Rugelj, Maria Lapina (eds.): Proceedings of SLET-2019 – International Scientific Conference Innovative Approaches to the Application of Digital Technologies in Education and Research, Stavropol – Dombay, Russia, 20-23 May 2019, published at https://ceur-ws.org/Vol-2494/

Abdurazakov, M., Volkova, S., Vasil'eva, P., Matveeva, E., & Tyutyunkova, M. (2020). Electronic educational resources as a means of digital education development. // CEUR Workshop Proceedings: "SLET 2020 – Proceedings of the International Scientific Conference on Innovative Approaches to the Application of Digital Technologies in Education". P p. 188–202. http://ceur-ws. org/Vol-2861/paper21.pdf

About the Western Ghats. (n.d.). WWF India. https://www.wwfindia.org/about_wwf/critical_regions/western_ghats2/about_the_western_ghats/

Adiyarta, K., Napitupulu, D., Syafrullah, M., Mahdiana, D., & Rusdah, R. (2020). Analysis of smart city indicators based on prisma: Systematic review. [IOP Publishing.]. *IOP Conference Series. Materials Science and Engineering*, *725*(1), 012113. DOI: 10.1088/1757-899X/725/1/012113

Admin. (2024, June 7). Artificial intelligence in fashion forecasting and trend analysis - Venuez.dk. Venuez.Dk - Your Strategic Partner In The Sports Clothing Industry. https://www.venuez.dk/artificial-intelligence-in-fashion-forecasting-and-trend-analysis/

Agarwal, P. (2019, March). Redefining banking and financial industry through the application of computational intelligence. In *2019 Advances in Science and Engineering Technology International Conferences (ASET)* (pp. 1-5). IEEE. DOI: 10.1109/ICASET.2019.8714305

Ahmad, S., Umirzakova, S., Mujtaba, G., Amin, M. S., & Whangbo, T. (2023). Education 5.0: requirements, enabling technologies, and future directions (arXiv:2307.15846). *arXiv*. https://doi.org//arXiv.2307.15846DOI: 10.48550

Ahmad, G., & Shahid, M. (2022, October). Towards portfolio selection in stock markets using grey wolf optimization approach. In *2022 International Conference on Data Analytics for Business and Industry (ICDABI)* (pp. 756-760). IEEE. DOI: 10.1109/ICDABI56818.2022.10041457

Ajzen, I. (1991). The theory of planned behavior. *Organizational Behavior and Human Decision Processes*, *50*(2), 179–211. DOI: 10.1016/0749-5978(91)90020-T

Ajzen, I. (2002). Perceived behavioral control, self-efficacy, locus of control, and the theory of planned behavior. *Journal of Applied Social Psychology*, *32*(4), 665–683. DOI: 10.1111/j.1559-1816.2002.tb00236.x

Ajzen, I., & Fishbein, M. (1980). *Understanding attitudes and predicting social behaviour*. Prentice-Hall.

Akram, S. V., Malik, P. K., Singh, R., Gehlot, A., Juyal, A., Ghafoor, K. Z., & Shrestha, S. (2022). Implementation of digitalized technologies for fashion industry 4.0: Opportunities and challenges. *Scientific Programming*, *2022*, 1–17. DOI: 10.1155/2022/7523246

Akudugu, M., Nkegbe, P., Wongnaa, C., & Millar, K. (2023). Technology adoption behaviors of farmers during crises: What are the key factors to consider? *Journal of Agriculture and Food Research*, *14*, 100694. DOI: 10.1016/j.jafr.2023.100694

Alaca, B. (2017). *Kindergarten teachers' perspectives on culturally responsive education*. University of Toronto.

Albright, W. F. (1998, August 23). *Ancient Middle East | History, cities, civilizations, & Religion*. Encyclopedia Britannica. https://www.britannica.com/place/ancient-Middle-East/Science-and-law

Alcocer, A., Cabrera, A., & García, E. (2024). La inteligencia artificial en la educación: desafíos éticos y perspectivas hacia una nueva enseñanza. *Latam Revista Latinoamericana De Ciencias Sociales Y Humanidades, 5*(6). DOI: 10.56712/latam.v5i6.3019

Alenezi, M., Mohamed, A., & Shaaban, T. (2023). Revolutionizing EFL special education: how ChatGPT is transforming the way teachers approach language learning. Innoeduca. *International Journal of AI technologies and Educational Innovation, 9*(2), 5–23. https://doi.org/DOI: 10.24310/innoeduca.2023.v9i2.16774

Alfred, T. (2014). The Akwesasne cultural restoration program: A Mohawk approach to land-based education. *Decolonization, 3*(3), 134–144.

Alina. (2023, June 28). How Fashion Analytics is Saving the Industry (2025). Retalon. https://retalon.com/blog/fashion-analytics

Al-khresheh, M. H., Mohamed, A. M., & Shaaban, T. S. (2024). A quasi-experimental study on the effectiveness of augmented reality technology on english vocabulary learning among early childhood pupils with learning disabilities. Innoeduca. *International Journal of Technology and Educational Innovation, 10*(2), 5–26. DOI: 10.24310/ijtei.102.2024.17823

Al-khresheh, M., Mohamed, A. M., & Asif, M. (2022). Teachers' Perspectives towards Online Professional Development Programs during the Period of COVID-19 Pandemic in the Saudi EFL Context. *FWU Journal of Social Sciences, 16*(2). Advance online publication. DOI: 10.51709/19951272/Summer2022/1

Al-khresheh, M., Mohamed, A., & Bin Ali, R. (2023). A Study on Writing Apprehension among Saudi Undergraduate EFL Students, *Language Teaching. Research Quarterly, 34*, 1–18. DOI: 10.32038/ltrq.2023.34.01

Anderson, L. W., Krathwohl, D. R., & Bloom, B. S. (2001). *A taxonomy for learning, teaching, and assessing: A revision of Bloom's taxonomy of educational objectives.* Longman.

Andres, B., Sempere-Ripoll, F., Esteso, A., & Alemany, M. M. E. (2022). Mapping between industry 5.0 and education 5.0. *EDULEARN22 Proceedings,* 2921–2926. https://doi.org/DOI: 10.21125/edulearn.2022.0739

Anitson (2023, June 5). How can fashion businesses use data analytics. WFX FASHION & APPAREL BLOG. https://www.worldfashionexchange.com/blog/how-data-analytics-is-revolutionizing-the-fashion-industry/

Antonova, A., & Dankov, Y. (2023). Smart services in education: Facilitating teachers to deliver personalized learning experiences. In *Data Science and Algorithms in Systems* (pp. 108–117). Springer International Publishing. DOI: 10.1007/978-3-031-21438-7_9

Anusree, & Umajyothi. (2013). A case study of the Educational and Socio-Economic status of the Kattunayakan community in Kerala. https://achuthamenonfoundation.org/images/publications/vol15-4.pdf

Anusree, G. I., & Umajyothi, V. (2021). A case study of the educational and socio-economic status of the Kattunayakan community in Kerala. *Journal of the Achutha Menon Centre for Health Science Studies, 15*(4), 1–10. https://achuthamenonfoundation.org/images/publications/vol15-4.pdf

Appadurai, A. (1996). *Modernity at large: Cultural dimensions of globalization* (Vol. 1). University of Minnesota Press.

Arhin, I., Yeboah, E., Liu, X., Liu, A., Chen, X., & Li, X. (2024). Integrating farmers' perception of sustainable agricultural technologies towards the development of sustainable tea production in China. *International Journal of Agricultural Sustainability, 22*(1), 2303886. Advance online publication. DOI: 10.1080/14735903.2024.2303886

Aririguzoh, S. (2022). Communication competencies, culture and SDGs: effective processes to cross-cultural communication. *Humanities and Social Sciences Communications, 9*(1), 1-11. https://doi.org/DOI: 10.1057/s41599-022-01109-4

Arrow, K. J., Forsythe, R., Gorham, M., Hahn, R., Hanson, R., Ledyard, J. O., Levmore, S., Litan, R., Milgrom, P., Nelson, F. D., Neumann, G. R., Ottaviani, M., Schelling, T. C., Shiller, R. J., Smith, V. L., Snowberg, E., Sunstein, C. R., Tetlock, P. C., Tetlock, P. E., & Zitzewitz, E. (2008). The promise of prediction markets. *Science, 320*(5878), 877–878. DOI: 10.1126/science.1157679 PMID: 18487176

Artemova, I. (2024). Bridging Motivation and AI in Education: An Activity Theory Perspective. Digital Education Review. .DOI: 10.1344/der.2024.45.59-67

Arzanlou, A., & Sardroud, J. M. (2024). Enhancing Project Performance: Particle Swarm Optimization for Optimal Budget Allocation and Maintenance Scheduling. *KSCE Journal of Civil Engineering, 28*(5), 1635–1644. DOI: 10.1007/s12205-024-1348-1

Asadi, Z., Abdekhoda, M., & Nadrian, H. (2019). Understanding and predicting teachers' intention to use cloud computing in smart education. *Interactive Technology and Smart Education, 17*(1), 14–27. DOI: 10.1108/ITSE-05-2019-0019

Ashraf, Z., Shamim, M. S., Noman, R., Shahid, M., Ahmad, F., & Sami, L. (2025). Constrained Portfolio Optimization Using Hybrid Nature-Inspired Metaheuristic Algorithm in Stock Market. In *Nature-Inspired Optimization Algorithms for Cyber-Physical Systems* (pp. 53–74). IGI Global Scientific Publishing.

Ayamga, M., Tekinerdogan, B., Kassahun, A., & Rambaldi, G. (2020). Developing a policy framework for adoption and management of drones for agriculture in Africa. *Technology Analysis and Strategic Management, 33*(8), 970–987. DOI: 10.1080/09537325.2020.1858047

Aziz, S., Dowling, M., Hammami, H., & Piepenbrink, A. (2022). Machine learning in finance: A topic modeling approach. *European Financial Management, 28*(3), 744–770. DOI: 10.1111/eufm.12326

Badakhshan, E., & Bahadori, R. (2024). A simulation-based optimization model for balancing economic profitability and working capital efficiency using system dynamics and genetic algorithms. *Decision Analytics Journal, 12*, 100498. DOI: 10.1016/j.dajour.2024.100498

Bajaj, R., & Sharma, V. (2018). Smart education with artificial intelligence based determination of learning styles. *Procedia Computer Science, 132*, 834–842. DOI: 10.1016/j.procs.2018.05.095

Baker, W. (2022). *Intercultural and transcultural awareness in language teaching*. Cambridge University Press. DOI: 10.1017/9781108874120

Bandura, A. (1977). Self-efficacy: Toward a unifying theory of behavioral change. *Psychological Review, 84*(2), 191–215. DOI: 10.1037/0033-295X.84.2.191 PMID: 847061

Bandura, A. (1986). *Social foundations of thought and action: A social cognitive theory*. Prentice-Hall.

Barathkumar, R., Selvanayaki, S., Deepa, N., Kannan, P., & Prahadeeswaran, M. (2024). Impact of drone technology on agriculture - farmers' perception analysis. *Plant Science Today, 11*(sp4). Advance online publication. DOI: 10.14719/pst.5934

Bartlett, C., Marshall, M., & Marshall, A. (2012). Two-eyed seeing and other lessons learned within a co-learning journey of bringing together Indigenous and mainstream knowledges and ways of knowing. *Journal of Environmental Studies and Sciences, 2*(4), 331–340. DOI: 10.1007/s13412-012-0086-8

Bearman, M., Ryan, J., & Ajjawi, R. (2023). Discourses of artificial intelligence in higher education: A critical literature review. *Higher Education, 86*(2), 369–385. DOI: 10.1007/s10734-022-00937-2

Beaulieu, D. (2019). A Survey and Assessment of Culturally Based Education Programs for Native American Students in the United States. https://www.semanticscholar.org/paper/A-Survey-and-Assessment-of-Culturally-Based-for-in-Beaulieu/c08108f30cf49353ac4f01a3c4cdf4accfb1e46b

Becker, G. S. (1964). *Human capital: A theoretical and empirical analysis, with special reference to education*. University of Chicago Press.

Behera, K. K., Mandal, P., & Mahapatra, D. (2006). Green Leaves for Diarrhoeal Diseases Used by the Tribals of Kenojhar and Mayurbhanj District of Orissa, India. Ethnobotanical Leaflets, 2006(1), Article 34. https://opensiuc.lib.siu.edu/ebl/vol2006/iss1/34

Belda-Medina, J. (2022). Promoting inclusiveness, creativity and critical thinking through digital storytelling among EFL teacher candidates. *International Journal of Inclusive Education, 26*(2), 109-123. https://doi.org/DOI: 10.1080/13603116.2021.2011440

Benkler, Y. (2006). *The wealth of networks: How social production transforms markets and freedom*. Yale University Press.

Bennett, M. J. (2022). Intercultural communication: A current perspective. In Basic Concepts of Intercultural Communication: Selected Readings (ed. Bennett, J.) (Intercultural Press, 1998).

Berlian, Z., & Huda, M. (2022). Reflecting culturally responsive and communicative teaching (CRCT) through partnership commitment. *Education Sciences, 12*(5), 295. https://doi.org/DOI: 10.3390/educsci12050295

Biggs, J., & Tang, C. (2007). *Teaching for quality learning at university: What the student does* (3rd ed.). McGraw-Hill Education.

Billett, S. (2006). Constituting the workplace curriculum. *Journal of Curriculum Studies, 38*(1), 31–48. DOI: 10.1080/00220270500153781

Bindhani, B. (2021). *Tribal education in India: Issues and challenges*. Concept Publishing Company.

Bindhani, B. K. (2021). Status and problems of educational scenario among the tribals in Koraput district, Odisha. *The Oriental Anthropologist a Bi-annual International Journal of the Science of Man, 21*(1), 158–163. DOI: 10.1177/0972558X21990627

Boote, D. N., & Beile, P. (2005). Scholars Before Researchers: On the Centrality of the Dissertation Literature Review in Research Preparation. Educational Researcher, 34(6), 3-15.

Bordovskaya, N. V. Pedagogy. Manual for High Schools / N.V. Bordovskaya, A.A. Rean. SPb.: Publishing house "Peter", 2000. 304 p.

Borg, M. (2004). The apprenticeship of observation. *ELT Journal, 58*(3), 274–276. DOI: 10.1093/elt/58.3.274

Bose, P. (2021). *Indigenous knowledge and sustainable development: Lessons from tribal communities*. Routledge.

Boud, D., & Falchikov, N. (2006). Aligning assessment with long-term learning. *Assessment & Evaluation in Higher Education, 31*(4), 399–413. DOI: 10.1080/02602930600679050

Boudouaia, A., Mouas, S., & Kouider, B. (2024). A study on ChatGPT-4 as an innovative approach to enhancing English as a foreign language writing learning. Journal of Educational Computing Research. https://doi.org/ Borenstein, M., Hedges, L. V., Higgins, J. P. T., & Rothstein, H. R. (2009). *Introduction to Meta-Analysis*. John Wiley & Sons. https://doi.org/DOI: 10.1177/0735633124124746

Brabazon, A., Kampouridis, M., & O'Neill, M. (2020). Applications of genetic programming to finance and economics: Past, present, future. *Genetic Programming and Evolvable Machines, 21*(1-2), 33–53. DOI: 10.1007/s10710-019-09359-z

Brahmanandam, T., & Babu, B. T. (2016). Educational status among the Scheduled Tribes: Issues and challenges. The NEHU Journal, 14(2), 55–70. https://nehu.ac.in/public/downloads/Journals/NEHU-Journal-July-Dec-2016-A5.pdf

Brayboy, B. M. J. (2005). Toward a Tribal Critical Race Theory in Education. *The Urban Review, 37*(5), 425–446. DOI: 10.1007/s11256-005-0018-y

Burke, A., & Hardware, S. (2015). Honouring ESL students' lived experiences in school learning with multiliteracies pedagogy. Language, Culture and Curriculum, 28(2), 143-157. https://doi.org/DOI: 10.1080/07908318.2015.1027214

Burke, P. (1998). *The European Renaissance: centres and peripheries*. Blackwell Publishers.

Butler, A. C., Marsh, E. J., Slavinsky, J. P., & Baraniuk, R. G. (2014). Integrating cognitive science and technology improves learning in a STEM classroom. *Educational Psychology Review, 26*(2), 331–340. DOI: 10.1007/s10648-014-9256-4

Byram, M. (2020). *Teaching and assessing intercultural communicative competence: Revisited*. Multilingual matters.

Caena, F., & Redecker, C. (2019). Aligning teacher competence frameworks to 21st century challenges: The case for the European Digital Competence Framework for Educators (Digcompedu). *European Journal of Education, 54*(3), 356–369. DOI: 10.1111/ejed.12345

Caley, L., Williams, S. J., Spernaes, I., Thomas, D., Behrens, D., & Willson, A. (2021). Frameworks for evaluating education programmes and work related learning: A scoping review. *Journal of Workplace Learning*, *33*(6), 486–501. DOI: 10.1108/JWL-09-2020-0157

Campbell, D. T. (1979). Assessing the impact of planned social change. *Evaluation and Program Planning*, *2*(1), 67–90. DOI: 10.1016/0149-7189(79)90048-X

Canicattì, M., & Vallone, M. (2024). Drones in vegetable crops: A systematic literature review. *Smart Agricultural Technology*, *7*, 100396. DOI: 10.1016/j.atech.2024.100396

Cao, A., Guo, L., & Li, H. (2025). Understanding farmer cooperatives' intention to adopt digital technology. *International Journal of Agricultural Sustainability*, *23*(1), 2464523. Advance online publication. DOI: 10.1080/14735903.2025.2464523

Cao, L. (2022). Ai in finance: Challenges, techniques, and opportunities. *ACM Computing Surveys*, *55*(3), 1–38. DOI: 10.1145/3502289

Carter, J., & Hollinsworth, D. (2017). Teaching Indigenous geography in a neo-colonial world. *Journal of Geography in Higher Education*, *41*(2), 174–187. DOI: 10.1080/03098265.2017.1290591

Castells, M. (2010). *The rise of the network society* (2nd ed.). Wiley-Blackwell.

Cavalcante, R. C., Brasileiro, R. C., Souza, V. L., Nobrega, J. P., & Oliveira, A. L. (2016). Computational intelligence and financial markets: A survey and future directions. *Expert Systems with Applications*, *55*, 194–211. DOI: 10.1016/j.eswa.2016.02.006

Cedeño, E., Quintero, A., Quiñónez, O., Zamora, M., & Prado, N. (2024). Análisis de tendencias y futuro de la inteligencia artificial en la educación superior: Perspectivas y desafíos. *Ciencia Latina Revista Científica Multidisciplinar*, *8*(1), 3061–3076. DOI: 10.37811/cl_rcm.v8i1.9637

Chakraborty, S., Hoque, S. M. A., & Kabir, S. M. F. (2020). Predicting fashion trend using runway images: Application of logistic regression in trend forecasting. *International Journal of Fashion Design, Technology and Education*, *13*(3), 376–386. DOI: 10.1080/17543266.2020.1829096

Chanal, P. M., Kakkasageri, M. S., & Manvi, S. K. S. (2021). Security and privacy in the internet of things: computational intelligent techniques-based approaches. In *Recent trends in computational intelligence enabled research* (pp. 111–127). Academic Press. DOI: 10.1016/B978-0-12-822844-9.00009-8

Chan, K., & Zary, N. (2019). Applications and Challenges of Implementing Artificial Intelligence in Medical Education: Integrative Review. *JMIR Medical Education, 5*(1), e13930. Advance online publication. DOI: 10.2196/13930 PMID: 31199295

Chen, L., Chen, P., & Lin, Z. (2020). Artificial Intelligence in Education: A Review. *IEEE Access : Practical Innovations, Open Solutions, 8*, 75264–75278. DOI: 10.1109/ACCESS.2020.2988510

Chen, X., & Long, Z. (2023). E-commerce enterprises financial risk prediction based on FA-PSO-LSTM neural network deep learning model. *Sustainability (Basel), 15*(7), 5882. DOI: 10.3390/su15075882

Chernyakova, N. S. (2013). Digitized embodiment of meanings as a subject of cultural interpretation. Cultural sciences in the perspective of "digital humanites". St. Petersburg. Materials of the International Conference - St. Petersburg: Asterion

Cheruku, J. K., & Katekar, V. (2023). Digitalisation of Agriculture in India: The case for doubling farmers' income. Indian Institute of Public Administration, 194-205.

Chirume, S. (2020). Sustainable professional development of primary school mathematics teachers in zimbabwe through philosophy of education 5.0: challenges and prospects. *Sumerianz Journal of Social Science*, 150–161. https://doi.org/DOI: 10.47752/sjss.312.150.161

Chisholm, I. M. (1998). Six elements for technology integration in multicultural classrooms. *Journal of Information Technology for Teacher Education, 7*(2), 247–268. DOI: 10.1080/14759399800200033

Chounta, I. A., Bardone, E., Raudsep, A., & Pedaste, M. (2022). Exploring teachers' perceptions of artificial intelligence as a tool to support their practice in Estonian K-12 education. *International Journal of Artificial Intelligence in Education, 32*(3), 725–755. DOI: 10.1007/s40593-021-00243-5

Cisternas-San Martín, N., Guzmán Muñoz, E., & Rivas Poblete, A. (2025). Inteligencia artificial como asistente para desarrollar un currículum de educación continua basado en competencias. *Revista de Investigación Educacional, 43*(43). Advance online publication. DOI: 10.6018/rie.619441

Colin, K. K., & Ursul, A. D. (2015). Information and culture. Introduction to Information Cultural Studies. Moscow: "Strategic - priorities"– 300 p.

Coll, C., & Rochera, M. J. (2022). *Aprendizaje significativo y contextos reales en la educación del siglo XXI*. Editorial UOC.

Compayré, G., & Payne, W. H. (2015). *The history of pedagogy*. Routledge. Daniela, L. Smart Pedagogy as a Driving Wheel for Technology-Enhanced Learning. Tech

Contreras, R. C., Xavier da Silva, V. T., Xavier da Silva, I. T., Viana, M. S., Santos, F. L. D., Zanin, R. B., Martins, E. F. O., & Guido, R. C. (2024). Genetic Algorithm for Feature Selection Applied to Financial Time Series Monotonicity Prediction: Experimental Cases in Cryptocurrencies and Brazilian Assets. *Entropy (Basel, Switzerland)*, *26*(3), 177. DOI: 10.3390/e26030177

Cooper, H. (2010). *Research Synthesis and Meta-Analysis: A Step-by-Step Approach*. Sage publications.

Core, T. D. H. (2016, September 22). Terre Des Hommes CORE Trust. https://tdhcore.org/6540-2/

Corntassel, J. (2012). Re-envisioning resurgence: Indigenous pathways to decolonization and sustainable self-determination. *Decolonization*, *1*(1), 86–101.

Cotrina-Aliaga, J., Vera-Flores, M., Ortiz-Cotrina, W., & Sosa-Celi, P. (2021). *Uso de la inteligencia artificial (ia) como estrategia en la educación superior*. Revista Iberoamericana De La Educación., DOI: 10.31876/ie.vi.81

Cranton, P. (2016). *Understanding and promoting transformative learning: A guide to theory and practice* (3rd ed.). Routledge.

Creswell, J. W., & Creswell, J. D. (2017). *Research design: Qualitative, quantitative, and mixed methods approaches*. Sage publications.

Cruz, C., Bejarano, J., Santana, X., & Villamarín, J. (2024). El impacto de la inteligencia artificial en la enseñanza de la economía y la administración: Tendencias, desafíos y oportunidades. *Revista Social Fronteriza*, *4*(2), e42239. DOI: 10.59814/resofro.2024.4(2)239

Dai, Z., Xiong, J., Zhao, L., & Zhu, X. (2023). Smart classroom learning environment preferences of higher education teachers and students in China: An ecological perspective. *Heliyon*, *9*(6), e16769. DOI: 10.1016/j.heliyon.2023.e16769 PMID: 37303554

Daniela, L. (2021). Smart Pedagogy as a Driving Wheel for Technology-Enhanced Learning. *Know Learn*, *26*(4), 711–718. DOI: 10.1007/s10758-021-09536-z

Daponte, P., De Vito, L., Glielmo, L., Iannelli, L., Liuzza, D., Picariello, F., & Silano, G. (2019). A review on the use of drones for precision agriculture. *IOP Conference Series. Earth and Environmental Science*, *275*(1), 012022. DOI: 10.1088/1755-1315/275/1/012022

Daripa, S. (2017). Tribal Education in India: Government Initiative and Challenges. *International Journal of Research in Social Sciences*, *7*(11), 2249–2496. https://www.ijmra.us/project%20doc/2017/IJRSS_OCTOBER2017/IJMRA-12325.pdf

Davis, F. D. (1989). Perceived usefulness, perceived ease of use, and user acceptance of information technology. *Management Information Systems Quarterly*, *13*(3), 319–340. DOI: 10.2307/249008

Deardorff, D. K. (2020). *Manual for developing intercultural competencies: Story circles.* UNESCO Publishing.

Debnam, K. J., Smith, L. H., Aguayo, D., Reinke, W. M., & Herman, K. C. (2023). Nominated exemplar teacher perceptions of culturally responsive practices in the classroom. *Teaching and Teacher Education, 125*, 104062. https://doi.org/DOI: 10.1016/j.tate.2023.104062

Deci, E. L., & Ryan, R. M. (2017). *Self-determination theory: Basic psychological needs in motivation, development, and wellness.* Guilford Press.

Delić, H., & Bećirović, S. (2016). Socratic method as an approach to teaching. *European Researcher. Series A*, (10), 511–517.

Demera, C. (2024, April 16). Harnessing the potential of price optimization with machine learning. Netguru. https://www.netguru.com/blog/price-optimization-with-machine-learning

Demir, K. A. (2021). Smart education framework. *Smart Learning Environments*, *8*(1), 1–36. DOI: 10.1186/s40561-021-00170-x PMID: 40477293

Desai, A. R. (2010). *Tribes and Indian society.* Rawat Publications.

Dewey, J., Small, A. W., & Dutton, S. T. (1897). *My pedagogic creed. and The demands of sociology upon pedagogy.* No Title.

Díaz-Parra, O., Fuentes-Penna, A., Barrera-Cámara, R. A., Trejo-Macotela, F. R., Ramos-Fernández, J. C., Ruiz-Vanoye, J. A., Zezzatti, A. O., & Rodríguez-Flores, J. (2022). Smart Education and future trends. *International Journal of Combinatorial Optimization Problems and Informatics*, *13*(1), 65.

Dimitrov, N., & Deardorff, D. (2023). Intercultural competence as the core to developing globally engaged teachers. *At school in the world: Developing globally engaged teachers*, 3-27.

Dmitrenko, N. Y., Voloshyna, O. V., Kizim, S. S., Mnyshenko, K. V., & Nahorniak, S. V. (2023). Smart education in the prospective teachers' training. *CTE Workshop Proceedings*, 10, 414–429. DOI: 10.55056/cte.568

Dooly, M., & Darvin, R. (2022). Intercultural communicative competence in the digital age: critical digital literacy and inquiry-based pedagogy. *Language and Intercultural Communication, 22*(3), 354-366. https://doi.org/DOI: 10.1080/14708477.2022.2063304

Dube, F., Nzimande, N., & Muzindutsi, P. F. (2023). Application of artificial neural networks in predicting financial distress in the JSE financial services and manufacturing companies. *Journal of Sustainable Finance & Investment, 13*(1), 723–743. DOI: 10.1080/20430795.2021.2017257

Dulhunty, A. (2022). The microcredit continuum: From 'smart economics' to holistic programming. *Journal of International Development, 34*(6), 1213–1226. DOI: 10.1002/jid.3625

Dutt, T. (2017). Education in Vedic and Post Vedic Era. *Journal of National Development, 30*(2), 159–166.

Egger, D. J., Gambella, C., Marecek, J., McFaddin, S., Mevissen, M., Raymond, R., Simonetto, A., Woerner, S., & Yndurain, E. (2020). Quantum computing for finance: State-of-the-art and future prospects. *IEEE Transactions on Quantum Engineering, 1*, 1–24. DOI: 10.1109/TQE.2020.3030314

Elek, C., Gubhaju, L., Lloyd-Johnsen, C., Eades, S., & Goldfeld, S. (2020). Can early childhood education programs support positive outcomes for indigenous children? A systematic review of the international literature. *Educational Research Review, 31*, 100363. DOI: 10.1016/j.edurev.2020.100363

Elyakov A. D. (2012). Homo Informaticus and the modern information environment. Philosophy and Society. Issue #3 (67)

Englund, R. K. (2004). Proto-cuneiform account-books and journals. Creating economic order: Record-keeping, standardization and the development of accounting in the ancient Near East, 23-46.

Eskelson, T. C. (2020). How and Why Formal Education Originated in the Emergence of Civilization. *Journal of Education and Learning, 9*(2), 29–47. DOI: 10.5539/jel.v9n2p29

Estrada-Araoz, E., Quispe-Aquise, J., Malaga-Yllpa, Y., Larico-Uchamaco, G., Pizarro-Osorio, G., Mendoza-Zuñiga, M., Velasquez-Bernal, A. C., Roque-Guizada, C. E., & Huamaní-Pérez, M. (2024). Role of artificial intelligence in education: Perspectives of peruvian basic education teachers. *Data & Metadata, 3*, 325. DOI: 10.56294/dm2024325

Evening Coaching Centers (ECC) – CTRD Trust. (2025). Ctrd.in. https://ctrd.in/evening-coaching-centers-ecc/

Eze, U. T., & Nnia, I. (2022). Integrating African Indigenous Knowledge In Nigeria's Formal Education System: It's Potential For Sustainable Development. Journal of Education and Practice. https://www.semanticscholar.org/paper/Integrating-African-Indigenous-Knowledge-In-Formal-Eze-Nnia/d20712b209d23c5775894d5726dd32da2240788b

Familoni, B., & Onyebuchi, N.Babajide Tolulope FamiloniNneamaka Chisom Onyebuchi. (2024). ADVANCEMENTS AND CHALLENGES IN AI INTEGRATION FOR TECHNICAL LITERACY: A SYSTEMATIC REVIEW. *Engineering Science & Technology Journal*, *5*(4), 1415–1430. DOI: 10.51594/estj.v5i4.1042

Farrokhi, F., & Mahmoudi-Hamidabad, A. (2012). Rethinking convenience sampling: Defining quality criteria. *Theory & practice in language studies,* *2*(4). https://doi.org/DOI: 10.4304/tpls.2.4.784-792

Fashion Law Journal. (2023, September 18). Fashion tech and data privacy: Legal challenges in wearable technology and smart fashion. Fashion Law Journal. https://fashionlawjournal.com/fashion-tech-and-data-privacy-legal-challenges-in-wearable-technology-and-smart-fashion/

Fawns, T. (2022). An entangled pedagogy: Looking beyond the pedagogy – technology dichotomy. *Postdigital Science and Education*, *4*(3), 711–728. DOI: 10.1007/s42438-022-00302-7 PMID: 40477872

Federal budget | The Official Portal of the UAE Government. (n.d.). https://u.ae/en/information-and-services/finance-and-investment/budget/federal-finance#:~:text=Federal%20Budget%202023&text=is%20as%20follows:-,AED%2023.9%20billion

Federal State Educational Standards. (2012). Order of the Ministry of Education and Science of Russia dated 05/17/2012 N 413 (as amended on 12/11/2020). https://fgos.ru/ Access mode: 02/27/2025.

Feng, X., Zhang, N., Yang, D., Lin, W., & Maulana, R. (2024). From awareness to action: Multicultural attitudes and differentiated instruction of teachers in Chinese teacher education programmes. *Learning Environments Research*, •••, 1–42. DOI: 10.1007/s10984-024-09518-9

Firat, M. (2023). How ChatGPT can transform autodidactic experiences and open education. *Open Education Faculty, Anadolu University.* http://doi.org/DOI: 10.31219/osf.io/9ge8m

Fishbein, M., & Ajzen, I. (1975). *Belief, Attitude, Intention, and Behavior: An Introduction to Theory and Research.* Addison-Wesley., DOI: 10.1017/CBO9781107415324.004

Floridi, L. (2014). *The fourth revolution: How the infosphere is reshaping human reality.* OUP.

Francis, A. J. J., Eccles, M. P. M., Johnston, M., Walker, A., Grimshaw, J., Foy, R., Kaner, E. F. S., Smith, L., Bonetti, D., & Francis, J. (2004). CONSTRUCTING QUESTIONNAIRES BASED ON THE THEORY OF PLANNED BEHAVIOUR A MANUAL for HEALTH SERVICES RESEARCHERS. In Direct. https://doi.org/DOI: 0-9540161-5-7

Francis, W. (1908). *Madras District Gazetteers: The Nilgiris.* Logos Press.

Frolov, A. A. (2017). Technology of intellectual education. Access mode: 04/15/2025. http://ncrao.rsvpu.ru/sites/default/files/library/tehnologiya_intellektualnogo_obrazovaniya.pdf

Fursov, I., Morozov, M., Kaploukhaya, N., Kovtun, E., Rivera-Castro, R., Gusev, G., & Burnaev, E. (2021, August). Adversarial attacks on deep models for financial transaction records. In *Proceedings of the 27th ACM SIGKDD Conference on Knowledge Discovery & Data Mining* (pp. 2868-2878). DOI: 10.1145/3447548.3467145

Gadgil, M., & Meher-Homji, V. M. (1986). Role of protected areas in conservation. In Chopra, V. L., & Khoshoo, T. N. (Eds.), *Conservation for Productive Agriculture* (pp. 143–159). Indian Council of Agricultural Research.

Gao, H., Kou, G., Liang, H., Zhang, H., Chao, X., Li, C. C., & Dong, Y. (2024). Machine learning in business and finance: A literature review and research opportunities. *Financial Innovation, 10*(1), 86. DOI: 10.1186/s40854-024-00629-z

Gao, J. (2024). *Research on the cultivation of new media literacy among college teachers from the perspective of smart education.* Media and Communication Research., DOI: 10.23977/mediacr.2024.050325

García, O., & Wei, L. (2014). *Translanguaging: Language, Bilingualism and Education.* Palgrave Macmillan. DOI: 10.1057/9781137385765

Gay, G. (2013). Teaching to and through cultural diversity. *Curriculum inquiry, 43*(1), 48-70. https://doi.org/DOI: 10.1111/curi.12002

Gee, J. P. (2021). *What Video Games Have to Teach Us About Learning and Literacy* (2nd ed.). St. Martin's Press.

George, A. (2005). In search of the é. dub. ba. a: The ancient Mesopotamian school in literature and reality.

George, D., & Mallery, P. (2003). *SPSS for Windows step by step*. Pearson A and B.

Ghonge, M. M., Bag, R., & Singh, A. (2020). Indian education: Ancient, medieval and modern. In *Education at the Intersection of Globalization and Technology*. IntechOpen.

GhorbanTanhaei, H., Boozary, P., Sheykhan, S., Rabiee, M., Rahmani, F., & Hosseini, I.GhorbanTanhaei. (2024). Predictive analytics in customer behavior: Anticipating trends and preferences. *Results in Control and Optimization*, *17*, 100462. DOI: 10.1016/j.rico.2024.100462

Gipps, C. (1994). *Beyond Testing: Towards a Theory of Educational Assessment* (1st ed.). Routledge.

Giudici, P., Centurelli, M., & Turchetta, S. (2024). Artificial Intelligence risk measurement. *Expert Systems with Applications*, *235*, 121220. DOI: 10.1016/j.eswa.2023.121220

Goforth, A. N., Nichols, L. M., Sun, J., Violante, A. E., Brooke, E., Kusumaningsih, S., Howlett, R., Hogenson, D., & Graham, N. (2022). Cultural Adaptation of an Educator Social–Emotional Learning Program to Support Indigenous Students. *School Psychology Review*, *53*(4), 1–17. DOI: 10.1080/2372966X.2022.2144091 PMID: 39100141

Gogas, P., & Papadimitriou, T. (2021). Machine learning in economics and finance. *Computational Economics*, *57*(1), 1–4. DOI: 10.1007/s10614-021-10094-w

Golinkoff, R. M., Hassinger-Das, B., Kucirkova, N., Myers, L., & Dore, R. A. (2025). The Explosion of EdTech: Can Its Promise Be Fulfilled? In Christakis, D. A., & Hale, L. (Eds.), *Handbook of Children and Screens*. Springer., DOI: 10.1007/978-3-031-69362-5_75

Gómez, W. (2023). La inteligencia artificial y su incidencia en la educación: transformando el aprendizaje para el siglo xxi. *Revista Internacional De Pedagogía E Innovación Educativa,* *3*(2), 217-230. DOI: 10.51660/ripie.v3i2.133

Gong, Y. (2024). The impact of big data on the fashion industry. Advances in Economics. *Management and Political Sciences*, *143*(1), 190–196. DOI: 10.54254/2754-1169/2024.GA18973

Gopalakrishnan, M. (1995). Gazetteers of India, Tamil Nadu State, The Nilgiris District. Madras Press.

Gorski, P. C., & Dalton, K. (2020). Striving for critical reflection in multicultural and social justice teacher education: Introducing a typology of reflection approaches. *Journal of Teacher Education, 71*(3), 357-368. https://doi.org/DOI: 10.1177/0022487119883545

Graczyk-Kucharska, M., Olszewski, R., Przybyła, J., Łuszkiewicz, J., Hojka, K., & Spychała, M. (2024). *Modelling Accelerating Acquisition of Teamwork Competences with Transversal Competences and Artificial Intelligence*. Krakow Review of Economics and Management/Zeszyty Naukowe Uniwersytetu Ekonomicznego w Krakowie., DOI: 10.15678/krem.13131

Graczyk-Kucharska, M., Özmen, A., Szafrański, M., Weber, G., Goliński, M., & Spychała, M. (2019). Knowledge accelerator by transversal competences and multivariate adaptive regression splines. *Central European Journal of Operations Research, 28*(2), 645–669. DOI: 10.1007/s10100-019-00636-x

Grigor'ev, S. G., & Grinshkyn, V. V. (2006). *Educational electronic editions and resources: study guide. manual for stud. pedagogical universities and students of the system of advanced training of educational workers*. KSU.

Hair, J. F., Black, W. C., Babin, B. J., Anderson, R. E., & Tatham, R. L. (2006). *'Multivariate data analysis*. Pearson Prentice Hall, Upper Saddle River.

Hair, J. F., Tatham, R. L., & Black, W. (1998). *Multivariate data analysis*. Prentice-Hall.

Hall, S. (2006). Cultural studies and its theoretical legacies. In *Stuart Hall* (pp. 272–285). Routledge.

Hannafin, M. J., & Hannafin, K. M. (2010). Cognition and student-centered, web-based learning: Issues and implications for research and theory. *Learning and instruction in the digital age*, 11-23.

Hardy, T. (1995, March). Multi-objective decision-making under uncertainty-fuzzy logic methods. In *10th Computing in Aerospace Conference* (p. 1000). DOI: 10.2514/6.1995-1000

Hare, J. (2011). "They tell a story and there's meaning behind that story": Indigenous knowledge and young indigenous children's literacy learning. *Journal of Early Childhood Literacy, 12*(4), 389–414. DOI: 10.1177/1468798411417378

Harreis, H., Koullias, T., Roberts, R., & Te, K. (2023, March 8). Holger Harreis. McKinsey & Company. https://www.mckinsey.com/industries/retail/our-insights/generative-ai-unlocking-the-future-of-fashion

Herrington, J., & Oliver, R. (2022). Authentic learning environments and tasks. *Educational Technology Research and Development*, *70*(1), 1–15. DOI: 10.1007/s11423-021-10041-9

Herskovits, M. F. (1955). Cultural anthropology. Oxford & IBH Publishing Co.

Heuritech. (2020, September 3). Trend forecasting with AI: Fashion's way forward. Heuritech. https://heuritech.com/trend-forecasting-fashion-ai/

Hoff, H. E. (2020). The evolution of intercultural communicative competence: Conceptualisations, critiques and consequences for 21st century classroom practice. *Intercultural Communication Education, 3*(2), 55-74. https://doi.org/DOI: 10.29140/ice.v3n2.264

Hoffman, D. L., Leong, P., Ka'aloa, R. P. H., & Paek, S. (2022). Teachers' Perspectives on Culturally-Relevant Computing: Principles and Processes. *TechTrends, 66*(3), 423-435. https://doi.org/DOI: 10.1007/s11528-022-00733-w

Hohenstein, J., Kizilcec, R. F., DiFranzo, D., Aghajari, Z., Mieczkowski, H., Levy, K., Naaman, M., Hancock, J., & Jung, M. F. (2023). Artificial intelligence in communication impacts language and social relationships. *Scientific Reports, 13*(1), 5487. DOI: 10.1038/s41598-023-30938-9 PMID: 37015964

Holland, J. H. (1975). *Adaptation in Natural and Artificial Systems*. University of Michigan Press.

Holmes, W., Bialik, M., & Fadel, C. (2021). *Artificial Intelligence in Education: Promises and Implications for Teaching and Learning*. https://curriculumredesign.org/wp-content/uploads/AI-in-Education.pdf

Holmes, W., Porayska-Pomsta, K., Holstein, K., Sutherland, E., Baker, T., Shum, S., Santos, O., Rodrigo, M., Cukurova, M., Bittencourt, I., & Koedinger, K. (2021). Ethics of AI in Education: Towards a Community-Wide Framework. *International Journal of Artificial Intelligence in Education, 32*(3), 504–526. DOI: 10.1007/s40593-021-00239-1

Howland, D. (2021, April 29). Algorithms vs. humans: Who's better at predicting fashion trends? Retail Dive. https://www.retaildive.com/news/algorithms-vs-humans-whos-better-at-predicting-fashion-trends/599050/

Hua, L., & Wang, S. (2019). *Antecedents of consumers' intention to purchase energy-efficient appliances: An empirical study based on the technology acceptance model and theory of planned Behavior*. Sustainability., DOI: 10.3390/su11102994

Huang, R., Tlili, A., Chang, T. W., Nascimbeni, F., & Burgos, D. (2022). Roles of AI in fostering self-directed learning and enhancing learner motivation. *Educational Technology Research and Development*, *70*(4), 1925–1943. DOI: 10.1007/s11423-022-10116-4

Hwang, G. J. (2014). Definition, framework and research issues of smart learning environments – a context-aware ubiquitous learning perspective. *Smart Learning Environments*, *1*(1), 4. DOI: 10.1186/s40561-014-0004-5

Idara, W. N. (2023, November 6). Adidas sales analysis: Unveiling insights through data-driven exploration and visualization. Medium. https://medium.com/@williamsnseidara/adidas-sales-analysis-unveiling-insights-through-data-driven-exploration-and-visualization-aa77383c6411

Indigenous People's Plan Kattunayakans. (n.d.). https://www.cepf.net/sites/default/files/56154-ipp-kattunayakan.pdf

Indrawati, A., N., & Amani, H. (2018). Indicators to measure smart economy: An Indonesian perspective. In *ICBIM'18: Proceedings of the 2nd International Conference on Business and Information Management* (pp. 173–179). Association for Computing Machinery. DOI: 10.1145/3278252.3278278

Ingridhaynesphd. (2024, March 7). Literacy Statistics 2024- 2025 (Where we are now). *National Literacy*. https://www.thenationalliteracyinstitute.com/post/literacy-statistics-2024-2025-where-we-are-now#:~:text=On%20average%2C%2079%25%20of%20U.S.,below%205th%2Dgrade%20level)

Iqbal, H. M. N., Parra-Saldivar, R., Zavala-Yoe, R., & Ramirez-Mendoza, R. A. (2020). Smart educational tools and learning management systems: Supportive framework. [IJIDeM]. *International Journal on Interactive Design and Manufacturing*, *14*(4), 1179–1193. DOI: 10.1007/s12008-020-00695-4

Iryani, I., & Yulianto, H. (2023). Artificial Intelligence (AI) of Financial in the VUCA Era: A Systematic Mapping Study. *Journal of Computer Networks. Architecture and High Performance Computing*, *5*(2), 398–413. DOI: 10.47709/cnahpc.v5i2.2201

Jackson, C. M., Mun, Y. Y., & Park, J. S. (2013). An empirical test of three mediation models for the relationship between personal innovativeness and user acceptance of technology. *Information & Management*, *50*(4), 154–161. DOI: 10.1016/j.im.2013.02.006

Jan, A., Sultana, I., & Adnan, M. (2020). Digital media and smart education in Pakistan: Challenges and prospects for the teachers in the age of E-learning. *Global Educational Studies Review*. https://www.academia.edu/download/79117620/4fgmQJvxwr.pdf

Jančiarová, B. (2024). PlatformE. PlatformE. https://www.platforme.com/post/data-driven-decision-making-in-fashion-how-platforme-is-enabling-brands-to-make-better-informed-decisions

Jayakumar, C., Sankaran, S., & Gangadharan, P. (2024). Looking inward, looking forward: Articulating alternatives to the current education system for adivasis, by Adivasis. *Contemporary Education Dialogue*, *21*(1), 108–124. DOI: 10.1177/09731849231212111

Jiao, Y., & Jakubowicz, J. (2017, December). Predicting stock movement direction with machine learning: An extensive study on S&P 500 stocks. In *2017 IEEE International Conference on Big Data (Big Data)* (pp. 4705-4713). IEEE.

Jim, J. R., Talukder, M. A. R., Malakar, P., Kabir, M. M., Nur, K., & Mridha, M. F. (2024). Recent advancements and challenges of nlp-based sentiment analysis: A state-of-the-art review. *Natural Language Processing Journal*, 100059.

Jmaiel, H. A., Abukhait, R. O., Mohamed, A. M., Shaaban, T. S., Al-khresheh, M. H., & AL-Qadri, A. H. (2025). The role of ChatGPT in enhancing EFL students' ESP writing skills: an experimental study of gender and major differences. Discover Education, 4(1), 1-19. https://doi.org/DOI: 10.1007/s44217-025-00700-6

Johnson, A. P. (2020). *Contextual Teaching and Learning: An Introduction and an Overview*. SAGE Publications.

Jones, L. (2007). *The student-centered classroom* (Vol. 32). Cambridge university press.

Joshi, N., Patel, N., Sharma, P., & Singh, A. (2020). Revolution of Indian agriculture through drone technology. *Research Today*, *2*(5), 174–176.

Kahn, H. E., & Agnew, M. (2017). Global learning through difference: Considerations for teaching, learning, and the internationalization of higher education. *Journal of Studies in International Education, 21*(1), 52-64. https://doi.org/DOI: 10.1177/102831531562202

Kamalov, F., Calonge, D., & Gurrib, I. (2023). New Era of Artificial Intelligence in Education: Towards a Sustainable Multifaceted Revolution. *Sustainability (Basel), 15*(16), 12451. DOI: 10.3390/su151612451

Kamble, P. N., Ajotikar, M., Suryawanshi, P., & Tarde, V. (2024). Awareness of drone technology application in agriculture amongst progressive farmers. *International Journal of Agriculture Extension and Social Development*, *7*(12), 307–310. DOI: 10.33545/26180723.2024.v7.i12e.1418

Kanti, A. (2015). Holistic Tribal Education: A Case Study of Kalinga Institute of Social Sciences in Odisha, India. https://www.semanticscholar.org/paper/Holistic-Tribal-Education%3A-A-Case-Study-of-Kalinga-Kanti/b510a16c24faa918bbcdb649befc18e5ac46563c

Karataş, S., Eckstein, K., Noack, P., Rubini, M., & Crocetti, E. (2023). Meeting in school: Cultural diversity approaches of teachers and intergroup contact among ethnic minority and majority adolescents. *Child Development, 94*(1), 237-253. https://doi.org/DOI: 10.1111/cdev.13854

Kaur, N., & Kaur, G. (2022). Educating the Tribals: Government Interventions. GHG Journal of Sixth Thought, 9(1). https://www.ghgkce.org/files/education/journal/j_6_3_2022.pdf

Kayalar, F. (2016). Cross-Cultural Comparison of Teachers' Views upon Integration and Use of technology in Classroom. *Turkish Online Journal of Educational AI technologies -TOJET, 15*(2), 11-19. https://eric.ed.gov/?id=EJ1096412

Keengwe, J. (2010). Fostering cross-cultural competence in preservice teachers through multicultural education experiences. *Early Childhood Education Journal, 38*(3), 197–204. DOI: 10.1007/s10643-010-0401-5

Kelman, H. C. (1958). Compliance, identification, and internalization: Three processes of attitude change. *The Journal of Conflict Resolution*, *2*(1), 51–60. DOI: 10.1177/002200275800200106

Khlaisang, J., Songkram, N., Huang, F., & Teo, T. (2021). Teachers' perception of the use of mobile technologies with smart applications to enhance students' thinking skills: A study among primary school teachers in Thailand. *Interactive Learning Environments*, *31*(8), 5037–5058. DOI: 10.1080/10494820.2021.1993933

Kim, D. (2020). Learning language, learning culture: Teaching language to the whole student. *ECNU Review of Education, 3*(3), 519-541. https://doi.org/DOI: 10.1177/2096531120936693

Kincheloe, J. L., & Horn, R. A., Jr. (2007). *The Praeger Handbook of Education and Psychology. Volume 1*. Praeger. Available from: Greenwood Publishing Group, Inc. PO Box 5926, Portsmouth, NH 03802-6926.

Kireet Joshi. (n.d.)_https://kireetjoshi.in/books/philosophy/philosophy-of-indian-pedagogy/philosophy-of-indian-pedagogy-2

Klerkx, L., Jakku, E., & Labarthe, P. (2019). A review of social science on digital agriculture. *NJAS Wageningen Journal of Life Sciences*, *90–91*(1), 1–16. DOI: 10.1016/j.njas.2019.100315

Klimova, B., & Chen, J. H. (2024). The impact of AI on enhancing students' intercultural communication competence at the university level: A review study. *Language Teaching Research Quarterly*, *43*, 102–120. DOI: 10.32038/ltrq.2024.43.06

Klimova, B., Pikhart, M., & Kacetl, J. (2023). Ethical issues of the use of AI-driven mobile apps for education. *Frontiers in Public Health*, *10*, 1118116. DOI: 10.3389/fpubh.2022.1118116 PMID: 36711343

Kochar, S. (2024, January 2). Top 10 technology trends reshaping the fashion industry. Techpacker Blog. https://techpacker.com/blog/design/top-09-fashion-technology-trends/

Kolb, D. A. (1984). *Experiential learning: Experience as the source of learning and development*. Prentice Hall.

Koren, M. (2024). The influence of social network on forecasting in fashion industry. *Journal ofTextile Science & Fashion Technology*, *11*(1). Advance online publication. DOI: 10.33552/JTSFT.2024.11.000752

Koretz, D. M. (2008). *Measuring up: What educational testing really tells us*. Harvard University Press. DOI: 10.4159/9780674039728

Korotenkova, V. V., & Korotenkov, Yu. G. (2021). *Methodological aspects of modern education monograph, M*. Pero.

Korotenkov, Yu. G. (2010). *Information educational environment of the basic school: a teaching aid // Yu. G.Korotenkov. – M*. Academy of IT.

Korpan, L. M. (2013). Graphic design and design education in the discourse of information society culture. St. Petersburg. Materials of the International Conference - St. Petersburg: Asterion.

Kostyk, Y., Tiuleniev, S., Goi, V., Kovalenko, O., & Pochernina, N. (2023). The National Model of the Smart Economy for Achieving the Goals of Innovative Development. *Revue d'Economie Financiere*, *21*, 622–632.

Krysiak-Adamczyk, A. (2024, August 16). Predicting customer behavior: Leveraging analytics for strategic insight. Survicate. https://survicate.com/blog/predicting-customer-behavior/

Kubota, R. (2020). Promoting and problematizing multi/plural approaches in language pedagogy. *Plurilingual pedagogies: Critical and creative endeavors for equitable language in education*, 303-321. https://doi.org/DOI: 10.1007/978-3-030-36983-5_14

Kumar, P., & Daniel, B. K. (2016). Integration of digital technologies in the classroom: A case study from India. *Education and Information Technologies, 21*(5), 1297–1316.

Kurani, A., Doshi, P., Vakharia, A., & Shah, M. (2023). A comprehensive comparative study of artificial neural network (ANN) and support vector machines (SVM) on stock forecasting. *Annals of Data Science, 10*(1), 183–208. DOI: 10.1007/s40745-021-00344-x

Kurbanova, F. K., Botirova, M., & Abdurahmonova, S. (2022). Promoting learning and developing cultural competence. *Academic research in educational sciences, 3*(3), 914-917.

Kuznetsov, A. A., Monakhov V. M., & Abdurazakov M. M. (2016a). Methodological recommendations for the teacher on designing the main educational program in informatics in accordance with the requirements of the second generation of the Federal State Educational Standards. (Text). *Informatics and education., 10*(279), 9–17.

Kuznetsov, A. A., Monakhov V. M., & Abdurazakov M. M. (2016b). Research activities of an informatics teacher in new didactic conditions for the functioning of the Federal State Educational Standard. (Text). *Informatics and education., 6*(275), 4–16.

Kwon, Y., & Lee, Z. (2024). A hybrid decision support system for adaptive trading strategies: Combining a rule-based expert system with a deep reinforcement learning strategy. *Decision Support Systems, 177*, 114100. DOI: 10.1016/j.dss.2023.114100

Labs, K. (2024, December 4). The role of AI & ML in trend forecasting! LinkedIn. https://www.linkedin.com/pulse/role-ai-ml-trend-forecasting-kreedalabs-pjrhf/

Ladson-Billings, G. (1995). Toward a theory of culturally relevant pedagogy. *American Educational Research Journal, 32*(3), 465–491. DOI: 10.3102/00028312032003465

Lai, J. Y., & Chen, C. C. (2011). Self-efficacy, effort, job performance, and job satisfaction: An exploratory study of the internet-enabled work environment. *Journal of International Management, 18*(1), 1–15.

Lau, W. S., & Shea, M. (2022). Empowering English learners in the classroom through culturally responsive social-emotional teaching practices. *Journal of Multilingual and Multicultural Development*, 1-18. \DOI: 10.1080/01434632.2022.2078337

Lauder, H., Young, M., Daniels, H., Balarin, M., & Lowe, J. (Eds.). (2012). *Educating for the knowledge economy? Critical perspectives.* Routledge. DOI: 10.4324/9780203817698

Lavryk, T. V., & Yaresko, E. V. (2015). The Development of the Foundations of Modern Pedagogy: Paradigmal and Methodological Aspects of Research. *European Journal of Contemporary Education, 12*(2), 150–157. DOI: 10.13187/ejced.2015.12.150

Lee, J. (2020). Informal digital learning of English and strategic competence for cross-cultural communication: Perception of varieties of English as a mediator. *ReCALL, 32*(1), 47–62. DOI: 10.1017/S0958344019000181

Leitão, M. (2024, July 29). Trend forecasting with big data and analytics. Izertis. https://www.izertis.com/en/-/blog/trend-big-data-analytics

Lema, R., Ruano, A., Rivas, N., Erazo, C., & Arias, R. (2024). Perspectivas de la educación con la inteligencia artificial a un cercano plazo. *Ciencia Latina Revista Científica Multidisciplinar, 8*(2), 5522–5536. DOI: 10.37811/cl_rcm.v8i2.10966

Levi-Faur, D. (2011). Regulation and regulatory governance. *In Handbook on the Politics of Regulation.*

Ley Orgánica (2020). de 29 de diciembre, por la que se modifica la Ley Orgánica 2/2006, de 3 de mayo, de Educación (LOMLOE). *Boletín Oficial del Estado*, nº 340, de 30 de diciembre de 2020, pp. 122868-122953. https://www.boe.es/boe/dias/2020/12/30/pdfs/BOE-A-2020-17264.pdf

Lifintsev, D., & Wellbrock, W. (2019). Cross-cultural communication in the digital age. *Estudos em Comunicação, 1*(28), 93-104. https://doi.org/DOI: 10.25768/fal.ec.n28.a05

Li, M. (2024). Using Prompt Engineering to Enhance STEM Education. *2024 IEEE Integrated STEM Education Conference (ISEC)*, 1-2. DOI: 10.1109/ISEC61299.2024.10664775

Lin, P., Chai, C., Jong, M., Dai, Y., Guo, Y., & Qin, J. (2021). Modeling the structural relationship among primary students' motivation to learn artificial intelligence. *omputers and Education. Artificial Intelligence, 2*, 100006. DOI: 10.1016/j.caeai.2020.100006

Liu, G., Ma, C., Bao, J., & Liu, Z. (2023). Toward a model of informal digital learning of English and intercultural competence: a large-scale structural equation modeling approach. *Computer Assisted Language Learning*, 1-25. https://doi.org/DOI: 10.1080/09588221.2023.2191652

Liu, J., Pang, Z., & Qi, L. (2020). Dynamic pricing and inventory management with demand learning: A bayesian approach. *Computers & Operations Research, 124*, 105078. DOI: 10.1016/j.cor.2020.105078 PMID: 32836690

Liu, L., & Liu, K. (2023). Can digital technology promote sustainable agriculture? *Cogent Food & Agriculture*, *9*(2), 2282234. Advance online publication. DOI: 10.1080/23311932.2023.2282234

Loayza, K. (2024). Transformando la educación básica: retos y perspectivas de la inteligencia artificial. *Vitalia, 5*(2), 01-17. DOI: 10.61368/r.s.d.h.v5i2.113

Long, J., & Lin, J. (2022). An empirical study on cultivating college students' cross-cultural communicative competence based on the artificial-intelligence English-teaching mode. *Frontiers in Psychology*, *13*, 976310. DOI: 10.3389/fpsyg.2022.976310 PMID: 36562067

López de Prado, M., Simonian, J., Fabozzi, F. A., & Fabozzi, F. J. (2024). Enhancing Markowitz's portfolio selection paradigm with machine learning. *Annals of Operations Research*, •••, 1–22.

Lorenzo Lledó, A. (2023). El modelo educativo inclusivo desde el diseño universal para el aprendizaje (DUA). En *Atención a la Diversidad en los Centros Educativos*. Universidad de Alicante.

Luan, H., Géczy, P., Lai, H., Gobert, J., Yang, S., Ogata, H., Baltes, J., Guerra, R., Li, P., & Tsai, C. (2020). Challenges and Future Directions of Big Data and Artificial Intelligence in Education. *Frontiers in Psychology*, *11*, 580820. Advance online publication. DOI: 10.3389/fpsyg.2020.580820 PMID: 33192896

Luckin, R. (2018). *Machine Learning and Human Intelligence: The Future of Education for the 21st Century*. UCL Institute of Education Press.

Luckin, R., Holmes, W., Griffiths, M., & Forcier, L. B. (2023). *Artificial Intelligence and Education: Promises and Implications for Teaching and Learning*. UNESCO Publishing.

Luiz, A. A. D. (1962). *The tribes of Kerala*. Bharatiya Adimjati Sevak Sangh.

Lundvall, B. Å., & Johnson, B. (1994). The learning economy. *Journal of Industry Studies*, *1*(2), 23–42. DOI: 10.1080/13662719400000002

Lu, W., & Liang, H. (2024). Research on strategies and measures for enhancing digital literacy of college teachers in the era of smart education. *2024 IEEE First International Conference on Data Intelligence and Innovative Application (DIIA)*, 1–3. DOI: 10.1109/DIIA62678.2024.10871896

Mac Fadden, I., García-Alonso, E. M., & López Meneses, E. (2024). Science Mapping of AI as an Educational Tool Exploring Digital Inequalities: A Sociological Perspective. *Multimodal Technologies and Interaction*, *8*(12), 106. DOI: 10.3390/mti8120106

Machi, L. A., & McEvoy, B. T. (2016). *The Literature Review: Six Steps to Success*. Corwin Press. DOI: 10.4135/9781071939031

Mack, G., Ritzel, C., Ammann, J., & Benni, N. E. (2024). Improving the understanding of farmers' non-compliance with agricultural policy regulations. *Journal of Rural Studies*, *106*, 103190. DOI: 10.1016/j.jrurstud.2023.103190

MACROTRENDS. (n.d.). *UAE literacy rate 1975-2025.* https://www.macrotrends.net/global-metrics/countries/ARE/uae/literacy-rate

Magazzo, S. (2022, June 20). 6 emerging technology trends in the fashion industry. Mondo Staffing Agency. https://mondo.com/insights/top-technology-trends-fashion-industry/

Mahasneh, H. (2024). Drones in agriculture: Real-world applications and impactful case studies. *Journal of Natural Science Review*, *2*(Special Issue), 643–656. DOI: 10.62810/jnsr.v2iSpecial.Issue.164

Mahler, C. (2024, October 30). Why Analytics is Essential for Fashion Inventory Forecasting. Prediko. https://www.prediko.io/blog/analytics-for-fashion-inventory-forecasting

Marcano, B., Rodríguez, H., & Lopez, L. (2024). Impacto de la inteligencia artificial en la mejora de la práctica docente. Actas IDI-UNICyT, 461-466. DOI: 10.47300/actasidi-unicyt-2023-71

Mariappan, K., & Zhou, D. (2019). *A threat of farmers' suicide and the opportunity in organic farming for sustainable agricultural development in India*. Sustainability., DOI: 10.3390/su11082400

Markowitz, H. (1952). Portfolio Selection. *The Journal of Finance*, *7*(1), 77–91.

Martín, A. C., Alario-Hoyos, C., & Kloos, C. D. (2019). Smart Education: A Review and Future Research Directions. *Proceedings*, *31*(1), 57.

McCarty, T. L., & Lee, T. S. (2014). Critical culturally sustaining/revitalizing pedagogy and Indigenous education sovereignty. *Harvard Educational Review*, *84*(1), 101–124. DOI: 10.17763/haer.84.1.q83746nl5pj34216

Meng, Q., Jia, J., & Zhang, Z. (2020). A framework of smart pedagogy based on the facilitating of high order thinking skills. *Interactive Technology and Smart Education, 17*(3), 251–266. DOI: 10.1108/ITSE-11-2019-0076

Menon, S., & Bawa, K. S. (1997). Applications of Geographic Information System, Remote-Sensing, and Landscape Ecology Approach to Biodiversity Conservation in the Western Ghats. *Current Science, 73*, 134–145. https://www.scirp.org/reference/referencespapers?referenceid=3010812

Metaactivity and metaknowledge. methodological advice. (2022). Access mode: 02/27/2025. https://metodsovet.ru/metadeyatelnost_metaznaniya

Mezirow, J. (1991). *Transformative dimensions of adult learning*. Jossey-Bass.

Mezirow, J. (Ed.). (2000). *Learning as transformation: Critical perspectives on a theory in progress*. Jossey-Bass.

Michalon, B., & Camacho-Zuñiga, C. (2023). ChatGPT, a brand-new tool to strengthen timeless competencies. *Frontiers in Education, 8*, 1251163. DOI: 10.3389/feduc.2023.1251163

Mieszajkina, E., & Ostapińska, K. (2024). Evaluation of the Impact of the BANI World on Small Business Innovation Activities. *European Research Studies, 27*(Special Issue B), 272–288. DOI: 10.35808/ersj/3489

Mitchell, T., & Steinmetz, G. (1999). Society, economy, and the state effect. *State/culture: State-formation after the cultural turn, 76*, 90.

Mohamed, A. (2024). Artificial intelligence in language education: Opportunities and challenges. *TESOL Quarterly, 58*(2), 210–230. DOI: 10.1002/tesq.5678

Mohamed, A., & Shaaban, T. (2023). Investigating College Students' Metacognitive Awareness in Enhancing ESP Writing Proficiency. *Journal of Teaching English for Specific and Academic Purposes, 11*(2), 387–401. DOI: 10.22190/JTESAP230420029M

Molur, S. (2009). Habitat and status assessment of mammals with special reference to rodents and bats in Western Ghats of Karnataka [PhD Thesis]. University of Mysore.

Molur, S., Smith, K. G., Daniel, B. A., Darwall, W. R. T., & Zoo Outreach Organisation. (2011). The status and distribution of freshwater biodiversity in the Western Ghats, India. IUCN.

Moore, G. C., & Benbasat, I. (1991). Development of an instrument to measure the perceptions of adopting an information technology innovation. *Information Systems Research, 2*(3), 192–222.

Moreton-Robinson, A. (2021). *Talkin' up to the white woman: Indigenous women and feminism*. University of Minnesota Press.

Morris, T., & Comeau, J. (2020). Portfolio creation using artificial neural networks and classification probabilities: A Canadian study. *Financial Markets and Portfolio Management*, *34*(2), 133–163. DOI: 10.1007/s11408-020-00350-8

Moybeka, A., Syariatin, N., Tatipang, D., Mushthoza, D., Dewi, N., & Tineh, S. (2023). Artificial Intelligence and English Classroom: The Implications of AI Toward EFL Students' Motivation. *Edumaspul: Jurnal Pendidikan*, *7*(2), 2444–2454. DOI: 10.33487/edumaspul.v7i2.6669

Muganda, B. W., & Kasamani, B. S. (2023, June). Parallel programming for portfolio optimization: A robo-advisor prototype using genetic algorithms with recurrent neural networks. In *2023 International Conference on Intelligent Computing, Communication, Networking and Services (ICCNS)* (pp. 167-176). IEEE. DOI: 10.1109/ICCNS58795.2023.10193396

Mujica-Sequera, R. (2024). Clasificación de las herramientas de la inteligencia artificial en la educación. *Revista Docentes 2 0, 17*(1), 31-40. DOI: 10.37843/rted.v17i1.513

Mukherjee, A., Chakraborty, D., Maity, A., Ray, M., & Kumari, S. (2017). How is drone going to impact Indian agriculture? *Biotech Articles*. Retrieved from http://www.biotecharticles.com

Muneer, V. (2015). Analysis of status of tribal students in selected aspects and functioning of tribal residential schools in Kerala. [Doctoral dissertation]. http://hdl.handle.net/10603/349029

Muñoz, G., Delgado, J., & Dávalos, J. (2024). Inteligencia artificial y gobernanza en la gestión académica y administrativa de la educación superior. *Revista Social Fronteriza*, *4*(6), e46508. DOI: 10.59814/resofro.2024.4(6)508

Myers, N., Mittermeier, R. A., Mittermeier, C. G., Da Fonseca, G. A. B., & Kent, J. (2000). Biodiversity hotspots for conservation priorities. *Nature*, *403*(6772), 853–858. DOI: 10.1038/35002501 PMID: 10706275

Nadeem, K., Ahmad, M., Javed, Z., & Habib, M. A. (2022). Development of a machine learning model for prediction of colour trends in fashion industry. *2022 International Conference on Frontiers of Information Technology (FIT)*, 296–301. DOI: 10.1109/FIT57066.2022.00061

Nag, D. (2024). When Technology Meets Tradition. https://doi.org/DOI: 10.1002/9781394200498.ch2

Nakata, M. (2007). *Disciplining the Savages: Savaging the Disciplines*. Aboriginal Studies Press.

Nanda, A. K., & Samanta, D. (2017). *Education and social change among Indian tribes*. Rawat Publications.

Nanda, A. K., & Samanta, S. (2017). Mainstreaming tribals through financial literacy – a review of literature. *International Journal of Social Economics, 45*(2), 437–444. DOI: 10.1108/IJSE-12-2016-0371

Natarajan, T. S. (1985). *Tribal habitats of Nilgiri District: A profile*. Tribal Research Centre.

Nayak, S. R., Kant, N., & Anjali, K. (2020). Strategy of using ICT in ODL to disseminate higher education in tribal communities: a case of MP, India. Asian Association of Open Universities Journal, 15(2), 189-206.

Nedović, L., & Devedžić, V. (2002). Expert systems in finance—A cross-section of the field. *Expert Systems with Applications, 23*(1), 49–66. DOI: 10.1016/S0957-4174(02)00027-1

Nikou, S. A., & Economides, A. A. (2017). Mobile-based assessment: Investigating the factors that influence behavioral intention to use. *Computers & Education, 109*, 56–73. DOI: 10.1016/j.compedu.2017.02.005

Noman, R., Shahid, M., & Alam, M. (2023, October). The Synergy of Blockchain and Artificial Intelligence in Fintech: Revolutionizing the Financial Landscape. In *2023 4th International Conference on Data Analytics for Business and Industry (ICDABI)* (pp. 425-430). IEEE.

Noor, F., & Noel, A. S. (2023). Perception of farmers regarding drones for pesticide spraying. *Asian Journal of Advances in Agricultural Research, 22*(3), 9–1. DOI: 10.9734/ajaar/2023/v22i3440

Ntoutsi, E., Fafalios, P., Gadiraju, U., Iosifidis, V., Nejdl, W., Vidal, M. E., Ruggieri, S., Turini, F., Papadopoulos, S., Krasanakis, E., Kompatsiaris, I., Kinder-Kurlanda, K., Wagner, C., Karimi, F., Fernandez, M., Alani, H., Berendt, B., Kruegel, T., Heinze, C., & Staab, S. (2020). Bias in data-driven artificial intelligence systems—An introductory survey. *Wiley Interdisciplinary Reviews. Data Mining and Knowledge Discovery, 10*(3), e1356. DOI: 10.1002/widm.1356

Nurmukhamedov, G. M. (2010). Multimedia textbook universal pedagogical teaching aid in modern education. *Informatics and education.*, (6), 29–34.

Nuryadi, M. H., & Widiatmaka, P. (2023). Strengthening civic literacy among students through digital literacy in society 5.0. [EduLearn]. *Journal of Education and Learning, 17*(2), 215–220. DOI: 10.11591/edulearn.v17i2.20746

Nwobodo, L. K. (2024). The importance of data analytics in modern fashion industry. Global Journal of Economic and Finance Research, 01(05). https://doi.org/01e5DOI: 10.55677/gjefr/04-2024-vol

O'Hara, S., & Pritchard, R. H. (2008). Meeting the challenge of diversity: Professional development for teacher educators. *Teacher Education Quarterly, 35*(1), 43-61. https://www.jstor.org/stable/23479030

Ofosu-Ampong, K. (2024). Beyond the hype: Exploring faculty perceptions and acceptability of AI in teaching practices. *Discover Education, 3*(1), 38. DOI: 10.1007/s44217-024-00128-4

Ole Henrik Magga. Nicolaisen, I., Trask, M., & Dunbar, R. (2005). Indigenous Children's Education and Indigenous Languages. ResearchGate. https://www.researchgate.net/publication/255571823_Indigenous_Children

Omer, W., Mohamed, A., & Al-khresheh, M. (2025). The Integration of English Language Skills in Business Education: Insights from Accounting in the Saudi Context. *Journal of Teaching in International Business* https://doi.org/DOI: 10.1080/08975930.2025.2452529

Omojola, S. (2025, February 18). Optimize supply chains using predictive fashion analytics. Woven Insights Blog. https://woveninsights.ai/site-blog/optimize-supply-chains-using-predictive-fashion-analytics/

Optimizing demand forecasting with historical data insights. (n.d.). EOXS- Where Steel Meets Technology. Retrieved March 17, 2025, from https://eoxs.com/new_blog/optimizing-demand-forecasting-with-historical-data-insights/

Ortiz, F. (1947). *Transculturation and Cuba*. University of Miami Press.

Ouko, K. O., Mukhebi, A. W., Obiero, K. O., & Opondo, F. A. (2022). Using the technology acceptance model to understand fish farmers' intention to use black soldier fly larvae meal in Nile tilapia production in Kenya. *All Life, 15*(1), 884–900. DOI: 10.1080/26895293.2022.2112765

Padilla, R. (2019). La llegada de la inteligencia artificial a la educación. *Revista De Investigación en Tecnologías De La Información, 7*(14), 260–270. DOI: 10.36825/RITI.07.14.022

Palos, R., & Maricutoiu, L. P. (2013). Teaching for Successful Intelligence Questionnaire (TSI-Q)-a new instrument developed for assessing teaching style. *Journal of Educational Sciences and Psychology, 3*(1).

Pang, X., Zhou, Y., Wang, P., Lin, W., & Chang, V. (2020). An innovative neural network approach for stock market prediction. *The Journal of Supercomputing, 76*(3), 2098–2118. DOI: 10.1007/s11227-017-2228-y

Parthasarathy, J. (2013). Kattunayakans of Nilgiris District, Tamil Nadu. Tribal Research Center (TRC) & Hill Area Development Programme (HADP), Government of Tamil Nadu.

Parthasarathy, J. (2013). *Education and Development Among The Tribes. Tribal Research Center (TRC) & Hill Area Development Programme (HADP).* Government of Tamil Nadu.

Patchipala, S.Surya Gangadhar Patchipala. (2023). Tackling data and model drift in AI: Strategies for maintaining accuracy during ML model inference. *International Journal of Science and Research Archive, 10*(2), 1198–1209. DOI: 10.30574/ijsra.2023.10.2.0855

Pedagogy: Innovative Teaching and Learning Strategies in Engineering Education, 2018 IEEE World Engineering Education Conference (EDUNINE), Buenos Aires, Argentina, pp. 1-6, DOI: 10.1109/EDUNINE.2018.8450962

Percherla, A. K. (2024). Dropshipping strategies in luxury and fashion retail - Unveiling the potential. *Journal of Business and Strategic Management, 9*(2), 52–60. DOI: 10.47941/jbsm.1896

Pérez, O., & González, N. (2024). Formación docente para el uso de la inteligencia artificial. *Ciencia Latina Revista Científica Multidisciplinar, 8*(5), 11772–11788. DOI: 10.37811/cl_rcm.v8i5.14594

Pham, L., & Tran, L. (2015). Understanding the symbolic capital of intercultural interactions: A case study of international students in Australia. *International Studies in Sociology of Education, 25*(3), 204-224. https://doi.org/DOI: 10.1080/09620214.2015.1069720

Phillips, L. A., Calantone, R., & Lee, M. T. (1994). International technology adoption: Behavior structure, demand certainty, and culture. *Journal of Business and Industrial Marketing, 9*(2), 12–19. DOI: 10.1108/08858629410059762

Piaget, J. (2016). Intellectual evolution from adolescence to adulthood1. In *Cognitive and moral development, academic achievement in adolescence* (pp. 1–12). Routledge.

Pingali, P. (2007). Agricultural growth and economic development: A view through the globalization lens. *Agricultural Economics*, *37*(s1), 1–12. DOI: 10.1111/j.1574-0862.2007.00231.x

Pintado, A., Prado, R., Peláez, C., & Aguilar, W. (2023). Inteligencia artificial y sostenibilidad: el compromiso de una institución de educación superior. *Magazine De Las Ciencias Revista De Investigación E Innovación, 8*(4), 12-28. DOI: 10.33262/rmc.v8i4.2954

Pisanets, N. G. (2020). Functional purpose of the information educational environment in the work of an educational institution. Access mode: 04/15/2025. https://www.prodlenka.org/metodicheskie-razrabotki/435430-funkcionalnoe-prednaznachenie-informacionnoj-

Popova, Y., & Popovs, S. (2022). Impact of smart economy on smart areas and mediation effect of national economy. *Sustainability (Basel)*, *14*(5), 2789. DOI: 10.3390/su14052789

Powell, W. W., & Snellman, K. (2004). The knowledge economy. *Annual Review of Sociology*, *30*(1), 199–220. DOI: 10.1146/annurev.soc.29.010202.100037

Prajapati, N. (2024, September 6). Predictive analytics: Enhancing supply chain management. Softweb Solutions. https://www.softwebsolutions.com/resources/role-of-predictive-analytics-in-supply-chain.html

Prakash, S., Muniammal, M. A., Maruthavanan, M., Natarajan, S., Thangavel, K., & Kumar, S. (2023). Education 5.0 revolutionizing learning for the future (Vol.1).

Prakasha, G. S. (2022). *Next-generation teacher competencies: competency balance scorecard, competency development cycle and strategies. in handbook of research on promoting global citizenship education*. IGI Global., DOI: 10.4018/978-1-7998-9542-8.ch014

productdatascrape. (2024, October 14). Scrape Zara product pricing data to understanding Zaras pricing strategy. IndiBlogHub. https://indibloghub.com/post/scrape-zara-product-pricing-data-to-understanding-zaras-pricing-strategy

Project, J. (n.d.). Kattunayakan in India. https://joshuaproject.net/people_groups/17305/IN

Pun, C. S., Wang, T., & Yan, Z. (2023). Data-Driven Distributionally Robust CVaR Portfolio Optimization Under A Regime-Switching Ambiguity Set. *Manufacturing & Service Operations Management*, *25*(5), 1779–1795. DOI: 10.1287/msom.2023.1229

Purcell, K., Buchanan, J., & Friedrich, L. (2024, April 14). *How teachers are using technology at home and in their classrooms*. Pew Research Center. https://www.pewresearch.org/internet/2013/02/28/how-teachers-are-using-technology-at-home-and-in-their-classrooms/

Pytel, K. (2020). Hybrid multi-evolutionary algorithm to solve optimization problems. *Applied Artificial Intelligence*, *34*(7), 550–563. DOI: 10.1080/08839514.2020.1730631

Qarkaxhija, J., Assanova, Z., Igenbayeva, R., Toxanova, S., Yensebayeva, G., Myrzanova, A., & Turlugulova, N. (2023). Development of the creative potential of future teachers based on smart education. [iJET]. *International Journal of Emerging Technologies in Learning*, *18*(24), 19–33. DOI: 10.3991/ijet.v18i24.43205

Raj, A. (2023, December 6). What is Supply Chain Predictive Analytics and how does it work? ThroughPut Inc. https://throughput.world/blog/predictive-analytics-in-supply-chain/

Raj, A. (2024, January 15). Top 5 demand forecasting methods in 2025. ThroughPut Inc. https://throughput.world/blog/demand-forecasting-methods

Rajagopalan, R. P., & Krishna, R. (2018). Drones: Guidelines, regulations, and policy gaps in India.

Rakovac Bekeš, E., & Galzina, V. (2023). Exploring the pedagogical use of ai-powered chatbots educational perceptions and practices. *International Convention on Information and Communication Technology, Electronics and Microelectronics.*

Ramachandran, B., Zvelebil, K. V., David, N. B., Buchanan, F., Dubois, J. A., & Aiyappan, A. Census of India, Tribal Research Center, Ooty, Keystone Foundation, & Seetha Kakkoth. (n.d.). Indigenous People's Plan Kattunayakans. https://www.cepf.net/sites/default/files/56154-ipp-kattunayakan.pdf

Ramakrishnan, R., Rajendrakumar, S., & Kothurkar, N. K. (2024). Regional sustainability of the Kattunayakan tribe in Kerala, India through the enhancement of agricultural, livestock, and livelihood options. *Agricultural Systems*, *217*, 103929. DOI: 10.1016/j.agsy.2024.103929

Ramaswamy, V., & Ozcan, K. (2014). *The co-creation paradigm*. Stanford University Press. DOI: 10.1515/9780804790758

Ramesh, B. R., Venugopal, P. D., Pélissier, R., Patil, S. V., Swaminath, M. H., & Couteron, P. (2010). Mesoscale patterns in the floristic composition of forests in the central western Ghats of Karnataka, India. *Biotropica*, *42*(4), 435–443. DOI: 10.1111/j.1744-7429.2009.00621.x

Rangasamy, S., Rajamohan, K., Janardhanan, A. K., & Manu, K. S. (2023). The role of financial analytics in decision-making for better firm performance. In *Analytics Enabled Decision Making* (pp. 139–169). Springer Nature Singapore. DOI: 10.1007/978-981-19-9658-0_7

Rani, A. L. K. A., Sharma, R., Verma, P., & Kumar, S. (2019). Drone: The green technology for future agriculture. *Harit Dhara*, *2*(1), 3–6.

Rao, Z., & Chen, H. (2020). Teachers' perceptions of difficulties in team teaching between local-and native-English-speaking teachers in EFL teaching. *Journal of Multilingual and Multicultural Development, 41*(4), 333-347. https://doi.org/DOI: 10.1080/01434632.2019.1620753

Raudenbush, S. W. (2004). What are value-added models estimating and what does this imply for statistical practice? *Journal of Educational and Behavioral Statistics*, *29*(1), 121–129. DOI: 10.3102/10769986029001121

Reddy, S. K., & Reddy, M. (2020). *Indigenous knowledge systems and sustainable practices: The case of the Kattunayakan tribe*. Indigenous Studies Press.

Redvers, J. (2020). "The land is a healer": Perspectives on land-based healing from Indigenous practitioners in northern Canada. *International Journal of Indigenous Health*, *15*(1), 90–107. DOI: 10.32799/ijih.v15i1.34046

Reporter, G. staff. (2023, October 1). 'You've got to be data-driven': The fashion forecasters using AI to predict the next trend. The Guardian. https://www.theguardian.com/technology/2023/oct/01/ai-artificial-intelligence-fashion-trend-forecasting-style

Rezaei, R., Safa, L., & Ganjkhanloo, M. M. (2020). Understanding farmers' ecological conservation behavior regarding the use of integrated pest management—An application of the technology acceptance model. *Global Ecology and Conservation*, *22*, e00941. DOI: 10.1016/j.gecco.2020.e00941

Rodríguez, A., Romero, M., Agreda, C., Saldarriaga, J., & Saldarriaga, M. (2024). Impacto de la inteligencia artificial en las prácticas educativas: percepciones y actitudes del profesorado. *Latam Revista Latinoamericana De Ciencias Sociales Y Humanidades, 5*(2). DOI: 10.56712/latam.v5i2.1933

Rodríguez, M., Rubio, A., Lingán, A., Rubio, D., Bocanegra, J., & Carrillo-Flores, J. (2023). Inteligencia artificial en la educación digital y los resultados de la valoración del aprendizaje. DOI: 10.31219/osf.io/c3pmd

Rogers, E. M. (1995). *Diffusion of innovations* (4th ed.). Free Press.

Rowe, R. K., & Shawanda, A. (2023). Indigenous Pedagogies and the Implications of EdTech, Data, and AI in the Classroom. In Perspectives on Indigenous Pedagogy in Education: Learning From One Another (pp. 215-247). IGI Global.

Rupavath, R. (2021). Poverty and Education: Attainments and challenges for tribal communities. *Contemporary Voice of Dalit, 15*(2), 230–247. DOI: 10.1177/2455328X211042718

Ryan, R. M., & Deci, E. L. (2020). Intrinsic and extrinsic motivations: Classic definitions and new directions. *Contemporary Educational Psychology, 61*, 101860. DOI: 10.1016/j.cedpsych.2020.101860 PMID: 10620381

Safna, H. (2025, February 25). How Predictive Inventory Analytics is Reducing Overstock & Stockouts for Retailers. Retailcloud. https://retailcloud.com/predictive-inventory-analytics/

Salinas, J., De Benito, B., & Pérez-Mateo, M. (2023). *Innovación educativa y desarrollo profesional docente en entornos digitales*. Narcea Ediciones.

Sánchez-Cartas, J. M., Tejero, A., & León, G. (2021). Algorithmic pricing and price gouging. Consequences of high-impact, low probability events. *Sustainability (Basel), 13*(5), 2542. DOI: 10.3390/su13052542

Sanchez-Roger, M., Oliver-Alfonso, M. D., & Sanchís-Pedregosa, C. (2019). Fuzzy logic and its uses in finance: A systematic review exploring its potential to deal with banking crises. *Mathematics, 7*(11), 1091. DOI: 10.3390/math7111091

Saravanan, S. P. (2024, November 18). Evening coaching classes enhance skills of tribal students in Sathyamangalam Tiger Reserve. The Hindu. https://www.thehindu.com/news/cities/Coimbatore/evening-coaching-classes-enhance-skills-of-tribal-students-in-sathyamangalam-tiger-reserve/article68881434.ece

Sarwari, A. Q., Javed, M. N., Mohd Adnan, H., & Abdul Wahab, M. N. (2024). Assessment of the impacts of artificial intelligence (AI) on intercultural communication among postgraduate students in a multicultural university environment. *Scientific Reports, 14*(1), 13849. DOI: 10.1038/s41598-024-63276-5 PMID: 38879546

Sasikumar, M. (1999). *Kattunaickans: An ethnographic report*. International School of Dravidian Linguistics.

Saulis, P. (2018). Indigenous knowledge and Western science [Video]. YouTube. https://www.youtube.com/watch?v=cwCfsAiERyk

Schleicher, A. (2024, March 18). Digital dreaming: the role of technology in expanding education access. https://www.unesco.org/en/articles/digital-dreaming-role-technology-expanding-education-access

Schön, D. A. (1983). *The reflective practitioner: How professionals think in action.* Basic Books.

Schön, D. A. (1987). *Educating the reflective practitioner: Toward a new design for teaching and learning in the professions*. Jossey-Bass.

Scriven, M. (1967). The Methodology of Evaluation. In Tyler, R. W., Gagne, R. M., & Scriven, M. (Eds.), *Perspectives of Curriculum Evaluation Chicago* (pp. 39–83). Rand McNally.

Seeling, P., Roberts, S., & Weible, J. (2022). Towards education 5.0: instruction with learners in the loop. *Proceedings of the 23rd Annual Conference on Information Technology Education*, 92–93. https://doi.org/DOI: 10.1145/3537674.3555793

Selvan, S. S., Wahid, A., Patel, A., Kumar, V., & Sahu, P. (2021). Challenges in Indian agriculture. *Agricultural Reviews (Karnal)*, *42*(1), 1–12. DOI: 10.18805/ag.R-2103

Setyono, B., & Widodo, H. P. (2019). The representation of multicultural values in the Indonesian Ministry of Education and Culture-Endorsed EFL textbook: a critical discourse analysis. *Intercultural Education, 30*(4), 383-397. https://doi.org/DOI: 10.1080/14675986.2019.1548102

Shang, L., Heckelei, T., Gerullis, M. K., Börner, J., & Rasch, S. (2021). Adoption and diffusion of digital farming technologies—Integrating farm-level evidence and system interaction. *Agricultural Systems*, *190*, 103074. DOI: 10.1016/j.agsy.2021.103074

Sharifzadeh, M. S., Damalas, C. A., Abdollahzadeh, G., & Ahmadi-Gorgi, H. (2017). Predicting adoption of biological control among Iranian rice farmers: An application of the extended technology acceptance model (TAM2). *Crop Protection (Guildford, Surrey)*, *96*, 88–96.

Shava, S. (2016). The Application/role of Indigenous Knowledges in Transforming the Formal education Curriculum: Cases from Southern Africa. In V. Msila & M. T. Gumbo (Eds.), Africanising the Curriculum: Indigenous Perspectives and Theories (pp. 121–137). AFRICAN SUN MeDIA.

Shewaramani, M. (2024, August 30). Ensuring data security and compliance in fashion retail with PIM. Credencys Solutions Inc. https://www.credencys.com/blog/ensuring-data-security-and-compliance-in-fashion-retail/

Shi, M., & Lewis, V. D. (2020, May 3). Using artificial intelligence to analyze fashion trends. arXiv.Org. https://arxiv.org/abs/2005.00986

Shirkhani, S., Mokayed, H., Saini, R., & Chai, H. Y. (2023). Study of ai-driven fashion recommender systems. *SN Computer Science*, *4*(5), 514. Advance online publication. DOI: 10.1007/s42979-023-01932-9

Siburian, J., & Mardiyanti, L. (2023). Prediction of Meta-Skills Based on Metcognition. *Jurnal Penelitian Pendidikan IPA*, *9*(12), 11053–11059. DOI: 10.29303/jppipa.v9i12.5683

Siddique, Z. (2024, August 14). Predictive analytics in fashion: Forecasting trends and optimizing inventory. Ceba Solutions. https://www.cebasolutions.com/articles/predictive-analytics-in-fashion-forecasting-trends-and-optimizing-inventory

Simpson, L. B. (2014). Land as pedagogy: Nishnaabeg intelligence and rebellious transformation. *Decolonization*, *3*(3).

Sivashinskaya, E. F. (2010). *Pedagogical systems and technologies: a course of lectures for students of pedagogical specialties of universities*. Ecoperspective.

Snyder, H. (2019). Literature review as a research methodology: An overview and guidelines. *Journal of Business Research*, *104*, 333–339. DOI: 10.1016/j.jbusres.2019.07.039

Sokolova, O. I. (2001). Modern information and educational environments. Access mode: 02/26/2025. https://search.rsl.ru/ru/record/01003220700

Sood, A., Bhardwaj, A. K., & Sharma, R. K. (2022). Towards sustainable agriculture: Key determinants of adopting artificial intelligence in agriculture. *Journal of Decision Systems*, *33*(4), 833–877. DOI: 10.1080/12460125.2022.2154419

Spaulding, W. (2022, December 28). *Education in the Middle Ages | Schools, History & Facts*. Study.com. https://study.com/academy/lesson/education-middle-ages-schools-history-facts.html

Spellman, K., Dillenbeck, J., Edwards, N. N., & Bohecker, L. (2022). Supporting marginalized students in counselor education and supervision programs. *Journal of Counselor Leadership and Advocacy, 9*(1), 32-44. https://doi.org/DOI: 10.1080/2326716X.2021.2007429

Sridevi, M., ManikyaArun, N., Sheshikala, M., & Sudarshan, E. (2020). Personalized fashion recommender system with image based neural networks. *IOP Conference Series. Materials Science and Engineering*, *981*, 022073. DOI: 10.1088/1757-899X/981/2/022073

StartUs Insights. (2022, September 6). Top 8 fashion technology trends in 2023. StartUs Insights. https://www.startus-insights.com/innovators-guide/fashion-technology-trends/

Sternberg, R. J., & Grigorenko, E. L. (1995). Styles of thinking in the school. *European Journal for High Ability*, *6*(2), 201–219. DOI: 10.1080/0937445940060211

Strathern, M. (1997). 'Improving ratings': Audit in the British University system. *European Review (Chichester, England)*, *5*(3), 305–321. DOI: 10.1002/(SICI)1234-981X(199707)5:3<305::AID-EURO184>3.0.CO;2-4

Studiopedia. (2014). System of information communications. Access mode: 02/27/2025. https://studopedia.ru/1_87877_sistema-informatsionnih-kommunikatsiy.html

Subramanian, K. S., Pazhanivelan, S., Srinivasan, G., Santhi, R., & Sathiah, N. (2021). Drones in insect pest management. *Frontiers in Agronomy*, *3*, 640885. DOI: 10.3389/fagro.2021.640885

Suchodolski, B. (1961). Education for the future and traditional pedagogy. *International Review of Education*, •••, 420–431.

Sugumaran, V., Geetha, T. V., Manjula, D., & Gopal, H. (2017). Guest editorial: Computational intelligence and applications. *Information Systems Frontiers*, *19*(5), 969–974. DOI: 10.1007/s10796-017-9794-7

Sundar, C. B. V., Asokhan, M., & Karthikeyan, C. (2023). Adoption of drones in agriculture: Social, economic, and personal factors. *International Journal of Environment and Climate Change*, *13*(8), 587–597. DOI: 10.9734/ijecc/2023/v13i81987

Sun, G. (2024). Prompt Engineering for Nurse Educators. *Nurse Educator*, *49*(6), 293–299. Advance online publication. DOI: 10.1097/NNE.0000000000001705 PMID: 38968442

Sun, L. (2024). Enhancing intercultural competence of Chinese English majors through AI-enabled Collaborative Online International Learning (COIL) in the digital era. *Education and Information Technologies*, •••, 1–33. DOI: 10.1007/s10639-024-13143-7

Sutrisman, H., Simanjuntak, R., Prihartanto, A., & Kusumo, B. (2024). The Impact of Using AI in Learning on Understanding of Material by Young Students. *International Journal of Educational Research*, *1*(3), 24–32. Advance online publication. DOI: 10.62951/ijer.v1i3.43

Suvittawat, A. (2024). Investigating farmers' perceptions of drone technology in Thailand: Exploring expectations, product quality, perceived value, and adoption in agriculture. *Agriculture*, *14*(12), 2183. DOI: 10.3390/agriculture14122183

Suvorova T. N. (2014). Didactic functions, opportunities and properties of educational resources. Standards and electronic monitoring in education. No. 2.Pp. 27–35.

Sytnyk, L., & Podlinyayeva, O. (2024). AI in education: main possibilities and challenges. In Proceedings of the 8th International Scientific and Practical Conference" International Scientific Discussion: Problems, Tasks and Prospects (pp. 569-579). DOI: 10.51582/interconf.19-20.05.2024.058

Szelei, N., Tinoca, L., & Pinho, A. S. (2020). Professional development for cultural diversity: The challenges of teacher learning in context. *Professional development in education, 46*(5), 780-796. https://doi.org/DOI: 10.1080/19415257.2019.1642233

Szyliowicz, J. S., Lauwerys, J. A., Bowen, J., Browning, R., Thomas, R. M., Vázquez, J. Z., Mukerji, S., Nakosteen, M. K., Chen, T. H., Naka, A., Swink, R. L., Lawson, R. F., Shimahara, N., Gelpi, E., Anweiler, O., Graham, H. F., Ipfling, H., Moumouni, A., Scanlon, D. G., . . . Arnove, R. F. (2025, January 19). *Education | Definition, Development, History, Types, & Facts*. Encyclopedia Britannica. https://www.britannica.com/topic/education/Education-in-the-earliest-civilizations

Takagi, C., Purnomo, S. H., & Kim, M. (2020). Adopting smart agriculture among organic farmers in Taiwan. *Asian Journal of Technology Innovation*, *1*(1), 1–16. DOI: 10.1080/19761597.2020.1797514

Tandon, R., & Tandon, S. (2020). Education 4.0: A new paradigm in transforming the future of education in India. *International Journal of Innovative Science. Engineering & Technology*, *7*(2), 32–54.

Tao, S., Li, Y., Dong, X., Gunasekaran, N., & Aziz, A. (2021). Smart educational learning strategies for teachers and students in the higher education system. *J. Multiple Valued Log. Soft Comput.*, *36*, 99–115.

Tapalova, E., Vorobyev, A., & Smirnov, S. (2022). The impact of contextual learning on students' motivation and performance: An empirical study. *International Journal of Educational Technology in Higher Education*, *19*(1), 1–18. DOI: 10.1186/s41239-022-00336-4 PMID: 35013716

Tapalova, O., Zhiyenbayeva, N., & Gura, D. (2022). Artificial Intelligence in Education: AIEd for Personalised Learning Pathways. *Electronic Journal of e-Learning*, *20*(5), 639–653. DOI: 10.34190/ejel.20.5.2597

Tariq, M. U., & Sergio, R. P. (2025). Ethical Considerations and Social Impact of AI and Computational Intelligence in Society 5.0. In *Open AI and Computational Intelligence for Society 5.0* (pp. 225-248). IGI Global Scientific Publishing.

Taskan, B., Junça-Silva, A., & Caetano, A. (2022). Clarifying the conceptual map of VUCA: A systematic review. *The International Journal of Organizational Analysis, 30*(7), 196–217. DOI: 10.1108/IJOA-02-2022-3136

Taylor, E. W. (2007). An update of transformative learning theory: A critical review of the empirical research (1999–2005). *International Journal of Lifelong Education, 26*(2), 173–191. DOI: 10.1080/02601370701219475

Taylor, E. W., & Cranton, P. (2012). *The handbook of transformative learning: Theory, research, and practice*. John Wiley & Sons.

Team, F. (n.d.). Education during the Renaissance period. *Fairgaze.* https://fairgaze.com/educationnews/education-during-the-renaissance-period.html

Team, S. C. B. (2023, April 13). Can analytics prevent stockouts and overstocks? SPS Commerce. https://www.spscommerce.com/blog/in-a-ship-from-anywhere-era-can-analytics-prevent-stock-outs/

Teo, T., Lee, C. B., & Chai, C. S. (2008). Understanding pre-service teachers' computer attitudes: Applying and extending the Technology Acceptance Model. *Journal of Computer Assisted Learning, 24*(2), 128–143. DOI: 10.1111/j.1365-2729.2007.00247.x

Thakkar, A., & Chaudhari, K. (2021). A comprehensive survey on portfolio optimization, stock price and trend prediction using particle swarm optimization. *Archives of Computational Methods in Engineering, 28*(4), 2133–2164. DOI: 10.1007/s11831-020-09448-8

The Fullstory Team. (2024, April 17). Predicting customer behavior with data. Fullstory. https://www.fullstory.com/blog/predicting-customer-behavior/

TimesTech. (2024, July 8). How AI is Revolutionising Personalised Fashion. TimesTech. https://timestech.in/how-ai-is-revolutionising-personalised-fashion/

Toral, S. L., Martínez-Torres, M. R., & Barrero, F. (2007). Perceived usefulness and ease of use of e-collaboration tools in engineering education. *Computers & Education, 49*(4), 1097–1113.

Torre, C., López, M., Basantes, D., & Chango, E. (2024). El papel de la inteligencia artificial en la adaptación curricular: Perspectivas para la mejora continua de la educación digital. *Revista Imaginario Social*, *7*(2). Advance online publication. DOI: 10.59155/is.v7i2.186

Torres-Peña, R., Peña-González, D., Chacuto-López, E., Ariza, E., & Vergara, D. (2024). Updating Calculus Teaching with AI: A Classroom Experience. *Education Sciences*, *14*(9), 1019. Advance online publication. DOI: 10.3390/educsci14091019

Tranfield, D., Denyer, D., & Smart, P. (2003). Towards a methodology for developing evidence-informed management knowledge by means of systematic review. *British Journal of Management*, *14*(3), 207–222. DOI: 10.1111/1467-8551.00375

Trang, H. (2024, October 22). Digital Transformation in Education: Benefits, Challenges and Solutions. Tokyo Tech Lab. https://tokyotechlab.com/blogs/digital-transformation-in-education

Trend forecasting with data analysis and AI. (n.d.). WGSN. Retrieved March 17, 2025, from https://www.wgsn.com/en/wgsn/data

Trigger, B. G. (1998). Writing systems: A case study in cultural evolution. *Norwegian Archaeological Review*, *31*(1), 39–62. DOI: 10.1080/00293652.1998.9965618

Tsang, A., Aubrey, S., & Yuan, R. (2023). Multiculturalism and multilingualism in higher education: intercultural activity participation and opportunities for language learning. *International Journal of Multilingualism*, 1-17. https://doi.org/DOI: 10.1080/14790718.2022.2164769

Tsybulsky, D., & Levin, I. (2019). Science teachers' worldviews in the age of the digital revolution: Structural and content analysis. *Teaching and Teacher Education*, *86*, 102921. DOI: 10.1016/j.tate.2019.102921

Tuttle, H. (2013). Technology-Based Curriculum Development in Native Language Learning: How Computer-Assisted Cultural Education Supports Contemporary Native Youths.

UNESCO World Heritage Centre. (n.d.). Western Ghats. https://whc.unesco.org/en/list/1342/

UNESCO. (2022). *Reimagining our futures together: A new social contract for education*. UNESCO Publishing.

UN-HJDM. (2024, October 7). AI in fashion: How AI fashion trends shape the future of style. JD Institute of Fashion Technology. https://www.jdinstitute.edu.in/ai-in-fashion-how-ai-fashion-trends-shape-the-future-of-style/

United Nations. (2015). Transforming our world: The 2030 Agenda for Sustainable Development. United Nations. https://sdgs.un.org/2030agenda

Uskov, V. L. (2022). Smart Pedagogy: Concepts, Components, Outcomes. In Uskov, V. L., Howlett, R. J., & Jain, L. C. (Eds.), *Smart Education and e-Learning - Smart Pedagogy. SEEL-22 2022. Smart Innovation, Systems and Technologies* (Vol. 305). Springer., DOI: 10.1007/978-981-19-3112-3_1

Uskov, V., Bakken, J. P., Aluri, L., Rachakonda, R., Rayala, N., & Uskova, M. (2018, March). Smart pedagogy: Innovative teaching and learning strategies in engineering education. In *2018 IEEE World Engineering Education Conference (EDUNINE)* (pp. 1-6). IEEE.

Vakaliuk, T., Osova, O., Antoniuk, D., & Novitska, I. (2020). Future Foreign Language Teachers' Readiness for Innovative Activity in the Context of Smart Education: *The Results of the Pedagogical Experiment. Universal Journal of Educational Research, 8*(12B), 8223–8232. DOI: 10.13189/ujer.2020.082626

Vavilala, N. (2023, May 3). Predictive analytics in supply chain. Admin. https://dilytics.com/predictive-analytics-in-supply-chain/

Vela, G., Naranjo, B., Quinte, R., Villafuerte, V., & Velasco, J. (2024). Inclusión de la inteligencia artificial en la docencia universitaria. *Latam Revista Latinoamericana De Ciencias Sociales Y Humanidades, 5*(1). DOI: 10.56712/latam.v5i1.1642

Venkatesh, V. (2000). Determinants of perceived ease of use: Integrating control, intrinsic motivation, and emotion into the technology acceptance model. *Information Systems Research, 11*(4), 342–365. DOI: 10.1287/isre.11.4.342.11872

Venkatesh, V., & Bala, H. (2008). Technology Acceptance Model 3 and a research agenda on interventions. *Decision Sciences, 39*(2), 273–315. DOI: 10.1111/j.1540-5915.2008.00192.x

Venkatesh, V., Morris, M. G., Davis, G. B., & Davis, F. D. (2003). User acceptance of information technology: Toward a unified view. *Management Information Systems Quarterly, 27*(3), 425–478. DOI: 10.2307/30036540

Vera, D., Gutiérrez, M., Chuquillan, D., & Carrión, D. (2024). Avances tecnológicos para la inclusión: El rol de la inteligencia artificial en la diversificación del aprendizaje escolar. *Revista Social Fronteriza, 4*(3), e43265. DOI: 10.59814/resofro.2024.4(3)265

Vera-Flores, K., Cerna-Vinueza, A., Cornejo, M., & Vergel-Parejo, E. (2024). Inteligencia artificial en el campo de la educación de lengua y literatura. *Mqrinvestigar, 8*(4), 5013–5037. DOI: 10.56048/MQR20225.8.4.2024.5013-5037

Verma, R. (2024, June). Revolutionizing Derivative Market Predictions with Hybrid EGARCH-ANN Models: Embracing Machine Learning for Better Forecasting. In *2024 15th International Conference on Computing Communication and Networking Technologies (ICCCNT)* (pp. 1-5). IEEE.

Verma, P., & Sinha, N. (2018). Integrating perceived economic well-being into the technology acceptance model: The case of mobile-based agricultural extension service. *Technological Forecasting and Social Change*, *126*, 207–216. DOI: 10.1016/j.techfore.2017.08.013

Vincent, D. R., Deepa, N., Elavarasan, D., Srinivasan, K., Chauhdary, S. H., & Iwendi, C. (2019). Sensors-driven AI-based agriculture recommendation model for assessing land suitability. *Sensors (Basel)*, *19*(17), 3667. DOI: 10.3390/s19173667 PMID: 31450772

Voorhees, R. A. (2001). Competency-based learning models: A necessary future. *New Directions for Institutional Research*, *2001*(110), 5–13. DOI: 10.1002/ir.7

Vorozhtsov, A. (2023, May 3). Machine Learning for price optimization. Habr. https://habr.com/en/articles/692270/

Vowel, C. (2016). *Indigenous Writes: A Guide to First Nations, Métis, and Inuit Issues in Canada*. Highwater Press.

Vui, C. S., Soon, G. K., On, C. K., Alfred, R., & Anthony, P. (2013, November). A review of stock market prediction with Artificial neural network (ANN). In *2013 IEEE international conference on control system, computing and engineering* (pp. 477-482). IEEE.

Wang, X., & Chen, Q. (2024). Stochastic optimal control for dynamic pricing and production in fashion retailing: An economic sustainability approach. Discrete Dynamics in Nature and Society, 2024(1). DOI: 10.1155/ddns/4854557

Washington, A. N. (2020, February). When twice as good isn't enough: The case for cultural competence in computing. In *Proceedings of the 51st ACM technical symposium on computer science education* (pp. 213-219). https://doi.org/DOI: 10.1145/3328778.3366792

Waters, C. N., Zalasiewicz, J., Summerhayes, C., Barnosky, A. D., Poirier, C., Gałuszka, A., Cearreta, A., Edgeworth, M., Ellis, E. C., Ellis, M., Jeandel, C., Leinfelder, R., McNeill, J. R., Richter, D. B., Steffen, W., Syvitski, J., Vidas, D., Wagreich, M., Williams, M., & Wolfe, A. P. (2016). The Anthropocene is functionally and stratigraphically distinct from the Holocene. *Science*, *351*(6269), aad2622. DOI: 10.1126/science.aad2622 PMID: 26744408

Waterston, S., & Choudhary, N. (2024). Demand trends decoded: Data science approach. Pickl.AI. https://www.pickl.ai/blog/decoding-demand-the-data-science-approach-to-forecasting-trends/

Weber, P., Carl, K. V., & Hinz, O. (2024). Applications of explainable artificial intelligence in finance—A systematic review of finance, information systems, and computer science literature. *Management Review Quarterly*, *74*(2), 867–907. DOI: 10.1007/s11301-023-00320-0

West, M. (2023). *An ed-tech tragedy? Educational technologies and school closures in the time of COVID-19*. UNESCO., DOI: 10.54675/LYGF2153

What are the data privacy concerns with AI in fashion customer relations in 2024? (2024, June 2). Blog.Stateless.Nyc. https://blog.stateless.nyc/what-are-the-data-privacy-concerns-with-ai-in-fashion-customer-relations-in-2024/

Williamson, B., & Eynon, R. (2020). Education data futures: Data-intensive technologies and the aims of education. *Learning, Media and Technology*, *45*(1), 36–51.

Wolfers, J., & Zitzewitz, E. (2004). Prediction markets. *The Journal of Economic Perspectives*, *18*(2), 107–126. DOI: 10.1257/0895330041371321

Xaxa, V. (2001). Protective Discrimination: Why Scheduled Tribes Lag behind Scheduled Castes. *Economic and Political Weekly*, *36*(29), 2765–2772. https://www.jstor.org/stable/4410888

Xaxa, V. (2005). Politics of Language, Religion and Identity: Tribes in India. *Economic and Political Weekly*, *40*(13), 1363–1370. https://www.jstor.org/stable/4416402

Xaxa, V. (2016). Tribes and Indian national identity. *The Brown Journal of World Affairs*, *23*(1), 223–237. https://www.jstor.org/stable/26534720

Xia, Y., Shin, S.-Y., & Kim, J.-C. (2024). Cross-Cultural Intelligent Language Learning System (CILS): Leveraging AI to Facilitate Language Learning Strategies in Cross-Cultural Communication. *Applied Sciences (Basel, Switzerland)*, *14*(13), 5651. DOI: 10.3390/app14135651

Xing, P. (2023). A practical study on the application of smart education platform for teaching and research by primary and secondary school teachers. *International Journal of New Developments in Education*, *5*(8). Advance online publication. DOI: 10.25236/IJNDE.2023.050814

Yalcin, M. K. (2024, December 24). Personalization in apparel e-commerce with predictive analytics. Vizio.AI. https://www.vizio.ai/blog/personalization-in-apparel-e-commerce-with-predictive-analytics

Yaman, I., & Dalkılıç, T. E. (2021). A hybrid approach to cardinality constraint portfolio selection problem based on nonlinear neural network and genetic algorithm. *Expert Systems with Applications*, *169*, 114517. DOI: 10.1016/j.eswa.2020.114517

Yang, R., Zhao, B., Liu, M., & Wu, Z. (2023). Teachers' professional development trends and reflections from the perspective of smart education. In Atlantis Highlights in Computer Sciences (pp. 1130–1135). *Atlantis Press International BV.*

Yang, T. (2024). Impact of Artificial Intelligence Software on English Learning Motivation and Achievement. *SHS Web of Conferences*. DOI: 10.1051/shsconf/202419302011

Yang, Y., & Montgomery, D. (2013). Gaps or bridges in multicultural teacher education: AQ study of attitudes toward student diversity. *Teaching and Teacher Education, 30*, 27-37. https://doi.org/DOI: 10.1016/j.tate.2012.10.003

Yang, R. (2022). A study on the model of online learning support services for teachers in the context of smart education. *Proceedings of the 14th International Conference on Education Technology and Computers*, 337–343. DOI: 10.1145/3572549.3572603

Yépez Álvarez, L. D., Jirón Febre, T. I., Rumbaut Rangel, D., & Jurado Martínez, G. (2024). El papel de la inteligencia artificial en la personalización de la educación. *Magazine de las Ciencias*, *9*(3), 31–40. DOI: 10.33262/rmc.v9i3.3160

Yi, Y. (2021). Establishing the concept of AI literacy. *Jahr–European Journal of Bioethics*, *12*(2), 353–368.

Young, L. E., & Paterson, B. L. (Eds.). (2007). *Teaching nursing: Developing a student-centered learning environment*. Lippincott Williams & Wilkins.

Yusuf, A., Pervin, N., Román-González, M., & Noor, N. M. (2024b). Generative AI in education and research: A systematic mapping review. *Review of Education*, *12*(2), e3489. DOI: 10.1002/rev3.3489

Zawacki-Richter, O., Marín, V. I., Bond, M., & Gouverneur, F. (2019). Systematic review of research on artificial intelligence applications in higher education: A framework for future research. *International Journal of Educational Technology in Higher Education*, *16*(1), 1–27. DOI: 10.1186/s41239-019-0171-0

Zengin, Y., Naktiyok, S., Kaygın, E., Kavak, O., & Topçuoğlu, E. (2021). An investigation upon industry 4.0 and society 5.0 within the context of sustainable development goals. *Sustainability (Basel)*, *13*(5), 5. Advance online publication. DOI: 10.3390/su13052682

Zenkina, S. V., Suvorova, T. N., & Nikolaev, M. V. (2015). Electronic educational resources as part of the information and educational environment: educational and methodological manual for students of pedagogical universities and students of the system of advanced training of educators. Moscow – Kirov: "Publishing house Raduga – PRESS", – 99 p.

Zenkina, S. V. (2007). *Information and communication environment focused on new educational results (monograph)*. Education.

Zenkina, S. V. (2010). *Methodology for the development and assessment of electronic educational resources: textbook. Manual for students of the advanced training system, educators and students ped. Universities*. Izvestia Publishing House.

Zhang, L. F., & Sternberg, R. J. (2006). The nature of intellectual styles. Mahwah, NJ: Lawrence Erlbaum.

Zhang, X., & Zhou, M. (2019). Interventions to promote learners' intercultural competence: A meta-analysis. *International journal of intercultural relations, 71*, 31-47. https://doi.org/DOI: 10.1016/j.ijintrel.2019.04.006

Zhang, H. (2020). Optimization of risk control in financial markets based on particle swarm optimization algorithm. *Journal of Computational and Applied Mathematics, 368*, 112530. DOI: 10.1016/j.cam.2019.112530

Zhang, K., & Aslan, A. B. (2021). AI technologies for education: Recent research & future directions. *Educational Technology Journal, 2*, 100025. DOI: 10.1016/j.caeai.2021.100025

Zhang, L. F. (2001). Approaches and thinking styles in teaching. *The Journal of Psychology, 135*(5), 547–561. DOI: 10.1080/00223980109603718 PMID: 11804007

Zhang, L. F. (2001). Do thinking styles contribute to academic achievement beyond self-rated abilities? *The Journal of Psychology, 135*(6), 621–637. DOI: 10.1080/00223980109603724 PMID: 11931003

Zhang, L., & Sternberg, R. J. (2005). A Threefold Model of Intellectual Styles. *Educational Psychology Review, 17*(1), 1–53. DOI: 10.1007/s10648-005-1635-4

Zhang, Y., Luo, Y., Zhang, X., & Zhao, J. (2019). How green human resource management can promote green employee behavior in China: A technology acceptance model perspective. *Sustainability (Basel), 11*(19), 5408. DOI: 10.3390/su11195408

Zheng, L., Niu, J., Zhong, L., & Gyasi, J. (2021). The effectiveness of artificial intelligence on learning achievement and learning perception: A meta-analysis. *Interactive Learning Environments, 31*(9), 5650–5664. DOI: 10.1080/10494820.2021.2015693

Zhilavskaya, I. V. (2013). Media education of youth: monograph / Moscow: MGGU, – 243 p.

Zhu, Z. T., Yu, M. H., & Riezebos, P. (2016). A research framework of smart education. *Smart learning environments*, 3, 1–17.

Zhu, M., & Zhan, S. (2022). Research and implementation of a trend prediction model based on trend similarity for the changing trends of fashion elements in clothing. *Proceedings of the 2022 6th International Conference on Electronic Information Technology and Computer Engineering*, 1460–1466. DOI: 10.1145/3573428.3573685

Zhu, Z. T., & He, B. (2012). Smart education: New frontier of educational informatization. *E-Education Research*, *12*, 1–13.

Ziatdinov, R., & Cilliers, J. (2021). Generation Alpha: Understanding the next cohort of university students. *European Journal of Contemporary Education*, *10*(3). Advance online publication. DOI: 10.13187/ejced.2021.3.783

ZYOD. (2023, November 1). AI and predictive analytics: Revolutionising the fashion industry. https://www.linkedin.com/pulse/ai-predictive-analytics-revolutionising-fashion-industry-zyod-zjkyf/

About the Contributors

Prakasha G S is an Associate Professor at the School of Education, Christ University, Bangalore, India. He has completed his MEd from the Regional Institute of Education, India. PhD from Bangalore University. He also completed an MSc (Chemistry) and MPhil in Education. He cleared the National Eligibility Test (NET) and was awarded a junior research fellowship (JRF). He holds IB Educator certificates in MYP and DP from DePaul University, Chicago, USA. He is a visiting research fellow, International University, Malaysia. He visited Germany, France, Switzerland, Canada, & USA. He led EdTech Project with Switzerland's Teacher-Ed University (hep, Vaud). He visited Miami University, University of Cincinnati-Ohio, University of Mount Union, Johns Hopkins University-Washington DC, Syracuse University, Binghamton University, Campbellsville University- Kentucky, New York University, & the Catholic University of America. His research areas are EdTech, TeacherEdu, & HigherEdu. He has published 50+ research articles indexed in SCOPUS and WOS databases. He wrote several book chapters, articles, & conference papers. He has presented papers worldwide, to name a few: Bangkok, Switzerland, Russia, University of Oxford-UK, & Binghamton University, USA. He is an active editorial board member for SAGE, SPRINGER NATURE, and Taylor & Francis. He received a research grant from Sophia University, Tokyo, JAPAN. At the Centre for Advanced Research and Training (CART), he has conducted many national-level research data analysis workshops at Christ University. He has trained PhD scholars, students, and faculty members in quantitative data analysis using SPSS. He has delivered lectures nationwide, including ICSSR- New Delhi, XLRI, Mount Carmel, Jyothi Nivas, Christ Nagar, Thiagarajar College, & St. Joseph's University. He conducts research activities that have national and international emphasis. He completed a Major Research PROJECT on Teacher & Teacher Educator Competencies. His PROJECT with NGO-GRAMA contributed a sustainable & replicable "Informal-Education-Model" to uplift rural education in India. He authored 6 books 1) Teaching Internship Models Around the World-

Routledge, 2) Educational Perspectives on Digital Technologies in Modeling & Management- IGI, 3) Digital Technologies- IGI: Insights in Education and Industry, 4) Diagnostic Testing, 5) Attention-deficit-behaviour, 6) Teacher-effectiveness by Lap Lambert, Germany.

Maria Lapina Deputy Director for International Affairs Institute of Digital Development North-Caucasus Federal University

Francisco David is a professor in the Universidad de Málaga, Spain. He is highly published researcher.

Juan J. Leiva-Olivencia is Full Professor in the Department of Didactics and Educational Organisation at the University of Málaga (Spain). His professional career is linked to research projects on educational innovation, ICT, inclusive education, interculturality and teacher training. He combines his teaching and research work with the role of evaluator for various international agencies for the evaluation of quality in higher education.

***

Vigneshwaran S A is an Assistant Professor of Social Work in the Department of Sociology and Social Work at Christ University. He was a Senior Research Fellow at the Department of Social Work in the School of Social Sciences and Humanities at the Central University of Tamil Nadu. He completed his Bachelor's degree in Zoology from The American College affiliated to Madurai Kamaraj University in 2006, followed by a Master's degree in Social Work (Medical & Psychiatry) from the same institution in 2009. Before joining academia, Vigneshwaran worked as a Community Mental Health Social Worker and Clinical Services Coordinator for the Shant Manas Trust from 2009 to 2018. In this role, he helped promote mental health awareness and treatment in rural areas of Madurai, Tamil Nadu, India. He also assisted social work and nursing students in their placement training and conducted skill development programs. Vigneshwaran's research interests include road safety, post-trauma care for road injury victims, social interventions for mental illness, early intervention in psychosis, cultural understanding of mental illness, and yoga and mental health. He has published several research articles in indexed journals and edited a book on Disaster Risk Reduction and Management in India. He has also presented papers at various national conferences and symposiums.

Mahfooz Alam is a Research Scholar in the Department of Computer Science, at Aligarh Muslim University (AMU), Aligarh, India. M. Alam has also served for six years as an Assistant Professor in the Department of Computer Science at

Al-Barkaat College of Graduate Studies (ABCGS), Aligarh, India. He received his M.Tech. in Computer Science and Engineering from Dr. A. P. J. Abdul Kalam Technical University, Lucknow, India. He also completed his B.C.A. and M.C.A. in Computer Applications from Indira Gandhi National Open University (IGNOU) in New Delhi, India. He has published many research papers in international conferences and international journals of repute (including IEEE Access & Xplore, Springer, Elsevier, Wiley, IGI Global, etc.). His research interest areas are Cloud, Fog and Edge Computing, Internet of Things, Machine Learning, Workflow Scheduling, and Load Balancing.

Robert Ashton-Bell, Associate Professor Dr R.L.Tavis Ashton-Bell has been the Head of the Department of Western Music at CHRIST University since 2019. He received his doctorate in Higher Education Music Teacher Education at Monash University, Australia focusing on learner-centred approaches in higher education music pedagogy using mathematics to create a visual models of harmonic stuctures. He holds membership to the International Society for Music Education and the International Society for Assessment in Music Education. Specialising in learner-centred music curriculum design, his broader research interests involve finding connections between music education and neuroscience, artificial intelligence, physics and mathematics.

Grace Anu Backia J is a student at CHRIST (Deemed to be University), pursuing her Masters in Social Work, specializing in Clinical and Community Practice. Her areas of interest include Human Rights, especially Dalit and Tribal Rights. She envisions working with the marginalised sections of the society in making justice a reality for them & in their empowerment.

Pragya Bhattacharjee is a student at Christ University, with a strong commitment to research and academic inquiry. Her work explores AI applications in oncology, focusing on diagnostic, predictive, and generative tools in cancer care. She has contributed to interdisciplinary studies on AI's impact on medical accuracy, efficiency, and accessibility. Beyond healthcare and technology, she has also authored a paper on oppression and intersectionality in modern society, examining systemic inequalities and their broader implications. Passionate about bridging research with real-world impact, she continues to explore opportunities at the intersection of business, technology, and social change.

Emmanuel Daniel is a committed academic and development professional with over thirteen years of experience in teaching, research, and community-based practice. His core area of expertise lies in child wellbeing, with a particular focus on education, life skills, and gender sensitisation. He has led several participatory research and development initiatives among tribal communities, especially in the Nilgiris and Assam, generating valuable insights for culturally sensitive and inclusive educational interventions. He has also undertaken key evaluation projects for the Government of Karnataka, focusing on schemes for Scheduled Castes and Scheduled Tribes. In addition to his academic role as Assistant Professor in the Department of Social Work at CHRIST (Deemed to be University), Bangalore, he consults with NGOs and development agencies in India and abroad, supporting projects in grant writing, programme design, and impact assessment. His work reflects a strong commitment to integrating academic knowledge with grassroots realities to advance child-focused and community-driven development.

Greta D'Souza is an Associate Professor and Head, School of Education, CHRIST (Deemed to be University). She supervises doctoral students and teaches the postgraduate and graduate programs in teacher education. Her areas of research and expertise include Adolescence Education for adolescents, pre-service teachers, and in-service teachers. Teaching and Training include Language skills, Soft skills, Vocational Training, Curriculum design and instruction. She equips professionals, teacher educators, pre-service and in-service teachers to reach out to institutions in different learning contexts to make learning effective, engaging and productive. She has visited universities in Hong Kong to study the best practices in Higher Education. She believes that students should be empowered to communicate effectively, hone leadership skills, take advantage of new opportunities and make a positive impact on society.

Carmen Rocío Fernández-Fernández, Graduated in Early Childhood Education and Pedagogy, both degrees obtained from the University of Granada (UGR). Holder of a Master's degree in Innovation and Research in Curriculum and Teacher Training from the UGR. Recipient of a research initiation grant awarded by the UGR, as well as a collaboration grant from the Spanish Ministry of Education and Vocational Training, carried out in the Department of Didactics and School Organization at the UGR. Currently holds an FPU (University Teacher Training) contract funded by the Ministry of Education and Vocational Training. Author of indexed scientific publications in high-impact journals and books, as well as participant and speaker at national and international conferences on topics such as: Innovation and Technology in Education, Educational Methodologies and Pedagogy, Inclusion and Diversity in the Classroom, Educational Assessment and Quality, Teacher Professional Development, and Educational Research.

Naadia Ghalib Sheriff is a dedicated student at Christ University with a strong passion for research. She has contributed to studies on AI applications in oncology, gender representation in Indian cinema, assistive technology for visually impaired, and predictive analytics in fashion industry. Committed to exploring innovative solutions and promoting positive change, she continues to pursue opportunities that bridge business, technology and social impact.

Akanksha Khanna works as an Assistant Professor in Management at Symbiosis International University, Dubai. With a PhD in Marketing, she has over 18 years of experience teaching and mentoring undergraduate and postgraduate students. A Fellow (FHEA) of Advance HE, UK, and an Associate CIPD (Level 5), she has been a dedicated researcher focused on quality work.

Gagan Kumar B R holds Ph.D. in Healthcare Artificial Intelligence from the Presidency University; a MBA in Information Systems from the Alliance University; and MS in Computer Science & Software Engineering from Woolf University. His academic and corporate accomplishments are numerous; including a distinguished record of endeavours in higher education, international initiatives, and exemplary testimony in developing and fostering strong relationships with the business community are the hallmarks of his work.

Manuel Enrique Lorenzo-Martín, Professor at the Department of Didactics and School Organisation at the University of Granada. PhD in Education from the University of Granada, his professional career has developed around formal teaching in different educational centres in ESO, Baccalaureate and Vocational Training. He is an active member of the A.R.E.A. Group of the University of Granada (HUM - 672). He has been an expert teacher trainer in LOMLOE, DUA and Learning Situations, as well as an External Educational Consultant for Editorial Edebé. Currently, his lines of research focus on Learning Situations, application of SAD, use of AI in formal education, attention to diversity and the application of active methodologies for the promotion of transversal competences in students.

Amr M. Mohamed is an Assistant Professor of Curriculum and EFL Instruction at North Private College of Nursing in Saudi Arabia. He earned his Ph.D. in Curriculum and Instruction from Ain Shams University in Egypt. His research interests encompass language education, curriculum development, and assessment, particularly in the context of EFL instruction. Dr. Mohamed has published numerous research articles in peer-reviewed journals indexed in Scopus and Web of Science. He is also actively engaged in academic and professional development initiatives

Mouas Samia is a researcher and an associate professor (MCA) at the department of English and Literature at Batna 2 University, ALGERIA. She received her Ph.D. and HDR in Didactics of the English Language from the same university in 2019 and 2021, respectively. Her most recent publications focus on teacher professional development, intercultural awareness, and online education. She is a member of the reviewing committees of some Algerian and international journals. Her expertise is primarily related to the socio-intercultural dimension of language, research methodology, and blended learning in EFL classes.

Sunil M.P. is an Assistant Professor at the School of Commerce, Finance and Accountancy, CHRIST (Deemed to be University), Bangalore Central Campus, with over a decade of teaching and research experience. He holds a Ph.D. in Strategic Human Resource Management and specialises in Quality Management, Human Resources, Technovation, and Organizational Behavior. Dr. Sunil has published over twenty research articles and holds multiple patents, underscoring his commitment to advancing knowledge in his field. His work includes impactful projects, such as an initiative to empower persons with disabilities for careers in finance. With a background in business, Dr Sunil also offers consultancy services to student startups, providing practical insights that fuel innovation and growth, bridging the gap between academic learning and industry needs.

Mukhidinov Magomed Gosengadzhievich. - PhD, Associate Professor of the Department of Information Technology and Economics, Dagestan State Pedagogical University, Makhachkala, 367003, Russia, Republic of Dagestan, Makhachkala, st. Yaragskogo 57 Teacher of Mathematics and Computer Science E-mail: muhidinov59@mail.ru

|Raghib Noman - Contributing Author|I am currently a research scholar in the Department of Commerce at Aligarh Muslim University, India. My research interests include financial technology, consumer behavior, and portfolio optimization. I am dedicated to contributing to the field of commerce through rigorous research and academic writing.

Palangov Abulfat Gulam Oglu, Doctor Pedagogical Sciences, Professor, Department of Computer Science, Azerbaijan State Pedagogical University (Baku, Azerbaijan), Teacher of Mathematics and Computer Science, Address: AZ1000, Azerbaijan, Baku, Uzeyir Hajibeyov Street, 68, abulfat1@gmail.com, abulfat59@mail.ru

Sanskriti Rawat (9 December 1996) is a research scholar at the School of Education, Christ University, Central Campus, Bangalore, India. Her areas of interest in research include educational psychology, educational technology, teacher education, and English language teaching and learning.

Carmen Rodríguez Jiménez, Lecturer in the Department of Didactics and School Organisation at the University of Granada, Melilla campus. Recipient of a Research Initiation grant awarded by the University of Granada, a Collaboration grant awarded by the Ministry of Education and a FPU grant awarded by the Ministry of Education. PhD in Education from the University of Granada. Active member of the A.R.E.A. Group of the University of Granada (HUM - 672). Currently, her lines of research focus on digital competence, digital safety, training of trainers and active methodologies.

Adrián Segura-Robles, Degree in Psychopedagogy and PhD in Educational Sciences. He is currently a lecturer in the Department of Didactics and School Organisation at the University of Granada. Specifically, he teaches at the Faculty of Education and Sport Sciences of Melilla in different university degrees and master's degrees. His research career is framed in the analysis of active methodologies in the different educational stages of the educational system although he also has different research in multicultural contexts. In recent years she has belonged to different research groups where she has been able to acquire the necessary skills to develop her investments.

Mohammad Shahid is a committed academician with several years of extensive experience in teaching, and administration at a leading academic institution. Currently, he is working as an Assistant Professor in the Department of Computer Science, Aligarh Muslim University, India. He has completed Ph.D., M. Tech. and MCA degrees from the School of Computer and Systems Sciences, Jawaharlal Nehru University (JNU), New Delhi. His research interests include Parallel and Distributed Computing, Cloud Computing, Evolutionary & Swarm Intelligence Algorithms, and combinatorial optimization problems. He has published numerous research papers in international conferences and journals of repute (including IEEE, Elsevier, Springer, InderScience, Wiley, Taylor & Francis, Emerald etc.). He was awarded a research startup grant-2017 from University Grants Commission (UGC), India. Recently, he has awarded a major research project form ICCSR, New Delhi, India project on portfolio optimization using evolutionary optimization. He has presented his work at numerous conferences both in India and abroad. He has won the best research paper award one in abroad 2nd International Conference on Technological Advancements in Computational Sciences (ICTACS 2022), Tashkent, Uzbekistan, 2022. He has also won the best paper award in International Conference on Reinventing Business Practices, Startups and Sustainability (ICRBSS) 2022, SRM University, India. He has held positions on numerous editorial boards, TPC, and organizing committees for various journals and conferences. He is also reviewer of many journals of publishers including, Elsevier, Springer, Wiley, GI Global, etc.. He is an IEEE Member.

Yara Shokry is a lecturer of curriculum, Instruction, and Educational technology at Port Said University, Egypt. She earned her PhD in Applied Linguistics, with a focus on Computer-Assisted Language Learning (CALL), from the University of York in 2021.Herresearch explores integrating AI and digital tools into language education. She is also an active member of the AI in Language Learning and Assessment Lab (AILLA) and the Pedagogical Innovation in Virtual and Online Teaching and Learning Lab (PIVOTAL), where her research focuses on innovative approaches to language education and leveraging technology to optimize learning outcomes.

Sruthi Suresh is a doctoral research scholar in Psychology at CHRIST (Deemed to be University) and an IB Psychology Educator and TOK Coordinator at Knowledgeum Academy, Bengaluru. In 2018, she completed her M.Sc. in Counseling Psychology and worked as an IB DP Psychology Faculty for the next two years. After joining the Ph.D. program in 2020, she has been an active scholar and engaged with the department as a Research Assistant, Teaching Assistant, Impact Assessment Research Team Lead, and Department Magazine Editor. She writes and presents widely on issues related to child and adolescent mental health, teacher education, and social emotional learning.

Pooja holds a PhD. in Social Work and is an academic with the Indian Institute of Management Rohtak. She is an independent researcher and consults monitoring and evaluation programs of NGOs.

Nisha Varghese, M.Phil., Ph.D., was a full-time Research Scholar in the Department of Computer Applications at Bharathiar University. She is currently working as an Assistant Professor in the Department of Computer Science at Christ University. Her areas of interest include Natural Language Processing (Computational Linguistics), Large Language Models, Machine Learning, Deep Learning, and Cyber Security. She has authored two books in the areas of NLP and Cyber Security. Previously, she served as an Assistant Professor at Cochin University and Calicut University in Kerala.

Abdullahi Yusuf is a Senior Lecturer at Sokoto State University. He specialized in Educational Technology and Educational Computing. He has held various positions, including the role of Head of Department in Science Education. His scholarly contributions extend across national and international platforms, with numerous journal articles to his credit. Of particular note is his recent publication, which investigates the multicultural perspectives of Generative AI.

Index

A

C

D

E

F

H

www.ingramcontent.com/pod-product-compliance
Lightning Source LLC
LaVergne TN
LVHW060003031026
849161LV00025B/1362